D0855041

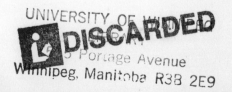

Ectomycorrhizae

THEIR ECOLOGY AND PHYSIOLOGY

PHYSIOLOGICAL ECOLOGY

A Series of Monographs, Texts, and Treatises

EDITED BY

T. T. KOZLOWSKI

University of Wisconsin
Madison, Wisconsin

T. T. KOZLOWSKI. Growth and Development of Trees, Volumes I and II — 1971

DANIEL HILLEL. Soil and Water: Physical Principles and Processes, 1971

J. LEVITT. Responses of Plants to Environmental Stresses, 1972

V. B. YOUNGNER AND C. M. MCKELL (Eds.). The Biology and Utilization of Grasses, 1972

T. T. KOZLOWSKI (Ed.). Seed Biology, Volumes I, II, and III — 1972

YOAV WAISEL. Biology of Halophytes, 1972

G. C. MARKS AND T. T. KOZLOWSKI (Eds.). Ectomycorrhizae: Their Ecology and Physiology, 1973

In Preparation

T. T. KOZLOWSKI (Ed.). Shedding of Plant Parts

Ectomycorrhizae

THEIR ECOLOGY AND PHYSIOLOGY

Edited by

G. C. MARKS

Forests Commission of Victoria
Treasury Place
Melbourne, Victoria, Australia

T. T. KOZLOWSKI

Russell Laboratory
University of Wisconsin
Madison, Wisconsin

ACADEMIC PRESS   New York and London   1973

A Subsidiary of Harcourt Brace Jovanovich, Publishers

ACADEMIC PRESS, INC.
111 Fifth Avenue, New York, New York 10003

United Kingdom Edition published by
ACADEMIC PRESS, INC. (LONDON) LTD.
24/28 Oval Road, London NW1

Library of Congress Cataloging in Publication Data

Marks, G   C
    Ectomycorrhizae.

    (physiological ecology)
    Includes bibliographies.
    1.   Mycorrhiza.   I.   Kozlowski, Theodore Thomas,
DATE   joint author.   II.   Title.
QK604.M35   581.5'24   72–9332
ISBN 0–12–472850–2

TO A. J. RIKER

Contents

1. Structure, Morphogenesis, and Ultrastructure of Ectomycorrhizae

G. C. Marks and R. C. Foster

2. Classification of Ectomycorrhizae

B. Zak

3. Distribution of Ectomycorrhizae in Native and Man-Made Forests

F. H. Meyer

4. Growth of Ectomycorrhizal Fungi around Seeds and Roots

G. D. Bowen and C. Theodorou

5. Mineral Nutrition of Ectomycorrhizae

G. D. Bowen

6. Carbohydrate Physiology of Ectomycorrhizae

Edward Hacskaylo

7. Hormonal Relationships in Mycorrhizal Development

V. Slankis

List of Contributors

Numbers in parentheses indicate the pages on which the authors' contributions begin.

G. D. Bowen, Division of Soils, Commonwealth Scientific and Industrial Research Organization, Glen Osmond, Adelaide, South Australia (107, 151)

R. C. Foster, Forest Products Laboratory, Division of Applied Chemistry, C.S.I.R.O. South Melbourne, Victoria, Australia (1)

Edward Hacskaylo, U.S.D.A. Forest Service, Forest Physiology Laboratory, Beltsville, Maryland (207)

G. C. Marks, Forests Commission of Victoria, Treasury Place, Melbourne, Victoria, Australia (1)

Donald H. Marx, U.S.D.A., Forest Service, Forestry Sciences Laboratory, Athens, Georgia (351)

F. H. Meyer, Institute for Landscape Management and Nature Conservation, Herrenharuser, West Germany (79)

Peitsa Mikola, Department of Silviculture, University of Helsinki, Finland (383)

Angelo Rambelli, Laboratory of Mycology, Institute of Botany, Roma State University, Rome, Italy (299)

V. Slankis, Ministry of Natural Resources, Research Branch, Southern Research Station, Maple, Ontario (231)

C. Theodorou, Division of Soils, Commonwealth Scientific and Industrial Research Organization, Glen Osmond, Adelaide, South Australia (107)

B. Zak, Forestry Sciences Laboratory, Pacific Northwest Forest and Range Experiment Station, U.S.D.A. Forest Services, Corvallis, Oregon (43)

Preface

This volume summarizes the present state of knowledge and opinion on the physiological ecology of ectomycorrhizae (which may be defined as symbiotic associations between nonpathogenic or weakly pathogenic fungi and living cells of roots). Although the book places considerable emphasis on forestry aspects of mycorrhizal problems, its wide ranging subject matter cuts across the boundaries of a number of traditional plant sciences. Thus, it will be of interest to a wide variety of researchers and teachers, especially agronomists, biochemists, foresters, horticulturists, mycologists, plant pathologists, soil scientists, plant ecologists, plant physiologists, and microbiologists. A short glossary of terms is included for use by those unfamiliar with mycological terminology.

The ecology of mycorrhizae is a somewhat neglected field of study, yet it is an area in which research can have its most significant impact on practical forestry problems. Not only could it assist directly in translating laboratory studies into improving forest yields but research findings may also help explain field observations. A vivid example of this was brought home to one of us (G.C.M.) when he discovered that a carefully balanced fertilizer mixture greatly reduced field resistance of *Eucalyptus* trees sensitive to feeder root rot caused by *Phytophthora cinnamomi* despite a very large increase in stand vigor, growth rate, etc. The information in D. H. Marx's chapter proved very useful when trying to explain this result.

Investigations into factors controlling genesis of mycorrhizae have provided a fertile field for controversy over the years. The formation of mycorrhizae is an important problem with heavy ecological overtones, and it must be clarified before greater practical use can be made of research findings. Some of the chapters in this book show that the origin of mycorrhizae is by no means resolved; it will be left to the student and researcher to evaluate the merits of the data presented. Finally, it is hoped that reviews, interpretations, and concepts proposed by authors of the various chapters will stimulate further work on microecology of forest tree roots because the information presented tends to highlight the need for additional research in this field and the lack of agreement on some fundamental questions.

In planning this book, invitations were extended to a group of leading modern investigators of the biology of mycorrhizae and their significance. We express our deep appreciation to each contributor for his scholarly work, patience, and attention to detail during the production process. The assistance of Mr. W. J. Davies and Mr. P. E. Marshall in preparation of the Subject Index is also acknowledged.

G. C. Marks
T. T. Kozlowski

Ectomycorrhizae

THEIR ECOLOGY AND PHYSIOLOGY

CHAPTER 1

Structure, Morphogenesis, and Ultrastructure of Ectomycorrhizae

G. C. MARKS[1] and R. C. FOSTER[2]

[1] Senior Forest Pathologist, Forests Commission, Victoria.
[2] Senior Research Scientist, Forest Products Laboratory, South Melbourne.

I. Introduction

The structure of mycorrhizae has been examined intermittently ever since the first observations were made by Unger in 1840. Frank (1885) provided one of the best of the early descriptions of mycorrhizal structure and he coined the name of this organ. Mycorrhizae have attracted the attention of foresters, mycologists, tree physiologists, biochemists, and biophysicists, and considerable information exists on their physiology and function.

In contrast to the wealth of information that has accumulated on structure and physiology of mycorrhizae, there is very little published information on the cytology of these associations. The majority of electron microscope studies of host/fungus interrelationships have been made on parasitic associations and have concentrated on fine structure of intracellular haustoria rather than intracellular hyphae. Ehrlich and Ehrlich (1963), Shaw and Manocha (1965), Manocha and Shaw (1967), and Van Dyke and Hooker (1969) looked at the Uredinales; McKeen *et al.* (1966) and Bracker (1968) looked at Erisiphales; and Peyton and Bowen (1963), Berlin and Bowen (1964), and Chou (1970) examined the Peronosporales. These studies provided useful data with which the cytology of symbiotic mycorrhizal associations may be compared. Like lichens, mycorrhizae are truly symbiotic and produce new, consistent morphological entities in which both organisms benefit from the association. Only recently, however, have they been examined at the level of the electron microscope, and the information gleaned gives new insight into the interrelationships of the two symbionts.

In this chapter we review information on the structure, cytology, and morphogenesis of mycorrhizae and describe factors in the environment that directly affect the latter phenomenon. In this manner a more meaningful picture of the organism can be presented. In order to achieve this, it will be necessary to briefly review the growth and development of primary tissues of the root and then follow root development once infection has taken place and a mycorrhiza is established.

II. Classification and Structure of Roots Based on External Appearance

Mycorrhizae form on unthickened roots of forest trees, and attempts have been made to classify these uninfected root tips on the basis of their growth characteristics. Noelle (1910) divided the root systems of conifers into long and short roots. Aldrich-Blake (1930) added a further category when he described the very fast-growing "pioneer" roots of *Pinus laricio* seedlings. These divisions were based primarily on growth rates of the roots, their method of branching, and, more recently, on appearance of the apical meristem (Wilcox, 1968). Although these categories are clearly defined, an individual root could change from one type to another when conditions in the environment are altered (Slankis, 1967; Wilcox, 1968). Separation into these categories is not always possible because they some-times intergrade as in broad-leaved trees (Chilvers and Pryor, 1965; Harley, 1969; Lyford and Wilson, 1964) and in some gymnosperm species (How, 1943; Zak, 1969). However, after infection and establishment of the mycorrhizal habit, the root assumes a different character which is quite distinctive (Clowes, 1950).

A. Morphogenesis of Long and Short Roots before Infection

In order to appreciate the events that take place in an infected root it is essential to compare the differences observed and the changes that occur in long and short roots prior to infection, since many of the changes in structure observed in mycorrhizae occur after infection. Separation of these two stages has not always been made in the literature on mycorrhizae.

B. Growth and Branching

The long root of conifers shows racemose branching and is capable of indefinite growth, unlike the dichotomously branched short root which grows for only a restricted period. The growth rates of the long roots are considerably greater (Fig. 1) and the duration of growth more prolonged than that observed in short roots (Hatch, 1937; Aldrich-Blake, 1930). However, it is worth observing that the growth rate of a long root can slow down to that of a short root, and the opposite situation often develops where a short root assumes the growth characteristics of a long root.

The structure of the apical meristem of an active long root reflects its increased capacity for growth. The meristem is conical in shape, and of larger diameter than that of the smaller, more hemispherically shaped meristem of the short root (Wilcox, 1968). The branches produced on the root are related to the activity of the apical meristem. In large, fast-growing long roots, lateral branching is often suppressed (Wilcox, 1968; Aldrich-Blake, 1930), while small-diameter, subordinate, mother roots may produce a relatively large number of lateral branches, which emerge from the cortex and grow vigorously. Wilcox (1968) stated that the dichotomous type of branching observed in the short roots of *Pinus resinosa* arose from the precocious growth of the laterals, with growth being suppressed in the apical meristems.

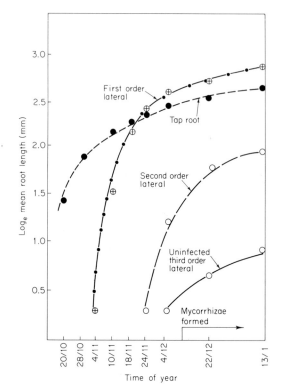

Fɪɢ. 1. Relationship between growth of tap roots, first-, second-, and third-order laterals of *Pinus radiata* and age of the seedlings. Note, the third-order laterals appear the latest, grow the slowest, and are the only roots that formed mycorrhizae.

During the period of active growth, the long root has a larger zone of white-colored tissue behind the apex than that seen in short roots. However, as growth slows down toward the end of the growing season, the size of this zone is reduced.

At the end of the growing season, a metacutized layer forms over the root apex. This consists of a layer of suberized cells that is continuous with the endodermis. Its chief function is to protect the delicate meristem during dormant periods. At the onset of a new growing period the apical meristem resumes activity, ruptures the metacutized layer, and the root begins to elongate once more (Wilcox, 1968; Aldrich-Blake, 1930).

C. Maturation of Root Tissues

In the young uninfected roots of conifers, deep-seated root hairs are sometimes produced (Aldrich-Blake, 1930), which are cast off as the root matures. Ideally, cells of the cortex are thin-walled, isodiametric or variable in shape. As maturation proceeds, these cells extend slightly and gradually turn brown in color (Aldrich-Blake, 1930; Wilcox, 1968; Marks, 1965), possibly because of the deposition of materials derived from the moribund cells. In older roots the primary cortex may persist for a period around the growing root as a collar of dead cells before it is sloughed off.

The fate of the endodermis is of particular interest. Deterioration of the cortex is associated with maturation of the endodermis and formation of suberin lamellae in cell walls. Aldrich-Blake (1930) recognized three stages of endodermal development in *P. laricio*, the embryonal condition in which the cells expanded, the primary condition in which the radial walls became suberized, and the secondary condition in which all the walls were thickened. In both *P. laricio* (Aldrich-Blake, 1930) and *P. lambertiana* (Marks, 1965) the larger, faster-growing roots had more living primary cortical tissue in the growing apex than roots that had stopped growing, while in the smaller, slower-growing roots, the zone of living, primary cortex was considerably reduced. Thus it appears that the rate of maturation of the endodermis is independent of growth rate of the apical meristem, and longevity of the primary cortex will depend on the interaction of these two events. There are no reliable data dealing with the question of whether longevity of individual cortical cells in long roots differs from that of cells in short roots. Judging from the rapidity with which tissues develop in fast-growing roots and the relatively slow growth and lack of secondary tissue in short shoots (Hatch, 1937), one might suspect that the cortical tissues of the latter survive for a longer period. When growth ceases the secondary endodermis extends up to the root apex (Aldrich-Blake, 1930; Hatch, 1937) and most of the primary cortical tissues die.

Maturation of tissues of the central stele will not be reviewed here, but it is worth noting that it follows the same pattern as that described for the endodermis. There are structural differences in the xylem elements of the long and short root (Wilcox, 1968) and these have been used to identify the growth potential of the two classes of root.

III. Process of Infection

A. METHODS OF INFECTION

It is worth recapitulating that a mycorrhiza is formed when a fungus (usually a soil basidiomycete) infects the living primary cortical cells of the root, and as Slankis (1963) observed, mycorrhizal formation will depend on specific physiological conditions in the root. An association will not form with either moribund or dead cells. There are no reports of intercellular infection occurring in a meristematic tissue, and it is safe to assume that no mycorrhiza will form in the apex of a root. Consequently, it is possible to demarcate a discrete zone where mycorrhizal formation can take place (Fig. 2) or, as Boullard (1957) noted, provide a gradient of resistance to infection. This zone lies behind the growing apex and in advance of the region where the primary cortex begins to deteriorate as the root matures. This zone will be referred to as the mycorrhizal infection zone (MIZ). The MIZ is not static, but moves acropetally as the root grows, and the size of the MIZ, its rate of acropetal movement, and time during which the cells of this region remain in the "receptive" conditions are influenced by factors affecting root growth and tissue morphogenesis. Consequently, when the process of mycorrhizal formation is considered, the morphogenic effects of the treatments used to analyze the process require close examination because of their possible effects on the MIZ.

There is an extensive literature on factors in the environment which affect mycorrhizal formation. It is difficult to establish whether they

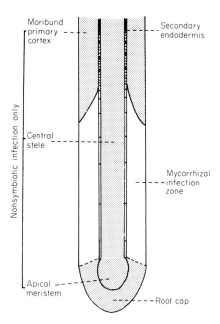

FIG. 2. Schematic drawing showing the regions of the root and the zones where mycorrhizal symbiosis can originate. Shaded areas represent regions where no true mycorrhizal infection occurs.

influence the process of infection or subsequent events. In order to obtain a better appreciation of these effects it is essential to follow the process of infection in a root. There are apparently two courses that may follow. One is shown by the pattern of primary infection seen in uninfected seedlings growing in infected soil. A weft of hyphae forms around the mother root about the time the first leaves appear (Morrison, 1956), and there is no intercellular penetration. New roots develop under this weft which expands and stretches to accommodate new growth. About the time the protoxylem vessels appear in the xylem, intercellular penetration occurs and a Hartig net is formed. Similar observations were made by Mihovic (1952) for *Quercus* in Russia, and by Clowes (1951) for *Fagus* in England. In the initial stages of infection, Hartig net formation occurs only in the outer cortical cells, but it gradually spreads inwards until it is stopped by the endodermis. Robertson (1954) observed mycorrhizal infection in seedlings grown at the edge of a pine forest at Brandon Park, England. He noticed that the outer cortical cells collapsed about the time a pseudoparenchymatous weft of hyphae formed around the young short root, and intercellular penetration did not occur. He concluded that Hartig net formation either preceded, or proceeded with, formation of the mantle. It is possible, however, that the rapid demise of the primary cortex before a mycorrhiza could form could have distorted the infection sequence or resulted from infection by more pathogenic fungi (Harley and Waid, 1955). "Secondary" infection can be described as infection of a new root surface by inoculum already established either on, or within the tissues of, older parts of the roots. This may occur either when regular growth takes place as trees develop or at the beginning of growth after a dormant period. Robertson (1954) observed that hyphae forming a Hartig net may either grow acropetally, infecting the newly produced cortex or that infection could take place from remnants of hyphal wefts on the root surface. Once infection is established the chances of emerging laterals being infected are greatly increased (Wilcox, 1968). However, Wilcox surmised that the anomolous distribution of mycorrhizae along a root, that is encountered so frequently, was caused by the great variation in the growth vigor of the emerging laterals. If the laterals emerged slowly the slow-growing fungus had time to "catch up" with the root and proceed to form a mycorrhiza. A vigorous root, on the other hand, escaped infection. Thus the process of mycorrhizal formation is reduced to very simple terms, depending predominantly on the rate of root growth and factors that affect it directly or indirectly. This observation has wide support in the literature (Mitchell *et al.*, 1937; Saljaev, 1958; Arnold, 1960; Anderson, 1967). In this context it is worth recapitulating a valid but often neglected criticism made by Robertson (1954) of Björkman's (1942) carbohydrate theory of mycorrhizal formation. "The effect of varying light intensity and unbalanced nitrogen manuring might be different on the fungus and the root, thus under any one environmental condition a balance may be achieved between the number of short roots and the number the fungus can reach and infect. . . ."

Foster and Marks (1966, 1967) described the process of intercellular penetration that occurs in infected roots. Unlike pathogenic organisms (Wood, 1967), the penetration process was not preceded either by plasmolysis or alteration in cytology of the host cells or by degeneration of tissues in advance of the fungus. Instead the walls of contiguous cells separated at the middle lamella, and there was no evidence of wall decomposition. When hyphae penetrated between host cells they appeared wedge-shaped, and, in cross sections, the hyphae in the inner regions of the cortex were small and densely packed and they had angular outlines. In the older, outer parts of the Hartig net, the angular outlines of the hyphae were lost and the cells became vacuolated. Thus it appeared in both *Pinus radiata* and *Pseudotsuga menziesii*, that the mycorrhizal fungus loosened the middle lamella region between contiguous cells and then wedged itself into the fissures so formed. The fungus then expanded, splitting the intercellular substance further. Thus the process of intercellular penetration is predominantly mechanical, and, on the basis of these observations, Foster and Marks (1966) suggested that when host cell expansion is rapid, intercellular penetration by a fungus is retarded. When root growth had slowed down and cortical cells attained maturity, penetration would be easier to accomplish. These cytological observations may explain why mycorrhizae do not form on rapidly growing roots, e.g., the long roots of most gymnosperms and well-fertilized trees with vigorous root systems, and why intercellular penetration only occurs after growth stops or slows down considerably (Fig. 1).

B. BARRIERS TO INFECTION

The process of successful fungal invasion is associated with overcoming many barriers. As mentioned, at the end of the growing season roots produce an impervious metacutized layer over the apex. Mycorrhizal fungi do not penetrate this barrier except where there are breaks in the layer (Wilcox, 1968), and acropetal advance of the fungus in the Hartig net is stopped by the same barrier (Marks, 1965; Wilcox, 1968). The formation of a metacutized layer sometimes results in a change in the mycorrhizal partnership, especially when the new root surface produced at the beginning of the next growth phase is infected by a more vigorous fungus (Marks and Foster, 1967; Kessler, 1966; Kelly, 1941). Foster and Marks (1966, 1967) observed that the tannin-filled outer cells of a mycorrhiza produced appreciable changes in morphology and cytology of the fungus as it passed through this barrier. Hillis *et al.* (1968) compared the phenolic compounds produced by *Pseudotsuga menziesii* and *Pinus radiata* and suggested that mycorrhizal formation in the former could be hindered by the variety of toxic phenolic compounds which the prospective mycorrhizal fungus would encounter during infection. Subsequent comparisons by Linnemann (1969) of the relative ability of firs (*Abies*) to form mycorrhizae appear to substantiate the work of Hillis *et al.* (1968). Other internal (mechanical) barriers to infection and penetration have been reported. Arnold (1960) found that empty hypodermal cells proved an effective barrier to Hartig net formation. Fungi have never been observed to penetrate the suberized layers of the endodermis or the peculiar, thickened cells produced in *Eucalyptus* roots in response to infection (Chilvers and Pryor, 1965). Failure to penetrate these barriers suggests that the fungus is only very weakly pathogenic despite reports that it could sometimes assume the role of a virulent pathogen (Jackson, 1947; Feliciani and Montefiore, 1954).

Further changes in root structure occur after infection has taken place and the tissues of the root undergo alteration, both in appearance and function (Clowes, 1951; MacDougal and Dufrenoy, 1944; Foster and Marks, 1966, 1967). Harley (1969) attempted to define this altered state. He considered that a mycorrhiza was one facet of an evolutionary state of root-inhabiting relationships. These relationships are primarily based on nutritive requirements of the fungus and the ability of the fungus to withstand the defensive mechanisms of the host. A Hartig net is the visual evidence of this close association. In practice, however, it is difficult to distinguish the Hartig net formed in a symbiotic relationship from intercellular infections by parasitic fungi. The most striking difference between the two lies in the ability of the former to prolong the life of adjacent cortical cells (Aldrich-Blake, 1930; Hatch, 1937; Robertson, 1954; Marks, 1965; Wilcox, 1968). The ability of a mycorrhizal fungus to prolong the life of the cortical cells is best observed where the cortex has been partially isolated by formation of a secondary endodermis and some of the cells are invested with mycorrhizal fungi. Parasitic fungi accelerate destruction of the cortex which shrivels and becomes discolored in the process. On the basis of these observations it appears more reasonable and accurate to define a mycorrhizal association as a partnership between a fungus and root in which intercellular penetration of the cortex occurs and the life span of the host cell is prolonged. This definition automatically excludes pseudomycorrhizae and infection of moribund cortical tissues by mycorrhizal fungi.

C. Factors Affecting Infection

Since there is little doubt that root growth and development are intimately associated with processes of mycorrhizal formation, it will be profitable to review briefly the morphogenic effects of some of the treatments used in experimental studies of mycorrhizal formation. This aspect of the problem has not been examined critically. It is however, beyond the scope of this review to examine all factors involved, and attention will be paid to those with well-documented effects.

Fertilizers rapidly stimulate initiation of new roots and increase growth rate of existing ones (Lyr and Hoffmann, 1967; Loub, 1963; Zottl, 1964; Kozlowski, 1971). In some instances the production of lateral roots is inhibited (Mitchell *et al.*, 1937) probably owing to the strong apical dominance of very fast-growing roots. Boron-deficient trees produce very poor root systems (Treshow, 1970), and moisture stresses produce similar results (Squire, 1972). Calcium deficiency produces aberrations in the cortical cells and retards lignification of xylem cells (Davies, 1949). Thus, reports on the role of nutrients and their deficiencies on mycorrhizal formation (Björkman, 1942; Morrison, 1956; Dumbroff, 1968) might require further investigation because of their direct effects on root structure.

Light and mycorrhizal formation have been so intimately associated that it is impossible to consider the problem of mycorrhizal morphogenesis without examining the direct and indirect effects of light on root development (Boullard, 1960). Reduction in illumination to 20% of full daylight greatly reduced the number of new roots produced (Hoffmann, 1967). In *Pinus densiflora* root length was independent of photoperiod and depended only on total light intensity (Kinugawa, 1965). It is generally agreed that reduction in illumination by 80–85% of daylight intensities virtually stops root growth (Elisson, 1968; Brown, 1955; Richardson, 1956; Anonymous, 1953), and there is evidence suggesting that in some species root growth which occurs immediately after reduction of shoot illumination is dependent on reserves of carbohydrate in the root (Richardson, 1953, 1956). Richardson also provides strong evidence that root growth requires a stimulus from the shoot. Girdling, decapitation, and defoliation consequently stop root growth. Thus the formation of new root surface for mycorrhizal colonization is directly affected by the interaction between shoot and light.

The indirect effects of light on mycorrhizae have provoked considerable controversy. Björkman (1942, 1944, 1949) pioneered work in this field and concluded that mycorrhizal formation required a surplus of carbohydrates in the roots. Although described elsewhere in this book in detail, it is worth recapitulating some aspects. Björkman stated that the inhibition of the mycorrhizal condition in roots was closely connected with a shortage of soluble carbohydrates in the roots. As evidence he showed that under conditions of greatly reduced illumination mycorrhizal formation was inhibited. When movement of photosynthate into the root was stopped by girdling the stem above ground level, mycorrhizal formation did not take place, irrespective of the level of illumination. There is little doubt that carbohydrates in the root are utilized by fungal hyphae of the Hartig net (Melin and Nilsson, 1959; Foster and Marks, 1966, 1967; Reid, 1968). However, the role played by "surplus carbohydrates" in mycorrhizal formation has been strongly disputed (Meyer, 1962; Schweers and Myer, 1970; Harley, 1969; Handley and Sanders, 1962; Richards and Wilson, 1963). When the evidence for and against Björkman's carbohydrate theory is weighed along with the striking morphogenic effects of the treatments he used in supporting his theory, it becomes increasingly difficult to accept, especially since girdling and reduced illumination greatly retard root growth, virtually stopping formation of new roots whether they become mycorrhizal or not.

Temperature affects mycorrhizal morphogenesis and structure in a manner somewhat similar to light intensity. Temperature has a direct effect on the rate of root growth and on production of new root tips (Barney, 1951). At temperatures of about 12°C root growth virtually stops (Lyr and Hoffmann, 1967), and it is reasonable to expect that mycorrhizal formation will follow similar trends. At 35°C, Barney (1951) observed that the rate of root growth decreased and maturation and suberization of the root tips increased greatly. The interaction between temperature and root maturation in turn could affect the infection process via formation of the MIZ. There is considerable variation in the reaction of individual roots to changes in temperature (Mullin, 1963). This may affect the overall reaction of a large root mass to temperature changes.

Slight changes in temperature have a direct effect on mycorrhizal structure. Redmond (1955) observed that either decreasing or increasing soil temperatures by 2°C resulted in changes in the thickness of the fungal mantle. Marx *et al.* (1970) found in controlled experiments that this type of reaction depended on the species of fungus. *Thelophora terrestris* produced mycorrhizae with *Pinus taeda* which were more sensitive to fluctuations in temperature than mycorrhizae produced by

Psiolithus tinctorius. Theodorou and Bowen (1971) observed a rapid decline in mycorrhizal production in *Pinus radiata* as temperature was decreased below 20°C, possibly owing to poor colonization at the lower temperatures.

Mycorrhizal formation and types of associations formed in the soil are influenced by soil moisture and aeration. Fassi (1967) reported that mycorrhizal formation was inhibited by heavy clay soils. The inhibition was attributed directly to failure of the roots to penetrate the soil. Mycorrhizal initiation could also have been influenced by poor aeration, when mycorrhizae are known to respire rapidly (Mikola, 1967), and are strongly aerobic (Harley, 1969). Water deficiency in the plant stops root growth. Root suberization is accelerated in dry soils, and this condition can sometimes lead to a change in the fungal partner (Marks and Foster, 1967), especially as some fungi such as *Cenococcum graniforme* are favored by dry soils (Meyer, 1964).

The role played by auxin in mycorrhizal formation is reviewed in Chapter 7, and it is not our intention to cover the subject here. However, the effect that auxin has on root growth and indirectly on formation of the MIZ will be mentioned briefly. One of the most obvious effects of auxin is that observed on growth rates and branching habit which are directly related to the concentration of external auxins (Slankis, 1958), and the frequency with which short roots are produced is determined by auxin applied to the root. Furthermore, the effects of auxin and light are directly related (Slankis, 1961), an important observation which requires considerably more attention than it has received in the past.

IV. Structure of Mycorrhizae

Detailed aspects of structure as pertaining to classification of mycorrhizae are dealt with in Chapter 2 and only the generalized features of mycorrhizal roots will be presented here. The mycorrhizal habit is much more widespread than believed initially. Mycorrhizae form on a variety of tree and shrub species. Mycorrhizal formation appears to be the rule rather than the exception among forest trees. A variety of fungi are associated with tree roots and extensive lists of associations have been compiled (Trappe, 1962). Most mycorrhizal fungi belong to the hymenomycetes and gasteromycetes. Only a few ascomycetous fungi have been reported (Sappa, 1940; Lihnell, 1942) forming mycorrhizae with trees.

The gross morphology and structural characteristics of mycorrhizae are sufficiently characteristic for them to be grouped into classes. In broadest terms, there are three classes, the ectomycorrhizae, ectendo-mycorrhizae, and endomycorrhizae. Two other types may be included but their niche in the classification system is disputed. These are the peritrophic associations and pseudomycorrhizae. It appears that there is some doubt as to the immutability of these groups. Laiho (1967) observed that young seedlings produced ectendomycorrhizae but when transplanted into the field the new associations were ectomycorrhizae. An even more disturbing aspect of these associations was illustrated by Marks *et al.* (1967) when they demonstrated that intracellular penetration of cortical cells (i.e., the ectendomycorrhizal condition) was common in aging mycorrhizae. Until further information on mycorrhizal development is forthcoming, these groupings must be considered arbitrary, although it appears they may reflect the host/fungal relationships that exist.

A. Peritrophic Associations

The term, peritrophic association, was coined by Jahn (1934) to describe fungi found on root surfaces. These include saprophytes, potential parasites, and mycorrhiza formers. It is possible that in the initial stages of mycorrhizal formation the thin weft of hyphae formed on the root surface prior to Hartig net formation could represent a special type of relationship between fungus and root. Foster and Marks (1967) found that a variety of fungal and bacterial forms can be seen closely associated with the outer hyphae of the mycorrhizal mantle. These associations could have special significance in mobilization of nutrients and nutrient absorption by roots.

B. Ectomycorrhizae

These represent one of the commonest forms of mycorrhizal association encountered and are the dominant forms in most forest trees. The fungus forms a mantle of hyphae covering the slow-growing, unsuberized parts of the roots, and it penetrates, intercellularly, to varying depths, into the host cortex. The hyphae never penetrate the endodermal tissues and the stele. Mycorrhizal roots lack root caps, possibly because the cells of this tissue have been incorporated into the tannin cells of the outer cortex and not sloughed off as in uninfected roots (Clowes, 1951; Chilvers, 1968).

Mycorrhizal roots differ in gross morphology from uninfected roots of similar age. They are thicker, more brittle, and usually colored differently. In *Abies, Fagus,* and *Eucalyptus* the roots are pinnately and racemosely branched (Harley, 1969; Trappe, 1967; Chilvers and Pryor,

1965), while in *Pinus* they are dichtomously branched. In roots in which dichtomy is repetitious and apical growth extremely slow, the mycorrhiza assumes the form of a nodule (Melin, 1923). The color observed is usually influenced by the tannin layer beneath the mantle. Marks (1965) described a mycorrhiza that had a pale, blue-green mantle when viewed in cross section and a dark, greenish-brown appearance when examined in reflected light. Thus the color of a mycorrhiza will vary with the method of examination and with time.

The mantle of mycorrhizae is a highly variable structure varying from loose wefts of hyphae covering the root to dense pseudoparenchyma with a firm surface. A variety of fungal structures with hairs, cystidia, and rhizomorphs may sometimes form in and on the outer mantle surface.

The rhizosphere of the mycorrhiza, or mycorrhizosphere, is discussed in detail (see Chapters 4 and 8). It is rich in bacteria, diatoms, actinomycetes, and other fungi. The bacteria are intimately associated with the hyphal cells of the mantle and probably form a nutritional relationship with the fungus (Foster and Marks, 1967). Pockets of lysed hyphae are sometimes seen in the outer mantle and these are packed with bacterial cells. Surface views of the mantle show that structurally different hyphae run along the surface. Thus the surface of the mantle provides a complex ecological niche for soil-inhabiting organisms. The role of the hyphae in nutrition, protection, and survival of the root is uncertain, and this is an area which could result in productive research.

The superficial layer of cells of the root lying adjacent to the mantle is frequently impregnated with dark-staining material consisting of tannins (MacDougal and Dufrenoy, 1946). The origin of this layer is uncertain. The hyphae from the mantle penetrate between the cells of this layer and then enter the intercellular spaces of the cortical cells. The hyphae pack closely in the intercellular space to form the Hartig net (named after Robert Hartig). The form and appearance of the Hartig net are characteristic, varying from a barely discernible net of fine hyphae that form bead-like chains in the intercellular spaces to coarse structures which stand out prominently. These features have been used to classify and identify mycorrhizae.

C. Pseudomycorrhizae

This term was first used by Melin (1917) to describe intracellular root infections in which the root did not produce the dichotomous branching characteristic of true mycorrhizae (Levisohn, 1954). Usually the root was invested with a fungal sheath which rarely became pseudoparenchymatous, and the intercellular hyphae were coarse and thick-walled. Intracellular penetration occurred, and the lumens of the cells were filled with hyphae. This condition was observed in gymnosperms with physiological disorders such as fused needles (Young, 1940) and in seedlings grown in old forest nurseries, raised under minimal light conditions or in soils which were grossly deficient in nutrients (Harley and Waid, 1955). However, pseudomycorrhizae have also been observed on trees growing in fertile soils (Marks, 1965), and intracellular penetration that is characteristic of this association can also occur in ectendomycorrhizae.

D. Ectendomycorrhizae

Like pseudomycorrhizal associations, these are characterized by a thin mantle, coarse Hartig net, and poor dichotomy of the infected root. Levisohn (1954) believed that these associations harmed the roots and restricted growth, while Laiho (1967) provided evidence that seedlings with ectendomycorrhizae could grow better than uninfected ones. Björkman (1949) suspected that in these associations the balance was tipped in favor of the fungus, and Marks and Foster (1967) observed that intracellular penetration could occur more frequently in moribund mycorrhizae where the cortical cells were collapsing. Thus, as Harley (1969) suggested, the differences that exist between ectomycorrhizae and ectendomycorrhizae may have been overemphasized because the range in the variability of mycorrhizal associations has not been investigated completely.

E. Mixed Mycorrhizal Associations

There is considerable evidence that mycorrhizal partnership can be formed between a single tree species and a number of different fungi (Zak and Marx, 1964). This is not altogether unexpected because not only can several species of basidiomycetes be used to form mycorrhizae in culture experiments, but also a number of types of mycorrhizae can be obtained from the one tree root under natural conditions. However, there are very few reports of more than one partnership forming on the same lateral root. Marks and Foster (1967) described situations where one mycorrhizal type was replaced by another when root growth was resumed after a dormant period. These observations were made possible by the distinctive appearance of the two mycorrhizal types examined. However, the frequency with which this occurred was low. Their observations suggest that a mycorrhiza is a stable entity and that the metacutized layer formed in the dormant root is a barrier to secondary infection of the new tissue, according to Robertson (1954).

F. Senescence in Mycorrhizae

Although mycorrhizal infection prolongs the life of roots, there is little information in the literature on the structural changes associated with aging and decay of mycorrhizae (Mikola and Laiho, 1962). Mycorrhizae are ephemeral structures that are discarded when secondary growth takes place. Marks *et al.* (1967) noted that as the mycorrhizae of *Pinus radiata* aged, the mantle shriveled and the roots darkened in color. This was associated with an increase in the amounts of dark-brown-colored deposits in cortical cells and an increase in intensity of staining in the tannin layer. In many cases, intracellular infection was observed, the central stele remaining intact, protected possibly by the endodermis. In the case of one unnamed blue-green species, the aging process was faster in the cortical cells immediately behind the apex, shriveling soon after infection took place. This condition was associated with intracellular infection, possibly owing to the greater pathogenicity of the fungal partner.

Some interesting conclusions emerge from this examination of the literature on mycorrhizal structure and morphogenesis. Harley (1969) made a plea for an integrated study of environmental factors, the role of hormones, light, and nutrients on the formation of mycorrhizae. He recognized the value and significance of Wilcox's work on mycorrhizal morphogenesis. The concept that mycorrhizal formation is dependent, in part, on the formation of a MIZ where optimal conditions occur for fungal penetration and establishment is but an extension of his ideas. It appears that this concept depends, *inter alia,* on subjecting *living* cortical tissue to maximum exposure to the fungal inoculum. Consequently, in roots that have stopped growing, and in very fast-growing roots, the MIZ will be either very restricted in size or exposed to infection for a much shorter period than in slow growing roots and those that show slow, secondary differentiation. On this basis, environmental factors that affect rates of root growth will not only have an indirect effect on host physiology, but also a direct effect on the infection process and hence mycorrhizal formation. This would help to explain the well-documented effects of nutrients, light, plant hormones, etc., on mycorrhizal formation. A simplified scheme for the effect of environmental factors on mycorrhizal formation based on this type of information is presented in Fig. 3. It will be noticed that once mycorrhizae are formed, they too will begin to influence the process of further mycorrhizal formation, especially in soils severely depleted of nutrients or in which nutrients are unavailable. It is unnecessary to emphasize that once a mycorrhizal condition is established and maintained at a satisfactory level by new root growth, improvement will be noticed in the shoots.

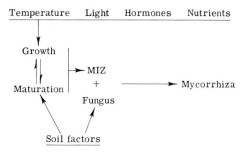

Fɪɢ. 3. Possible relationships of some of the factors that influence the formation of the mycorrhizal infection zone (MIZ).

V. The Ultrastructure of Ectomycorrhizae

Chilvers (1968) examined an angiosperm association in *Eucalyptus,* and Foster and Marks (1966, 1967), Scannerini (1968), and Hofsten (1969) studied gymnosperm associations (*Pinus* sp.). Chilvers (1968) published one low-power composite electron micrograph illustrating the general interrelationships of the fungal and angiosperm associates but he was unable to demonstrate the fine cytoplasmic details of either symbiont. Scannerini (1968) described the ultrastructure of a mycorrhizal association between the ascomycete *Tuber albidum* and *Pinus strobus* and showed that the cytological details were very similar to the white-type basidiomycetous association of *Pinus radiata* described by Foster and Marks (1966, 1967). Hofsten (1969) looked at a basidiomycetous association of *Pinus sylvestris.* This work was directly comparable with that of Foster and Marks on *Pinus radiata* and *Pseudotsuga menziesii.* The identity of the basidiomycetous fungi of Foster and Marks and of Hofsten are not known.

Despite the use of different host species, of different ages, and of material collected at different times of the year, and the fact that both basidiomycetous and ascomycetous associations have been examined, there is remarkable agreement as to the general cytological details of ectomycorrhizal associations. These will be discussed under four headings, the rhizosphere, the mantle, the "tannin layer," and the Hartig net.

A. The Rhizosphere

Neither Hofsten (1969) nor Scannerini (1968) dealt with the rhizosphere, although Foster and Marks (1966, 1967) and Marx (1969) had shown how closely rhizosphere and mantle morphology and physiology are related (Figs. 4–11). Foster and Marks (1967) published electron micrographs (Fig. 4) and micrographs of carbon replicas (Figs. 5

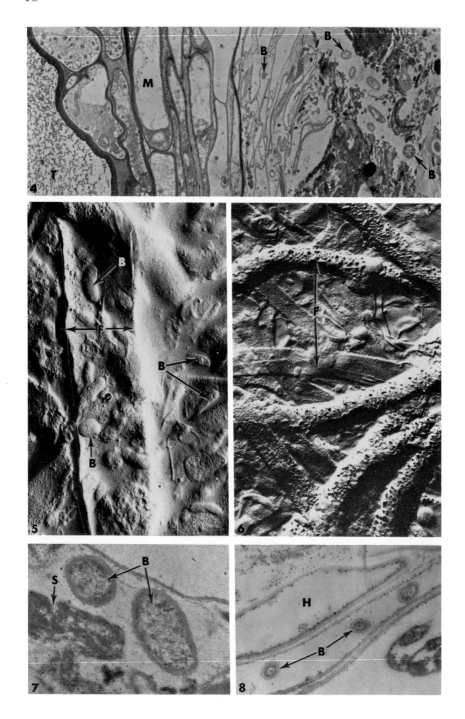

and 6) of the outer layers of short mycorrhizal roots of *Pinus radiata* which had been gently washed and showed that bacteria were closely associated with the hyphae of the outer mantle, whether it was loosely interwoven or pseudoparenchymatous. The bacteria occurred singly or in groups, both in the adjacent soil (Figs. 7 and 8) and adpressed to the surface of hyphae. In ultrathin sections through the rhizosphere (Fig. 9) the bacteria appeared to be separated from each other, from soil particles, and from the mantle hyphae by an electron-transparent slime layer. The bacteria common to the various mycorrhizal associations could be distinguished morphologically. Those from the white-type mycorrhizae were usually ovoid, 3×0.6 nm, whereas those from red-type mycorrhizae were more elongated. Oswald and Ferchau (1966) showed that particular bacterial forms may be associated with mycorrhizal roots. The details of the mycorrhizosphere are discussed in Chapter 8 and will not be dealt with here. Only the structural details will be described to provide a picture of the relationships which exist between the soil particles, bacteria, and mycorrhizal fungi. Zones of bacteria are formed around the root. In *Pinus radiata* the bacteria may reach a concentration of 200×10^9 per cm^3 in the most densely populated zone between 8 and 12 nm from the host surface (Foster and Marks, 1967). Groups of bacteria are often associated with lysed hyphae of the outer mantle (Foster and Marks, 1966, 1967) but their lytic action on both mantle and roots is not well established except in a few instances (Wood, 1967).

Marx (1969) showed that bacterial secretions play an essential part in the life history of root pathogens, and Zak (1964) suggested that their secretions may act as a chemical barrier to some root pathogens. The presence of rhizosphere bacteria could enhance the uptake of phosphorus by the host (Bowen and Rovira, 1966). Nitrogen accretion, which may occur in conifer forests (Stephenson, 1959), could also be

---

FIGS. 4–8. RHIZOSPHERE.

FIG. 4. Section of red-type mycorrhiza of *Pinus radiata* showing tannin cell (T) of host on the left, living mantle hyphae (M) with glycogen granules, crushed hyphae with no contents, and numerous bacteria (B) between the soil particles. Glutaraldehyde/OsO₄ fixation. Reynolds lead stain.

FIG. 5. Indirect metal shadowed carbon replica of surface of white mycorrhiza of *Pinus radiata*. Numerous bacteria (B) are closely associated with the fungal hyphae (F).

FIG. 6. Metal shadowed indirect carbon replica of black mycorrhiza of *Pinus radiata* showing hyphae (F) with characteristic surface.

FIG. 7. Bacteria (B) from a red-type *Pinus radiata* mycorrhiza rhizosphere; soil (S) particles are visible.

FIG. 8. Bacteria (B) inside empty hyphae of mantle; Hartig net (H) is visible.

associated with activity of bacteria in the mycorrhizosphere (Richards and Voigt, 1964; Hewett, 1966) rather than with activity of mycorrhizal fungi or with host roots (Harley, 1969). Hassouma and Wareing (1964) showed that rhizosphere bacteria can fix nitrogen in the presence of mannitol, and Lewis and Harley (1965a) observed that this is one of the carbohydrates secreted by beech mycorrhizae. This evidence suggests that bacteria may play an active role in the mycorrhizal complex of higher plants as they do on their leaves (Wichner and Libbert, 1968).

B. The Mantle

Only Scannerini and Foster and Marks have provided detailed descriptions of fine structure of the mantle (Figs. 12–15). The cytological details of different types of mantle in both *Pinus radiata* (Figs. 12 and 13) and *Pseudotsuga menziesii* (Figs. 14 and 15) are sufficiently characteristic to allow them to be distinguished at the ultrastructural level. In general, the fine structure of the mantle is similar to that of basidiomycetous hyphae described in the literature, or as in the case of *Tuber albidum,* to hyphae from the carpophore of the same species (Scannerini, 1968). Foster and Marks (1967) showed that the cell walls of the mantle hyphae were covered with an amorphous layer which on mild maceration revealed two layers of microfibrils, the inner one being more organized than the outer as is common in fungi of many groups (Aronson and Preston, 1960; Fuller and Barshad, 1960; Hawker, 1965). In basidiomycetous mycorrhizae the hyphae showed the typical dolipore septae (Moore and McAlear, 1962; Bracker and Butler, 1963) in both the mantle (Foster and Marks, 1966, 1967) and in the Hartig net (Foster and Marks, 1966; Hofsten, 1969). In *Pinus radiata* and *Pseudotsuga menziesii* the hyphae of the outer mantle were usually devoid of cytoplasmic contents and were sometimes partially collapsed and transversely stretched, whereas those of the inner mantle were more rounded and richly cytoplasmic. In both species the carbohydrate reserves were glycogen and occasional oil droplets. The number of glycogen granules per cell increased as the Hartig net was approached. The cells of the outer mantle had no granules while those of the inner mantle were almost filled. In *Tuber albidum* mantles, Scannerini distinguished two types of hyphae which he showed also occurred in the carpophore. These were hyphae characteristic of mycelia undergoing rapid cytoplasmic synthesis and

---

Fig. 9. RHIZOSPHERE. The rhizosphere of a red type *Pinus radiata* mycorrhiza showing the tannin-filled cells of the host on the left, the living cells of the fungal mantle (M) enclosed by rather transversely stretched dead hyphae. Outside this is a layer of soil very rich in bacteria. Arrows indicate bacteria.

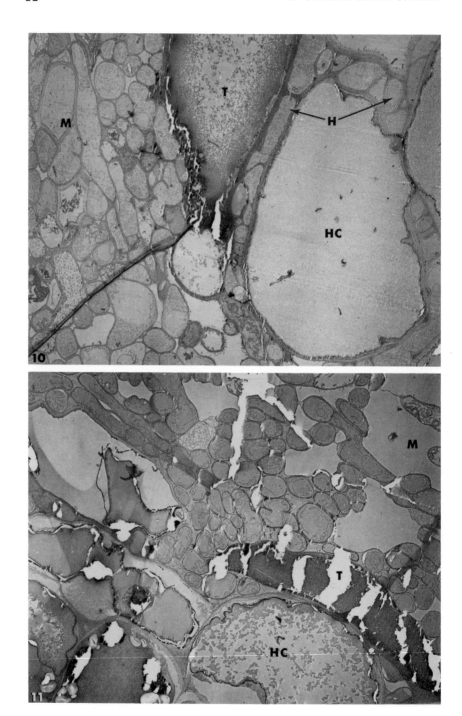

hyphae specialized for storage and transport of reserves. Although this clear functional differentiation was not observed in basidiomycete mycorrhizae, particular lobes of some hyphae were packed with glycogen granules, and they probably correspond to the hyphae *sia forme a funzione di conduzione di riserva* of Scannerini.

In individual cells of basidiomycete mycorrhizae two nuclei may be seen, confirming the dicaryotic nature of the mycelium (Foster and Marks, 1966). Occasionally, nuclei in the process of division and of migration through simple septae were observed. These were perhaps associated with formation of clamp connections which were frequently visible in hyphae isolated from the mantle (Foster and Marks, 1967). The hyphae contained mitochondria and profiles of endoplasmic reticulum, but lomasomes were not as common in the mantle as in the Hartig net. In some pseudoparenchymatous mantles, vesicles were secreted through the plasmalemmas and cell walls of mantle hyphae but the nature of their contents is unknown.

The inner mantle was characterized by more closely interwoven hyphae (Foster and Marks, 1966) and an increase in the number of cytoplasmic organelles as noted by Scannerini (1968) and an increase in the concentration of glycogen granules (Foster and Marks, 1966). In some cases, hyphae close to the tannin layer were immersed in an electron-dense, amorphous material similar in appearance to contents of host cells of the tannin layer. This is also common in the ectendotropic mycorrhizae of *Araucaria cunninghamii* (Foster, 1972). However, these hyphae had the normal cytology and morphology of the mantle. The increase in concentration of glycogen granules in the inner regions of the mantle suggests that the carbohydrates were derived from the host rather than from the leaf litter as suggested by Young (1940).

C. The "Tannin Layer"

The function and origin of the tannin layer has been the center of some controversy. This consists of a layer of cells which are naturally a dark-brown color and which are densely stained after normal methods of fixation for electron microscopy (Figs. 16 and 18). Foster and Marks (1966) showed that uninfected roots contained well-developed tannin-

---

Figs. 10–11. *Pseudotsuga menziesii* MYCORRHIZAE.

Fig. 10. Section of a mycorrhiza from a healthy nursery seedling showing well-developed Hartig net (H), host cell of Hartig net (HC), isolated tannin cell (T), and the mantle cells (M) filled with glycogen granules.

Fig. 11. Section of yellow stunted seedling growing adjacent to that in Fig. 10 showing absence of Hartig net and mantle hyphae (M) devoid of glycogen. Tannin cell (T) and host cell (HC) are visible.

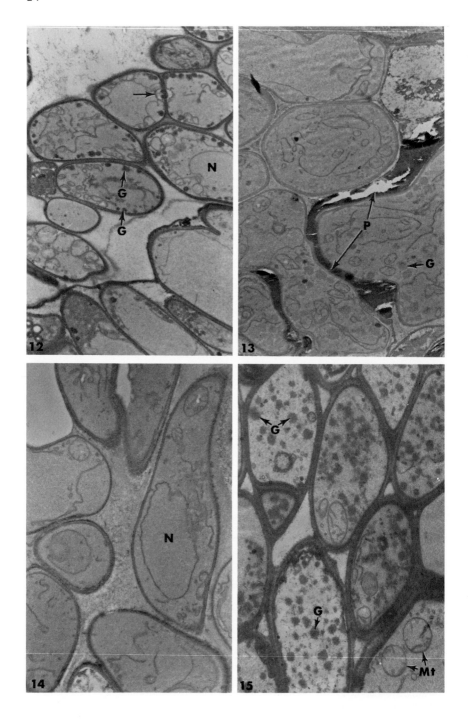

iferous vacuoles, especially in the outer cortex and epidermal cells, and that these cells retained their cytoplasmic contents throughout the life of the cortex. Such tanniniferous cells occurred even in the radicle of the ungerminated embryo as well as in sterile roots (Foster, 1972), therefore, as Scannerini correctly observed, they are not the result of fungal infection. However, in mycorrhizae there is a well-developed layer of brown-colored cells, usually only one or two cells thick which take up both OsO_4 and $KMnO_4$ to a remarkable degree (Foster and Marks, 1966; Scannerini, 1968). Owing to its hardness after fixation and dehydration, this tannin layer presents severe technical difficulties in embedding and sectioning specimens for electron microscopy (Foster and Marks, 1966, 1967; Chilvers, 1968; Hofsten, 1969). These cells contained no recognizable organelles in mature mycorrhizae and are considered in angiosperm mycorrhizae to be root cap cells which have become included in the mantle (Clowes, 1954; Chilvers, 1968). Chilvers stated that hyphae enter these cells so that their electron-dense contents become subdivided. He considered that the "tannin-like" materials of these cells originated in the same way as the tannins of the inner cortex. However, his electron micrographs and those of Foster and Marks (1966) showed that the physical organization of the electron-dense deposits in the "tanniniferous cells" and those of the cortical cells may be quite different. Unfortunately, so many substances take up the normal electron-dense materials used in electron microscopy (Foster *et al.*, 1971), and chemical analyses of mycorrhizae such as those of Hillis *et al.* (1968) do not differentiate between the contents of the various tissues of the association. It is possible therefore that in angiosperm mycorrhizae the electron-dense cells, through which hyphae are able to penetrate, are indeed root cap cells which are filled with polyuronides and polysaccharides such as hemicelluloses rather than polyphenols. Polyuronides and hemicelluloses are known to be osmiophilic

---

Figs. 12–15. MANTLE.

Fig. 12. Hyphae from white mycorrhiza of *Pinus radiata* which are losely interwoven. The cells contain glycogen granules (G). Nuclei (N) are visible. Dolipore septae separate the cells (arrow). $KMnO_4$ fixation. Reynolds lead stain.

Fig. 13. Cells from the inner mantle of white-type *Pinus radiata* mycorrhizae. The cells are immersed in a medium which is strongly reactive with heavy metal stains. Glycogen granules (G) abound, and polyphenol deposits (P) are visible. $KMnO_4$ fixation.

Fig. 14. Hyphae from a stunted *P. menziesii* seedling. The cell walls are thin and glycogen granules are absent. Nuclei (N) are visible. $KMnO_4$ fixation. Lead stain.

Fig. 15. Similar area of healthy *P. menziesii* seedling. The cell walls are thicker, and the hyphae contain abundant glycogen granules (G). Mitochondria (Mt) are visible. $KMnO_4$ fixation. Lead stain.

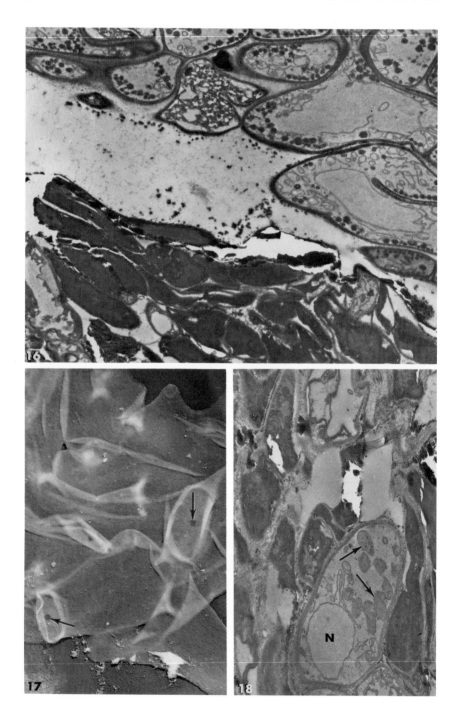

and to be secreted in large amounts by root cap cells (Mollenhauer *et al.*, 1961; Northcote and Pickett-Heaps, 1966). These cells may therefore be quite different in origin and contents from the cells filled with electron-dense materials found elsewhere in mycorrhizae. In mycorrhizae formed by *Tuber albidum* on *Pinus strobus* Scannerini observed that the arrangement and contents of the tannin layer suggested a hypersensitive reaction of the host. In stunted seedlings from nursery beds of *Pseudotsuga menziesii* there was a layer, several cells thick, of electron-dense cells through which there was no penetration of hyphae, and in these cases, there was no glycogen in the mantle hyphae, the fungal cells had thin walls, and there was no Hartig net (Fig. 14) (Foster and Marks, 1972). This was in marked contrast to healthy seedlings (Fig. 15) growing alongside the stunted ones of the same species, in which the electron-dense cells were widely scattered and fungal hyphae readily entered the cortex of the host to form a well-developed Hartig net. In these cases, there was transfer of carbohydrate between host and fungus, and the mantle cells were filled with glycogen.

Foster and Marks (1967) showed that mycorrhizal hyphae, isolated by gently macerating the tannin layer, were wide, distorted, and lobed. The cells were smaller than those of the mantle, owing to frequent septation. However, although the microfibrillar arrangements in the cell wall were normal, some of the hyphae appeared distorted and crushed and were filled with electron-dense material. These moribund cells were separated from quite normal ones by dolipore septae. The normal cells also contained some unusual organelles (Fig. 18), probably modified lomasomes, as well as more frequent mitochondria and endoplasmic reticulum, and they were filled with glycogen. Foster and Marks suggested that these morphological and cytological effects were due to the toxic nature of the polyphenols released by the host. Scannerini (1968) however, showed that in the ascomycete association on *Pinus strobus*, although there were cells containing electron-dense deposits in the tannin layer, there also were similar lobed hyphae in the carpophore, so that their structure fell within the normal range of morphological variation, and he considered they had no particular significance in the

---

Figs. 16–18. TANNIN ZONE.

Fig. 16. Mantle-tannin transition zone in white-type *Pinus radiata* mycorrhiza. Large glycogen-filled hyphae of inner mantle at top and the electron-dense smaller hyphae of the tannin zone below. $KMnO_4$ fixation. Lead stain.

Fig. 17. Whole-mounted, lightly extracted hyphae from tannin zone. The hyphae are lobed. Note central pore in septae (arrow). Metal shadowed.

Fig. 18. Detail of tannin zone hyphae. Nucleus (N) is visible. Occasional hyphae are less densely stained. Note unusual organelles (arrows). $KMnO_4$ lead stain.

tannin layer. Scannerini showed that these electron-dense cells differed from moribund cells in the carpophore in that in the tannin layer necrotic cells became completely impregnated with electron-dense materials, while those of the carpophore merely lost their cytoplasmic contents. This suggested that an unusual process takes place in the tannin layer during cell necrosis.

Hofsten (1969) implied that the presence of these electron-dense materials in the hyphae of the tannin layer indicated that these associations were not truly symbiotic. However, these unusual cells were not present in mycorrhizae of healthy *Pseudotsuga menziesii* trees and were relatively uncommon in mycorrhizae of forest trees. The formation of a Hartig net and evidence for carbohydrate transfer involving minimal cytoplasmic reaction between host and fungus strongly suggest that the roots examined by Foster and Marks were truly symbiotic mycorrhizal associations. Hess (1969) showed that it is characteristic of some fungi to accumulate electron-dense materials, so that the presence of these materials may have no special significance.

Hillis *et al.* (1968) found that the mycorrhizae of *Pinus radiata* and *Pseudotsuga menziesii* contain both polyphenols and resins as well as other compounds known to be fungistatic. It is also known that polyphenols leak out of healthy, sterile *Pinus radiata* roots grown on agar. Foster and Marks (1966, 1967) suggested that tannins secreted by the host may act as a biological screen that selects only fungal species which can tolerate these compounds and form a symbiotic relationship with the host. Polyphenolic compounds are well known to be toxic to fungi (Wood, 1967), and in some hosts differences between disease-resistant and disease-susceptible varieties can be traced to differences in polyphenol production. Polyphenols exert their protective function by selective inhibition of fungal enzymes (Mandels and Reese, 1963) or, in the case of the hypersensitive reaction, the host cells having secreted polyphenols die, thus isolating the pathogens from further potential nutrient sources in the host. Foster and Marks (1966, 1967) showed that fungal hyphae from the tannin region differed in both morphology and cytology from those of the Hartig net in the same section of root and that these changes were confined to the tannin layer. This suggests that these abnormalities were due to free polyphenols released from the tannin cells of the host.

In addition to the electron dense and normal hyphae, Scannerini again recognized hyphae rich in cytoplasm and others modified for storage and transport of reserves. This differentiation may be an ascomycete characteristic which is not so well developed in basidiomycetes. Scannerini also observed that the hyphae nearer the host cells were more richly cytoplasmic than those of the outer mantle. This is also the case in basidiomycetous mycorrhizae, although the effect is not so easy to observe owing to the large amounts of glycogen present. There may be important seasonal changes in the amount of storage materials which could account for the differences observed in the various associations which have been examined.

D. The Hartig Net

The presence of a well-developed Hartig net is assumed to be the distinguishing feature of true mycorrhizae. Associations in which the fungal mantle is formed without intercellular penetration are considered to be asymbiotic. The reasons for these assumptions are based on the observations that this is the region where the host and fungus come into close contact and where nutrient exchange takes place between the two symbionts. In all the mycorrhizal associations of *Pinus radiata, Pseudotsuga menziesii*, and *Pinus strobus* described in the literature, the fungal hyphae are confined to the intercellular spaces. Only in old ectomycorrhizae and in the zone at the base of the root where the host cortex is breaking down did fungal hyphae enter the cortical cells of the host. They never penetrated as far as the endodermis which was invariably filled with tannins. Hofsten (1969) reported that haustoria entered some cells of the cortex of *Pinus sylvestris* but these mycorrhizae were collected from old trees and at a different season of the year from those of Foster and Marks. Hofsten rightly stressed that the precise nature of the association may vary with seasonal physiological changes in the host. In basidiomycete associations there is no zone of apposition between the cells of the host and the fungus, a characteristic feature of parasitic associations, and the cells of the host are relatively unmodified, cytologically. They may be considerably enlarged in angiosperm mycorrhizae (Harley, 1969; Chilvers and Pryor, 1965). In some cases, however, Foster and Marks (1966) reported that the outer cortical cells appeared to have enlarged nuclei.

In most cells of *Pinus radiata* (Fig. 19) and *Pseudotsuga menziesii* (Fig. 20) there was a thin layer of cytoplasm with a few mitochondria, Golgi complexes, and profiles of endoplasmic reticulum typical of cortical cells of uninfected roots (Foster and Marks, 1966). The tannin deposits were limited to a thin layer lining the tonoplast (Figs. 19 and 20). Similar situations occurred in uninfected roots and in cortical cells of mycorrhizae remote from fungal hyphae (Foster and Marks, 1966, 1967; Scannerini, 1968). Thus the host cells appeared to be relatively unaffected by the presence of the fungus. In *Pinus sylvestris*, Hofsten (1969) also reported that the host cells contained few tannin par-

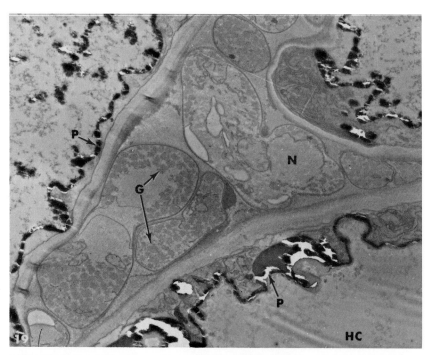

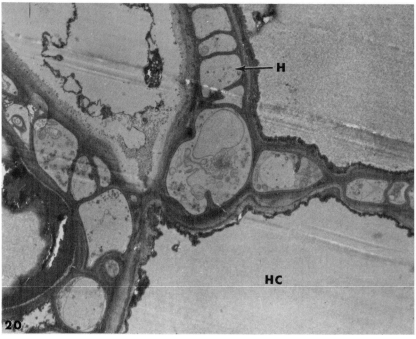

ticles, but the cytoplasm had the appearance of high metabolic activity as indicated by the presence of many mitochondria and plastids.

The fungal partner in basidiomycete mycorrhizae appeared to be very active. In *Pinus radiata* the hyphae were small, isodiametric, and non-vacuolate in the inner, recently penetrated areas of the cortex, and they contained abundant glycogen. They fitted closely against each other and against the host cell walls. In contrast, in the outer cortex the fungal hyphae were large, highly vacuolated, and round in cross section with large intercellular spaces. They contained numerous organelles, but mitochondria were not conspicuously abundant so that the cells of the Hartig net were similar to those of the mantle in this respect. *Pseudotsuga menziesii* mycorrhizae were characterized by regularity of the transverse septae of the fungal hyphae so that the Hartig net assumed a ladder-like appearance (Fig. 21). These septae were produced by fusion of lateral outgrowths from opposite sides of the hyphae and numerous intermediate stages were observed (Fig. 22). The hyphae of the Hartig net in both *Pinus* and *Pseudotsuga* were filled with glycogen granules although there were occasional oil droplets. The host cells adjacent to the fungal hyphae had small starch grains. This was in marked contrast to the large starch grains seen in the cells remote from fungal hyphae and those in similar positions in uninfected roots (Foster and Marks, 1966; Scannerini, 1968). Scannerini (1968), however, reported that there were no starch grains in the cells of the outer cortex associated with *Tuber albidum,* and he maintained that these outer cortical cells never contained starch. Hofsten (1969), on the other hand, showed that in *Pinus sylvestris* as in *Pinus radiata* the host cells in the Hartig net contained many plastids filled with starch. Both Foster and Marks (1966, 1967) and Hofsten (1969) showed that dolipores occur in the septae of hyphae in the Hartig net, confirming the basidiomycetous nature of the hyphae in this region of the mycorrhiza.

Foster and Marks (1967) showed that when the Hartig net region of *Pinus radiata* is gently macerated, intact host cells may be obtained with the fungal hyphae still adhering to them. In these cases the host cells revealed the underlying microfibrillar network, whereas the fungal cells did not. This suggested that hyphal penetration into the host cortex is initiated by enzymatic degradation of the middle lamella substances

---

FIGS. 19–20. HARTIG NET.

FIG. 19. *Pinus radiata* Hartig net showing host cells (HC) filled with glycogen (G). The host cells show tannin deposits in the vacuoles characteristic of outer cortical cells of these roots. Polyphenol deposits (P) and nuclei (N) are visible.

FIG. 20. *P. menziesii.* The Hartig net (H) shows a characteristic ladder-like appearance. The hyphae contain few glycogen granules and the host cells (HC) show little cytoplasmic detail. Glutaraldehyde/KMnO₄ fixation. Lead stain.

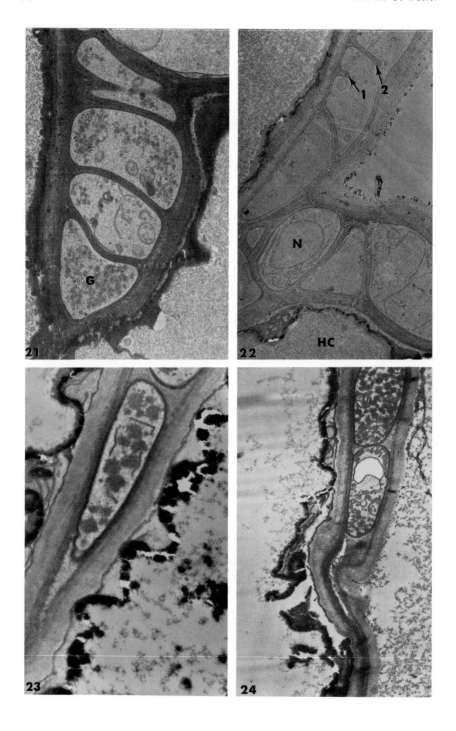

of the host cell walls by secretions from the wedge-shaped hyphae. Hofsten (1969) agrees that this probably occurs in *P. sylvestris* also. Foster and Marks further suggested that penetration is then aided by osmotic swelling as the cells become vacuolated and rounded off so that the newly penetrated cells push the host cells apart mechanically (Figs. 23 and 24). Hofsten agreed that penetration is both mechanical and enzymatic.

Hofsten (1969) elegantly illustrated the netlike structure of this region where the cortical cells were completely enveloped by the fungal cells. It is noteworthy, as both Hofsten and Foster and Marks found, that the cortical cells remained alive for some time, confirming several reports that mycorrhizal fungi prolong the life of the cortical cells.

Foster and Marks (1966, 1967) interpreted the disappearance of starch from plastids of the host cells, as indicated by their shrunken appearance and reduced size of the grains in otherwise cytologically unchanged cells of the host, as being evidence of translocation of carbohydrates from the host to the fungus. Scannerini, however, proposed that the host cells in contact with hyphae never produce starch. He found that in the *Pinus strobus/Tuber albidum* associations, the cell walls of the host broke down to give an electron-dense "involving layer" which he considered to be composed of polyuronides. He believed that it was this material and possibly glycoproteins secreted by the host which constituted the main carbon source of the fungal associate rather than hexoses derived from hydrolysis of host starch. Lewis and Harley (1965a,b,c) showed that hexoses were the usual carbon sources for the fungal components of *Fagus* mycorrhizae, and that glucose and fructose were secreted by the host and were converted to trehalose and mannitol, sugars which are not available to host tissues, by the fungus. The fungal hyphae therefore constituted a carbohydrate sink which led to unidirectional transport of carbohydrates to the fungus. In Hofsten's material, normal starch grains occurred in the host cells. The reasons for this are uncertain

---

Figs. 21–24. HARTIG NET.

Fig. 21. *P. menziesii*. Wedge-shaped hyphae pushing between host cells. The hyphae contain many glycogen granules (G). Mitochondria and endoplasmic reticulum are present.

Fig. 22. *P. menziesii*. The ladder-like appearance apparently caused by subdivision of hyphae by septae growing across the hyphae. Two stages are labeled 1 and 2. Host cell (HC) and nuclei (N) are visible.

Fig. 23. Possible mechanism of penetration into host tissue. Wedge-shaped glycogen-rich hyphae in the middle lamella zone of host. Note apparent lysis of middle lamella ahead of the hypha, *Pinus radiata*. KMnO₄ fixation. Lead stain.

Fig. 24. The hyphae appear to vacuolate, round off, and appear to push the host cells apart. *Pinus radiata*. KMnO₄ fixation. Lead stain.

but it could be a seasonal phenomenon or might be associated with the age of the trees. Horton and Keen (1966) have suggested that the secretion of sugars by the host may repress synthesis of cellulases in fungal pathogens so that as long as the host supplied carbohydrate to the fungus it would be restricted to the intercellular spaces of the host. The role of host polyphenols in this regard cannot be overlooked, especially when it is known that activity of cellulase is more sensitive than that of pectinase to inhibition by the types of polyphenols known to be secreted by pine roots (Lyr, 1961; Hillis *et al.*, 1968).

Several well-established ultrastructural changes in both the host and fungus have been described as characteristic of parasitic associations. The haustoria produced by the fungus are rich in mitochondria (Van Dyke and Hooker, 1969; Ehrlich and Ehrlich, 1963; McKeen *et al.*, 1966; Berlin and Bowen, 1964; Manocha and Shaw, 1967; Bracker, 1967; Chou, 1970) while in the host cells there is normally an increase in endoplasmic reticulum (Berlin and Bowen, 1964; Shaw and Manocha, 1965; Van Dyke and Hooker, 1969; Hess, 1969) and in the number of ribosomes (Shaw and Manocha, 1965; Van Dyke and Hooker, 1969). The number of lomasomes in the association also increases in parasitic associations (Berlin and Bowen, 1964; Manocha and Shaw, 1967; Van Dyke and Hooker, 1969; Ehrlich *et al.*, 1968). Frequently host cells become plasmolysed. The most characteristic features of parasitic associations, however, is the secretion of an electron-dense "zone of apposition" or "involving layer" between the host and fungal tissues (Payton and Bowen, 1963; Berlin and Bowen, 1964; Shaw and Manocha, 1965; Ehrlich *et al.*, 1968; McKeen *et al.*, 1969; Chou, 1970). None of these features were observed in the mycorrhizae of *Pinus radiata* or *Pseudotsuga menziesii* from forest trees, and Hofsten (1969) only mentions an increase in cytoplasmic organelles in *Pinus strobus.*

Foster and Marks (1966) and Hofsten (1969) independently discovered that although there was some slight lysis of the host cell wall, this was restricted to the middle lamella substances in basidiomycete associations. The massive accumulation of electron-dense material that accompanied the alteration in the staining properties of both the host cell walls and those of the fungus observed by Scannerini did not occur in the basidiomycete association. Furthermore, their observations that in basidiomycete associations the majority of the host cells remained alive and retained normal cytoplasmic organization indicated that the fungus and host coexisted with minimal physiological disturbance. This suggests a mutual tolerance typical of a symbiotic relationship. In the ascomycetous association described by Scannerini, the breakdown of the cell walls of the host, the death of the great majority of the host cells of the Hartig net, and the secretion of an electron-dense "involving layer" between the host and fungus, together with the fact that the host cells are said to die in advance of fungal penetration, strongly suggest that either a pathological relation existed between the fungus and the cortical cells or that the fungus had invaded moribund cortical tissues.

The work on the fine structure of mycorrhizae, beset as it is by the technical difficulties caused by the tannin layer, is only beginning to yield results, and it is not possible to provide generalizations from the few studies made so far. It is noteworthy that these independent investigations have yielded such similar results, bearing in mind the diversity of species, the difference in ages of the hosts, and the possible effects of climate, season, and fixation methods. Obviously there is need for more work on the interrelationships in the mycorrhizosphere, the role of phenolic compounds, the methods of intercellular penetration, and the survival of cortical cells, surrounded as they are by fungal hyphae. There is also need, as Scannerini points out, for fine structural studies to be undertaken with known fungal symbionts at various stages in the establishment of the mycorrhizal condition and at different seasons of the year.

References

Aldrich-Blake, R. N. (1930). The root system of the Corsican pine. *Oxford Forest. Mem.* **12**, 1.

Anderson, J. (1967). Investigations on mycorrhizal symbiosis in some Eucalypts in Central Italy. *Pubbl. Cent. Sper. Agr. Forest., Rome* **9**, 81 (translation).

Anonymous. (1953). Die-back of yellow-birch and mycorrhiza. *Rep. Sci. Serv. Dep. Agr. Can* **1952/53**, 11.

Arnold, B. C. (1960). Mycorrhizal infection of germinating seedlings of *Nothofagus solandri* var. *cliffortioides* (Hooke f.) Poole. *Pac. Sci.* **14**, 248.

Aroson, J. M., and Preston, R. D. (1960). An electron microscopic and X-ray analysis of the walls of selected lower phycomycetes. *Proc. Roy. Soc., Ser. B* **152**, 346.

Barney, C. W. (1951). Effects of soil temperature and light intensity on root growth of loblolly pine seedlings. *Plant Physiol.* **26**, 146.

Berlin, J. D., and Bowen, C. C. (1964). The host parasite interface of *Albugo candida* on *Raphanus sativus*. *Amer. J. Bot.* **51**, 445.

Björkman, E. (1942). Conditions favouring the formation of mycorrhizae in pine and spruce. *Symb. Bot. Upsal.* **6**, 1.

Björkman, E. (1944). The effect of strangulation on the formation of mycorrhiza in pine. *Sv. Bot. Tidskr.* **38**, 1.

Björkman, E. (1949). The ecological significance of the ectotrophic mycorrhizal associations in forest trees. *Sv. Bot. Tidskr.* **43**, 223.

Boullard, B. (1957). Etude des mycorrhizes dans le genre *Cedrus*. Première contribution. *Bull. Soc. Mycol. Fr.* **73**, 225.

Boullard, B. (1960). La lumière et les mycorrhizes. *Annee Biol.* [3] 36, 231.

Bowen, G. D., and Rovira, A. D. (1966). Microbial factors in short term phosphate uptake studies with plant roots. *Nature (London)* 211, 665.

Bracker, C. E. (1967). Ultrastructure of fungi. *Annu. Rev. Phytopathol.* 5, 343.

Bracker, C. E. (1968). Ultrastructure of the haustorial apparatus of *Erisiphe graminis* and its relationship to the epidermal cell of barley. *Phytopathology* 58, 12.

Bracker, C. E., and Butler, E. E. (1963). The ultrastructure and development of septa in the hyphae of *Rhizoctonia solani*. *Mycologia* 55, 35.

Brown, J. M. B. (1955). Ecological investigations: Shade and growth of oak seedlings. *Rep. Forest Res.* p. 24.

Chilvers, G. A. (1968). Low power electron microscopy of the root cap region of eucalypt mycorrhizas. *New Phytol.* 67, 663.

Chilvers, G. A., and Pryor, L. D. (1965). The structure of eucalypt mycorrhizas. *Aust. J. Bot.* 13, 245.

Chou, C. K. (1970). An electron microscope study of host penetration and early stages of haustorium formation of *Perenospora parasitica* (Fr.) Tul. on cabbage cotyledons. *Ann. Bot. (London)* [N.S.] 34, 189.

Clowes, F. A. L. (1950). Root apical meristems of *Fagus sylvatica*. *New Phytol.* 49, 248.

Clowes, F. A. L. (1951). The structure of mycorrhizal roots of *Fagus sylvatica*. *New Phytol.* 50, 1.

Clowes, F. A. L. (1954). The root cap of ectotrophic mycorrhizas. *New Phytol.* 53, 525.

Davis, D. E. (1949). Some effects of calcium deficiency on the anatomy of *Pinus taeda*. *Amer. J. Bot.* 36, 276.

Dumbroff, E. B. (1968). Some observations on the effect of nutrient supply on mycorrhizal development of pine. *Plant Soil* 28, 463.

Ehrlich, H. G., and Ehrlich, M. A. (1963). Electron microscopy of the sheath surrounding the haustorium of *Erysiphe graminis*. *Phytopathology* 53, 1378.

Ehrlich, M. A., Schafer, J. F., and Ehrlich, H. G. (1968). Lomasomes in wheat leaves infected by *Puccinia graminis* and *P. recondita*. *Can. J. Bot.* 46, 17.

Elisson, L. (1968). Dependence of root growth on photosynthesis in *Populus tremula*. *Physiol. Plant.* 21, 806.

Fassi, B. (1967). Recherches sur les mycorrhizes dan les pépinières de resineux en sols agricoles. *Proc., Int. Union Forest Res. Organ., 14th, 1967* Sect. 24, p. 191.

Feliciani, A., and Montifiori, R. (1954). Allivamento in vivaio di *Pseudotsuga douglasi* e colonizzione microrrizia delle piante. *Monti e Boschi* 5, 215.

Foster, R. C. (1972). Unpublished observations.

Foster, R. C., and Marks, G. C. (1966). The fine structure of the mycorrhizas of *Pinus radiata* D. Don. *Aust. J. Biol. Sci.* 19, 1027.

Foster, R. C., and Marks, G. C. (1972). Unpublished data.

Foster, R. C., and Marks, G. C. (1967). Observations on the mycorrhizas of forest trees. II. The rhizosphere of *Pinus radiata* D. Don. *Aust. J. Biol. Sci.* 20, 915.

Foster, R. C., Bland, D., and Logan, A. (1971). The mechanism of permanganate and osmium tetroxide fixation and distribution of lignin in the cell wall of *Pinus radiata*. *Holzforsch. Helzverwert.* 25, 137.

Frank, A. B. (1885). Ueber die auf Wurzelsymbiose beruhende Ernährung gewisser Bäume durch unterirdische Pilze. *Ber. Deut. Bot. Ges.* 3, 128.

Fuller, M. S., and Barshad, I. (1960). Chitin and cellulose in the cell walls of *Rhizidiomyces* sp. *Amer. J. Bot.* **47**, 105.

Handley, W. R. C., and Sanders, C. J. (1962). The concentration of easily soluble reducing substances in roots and the formation of ectotrophic mycorrhizal associations—a re-examination of Björkman's hypothesis. *Plant Soil* **16**, 42.

Harley, J. L. (1969). "The Biology of Mycorrhiza," 2nd ed. Leonard Hill, London, 334p.

Harley, J. L., and Waid, J. S. (1955). The effect of light on the roots of beech and its surface population. *Plant Soil* **7**, 96.

Hassouna, M. G., and Wareing, P. F. (1964). Possible role of rhizosphere in the nitrogen metabolism of *Ammophila arenaria*. *Nature* (*London*) **202**, 467.

Hatch, A. B. (1937). The physical basis of mycotrophy in the genus *Pinus*. *Black Rock Forest Bull.* **6**.

Hawker, L. E. (1965). The fine structure of fungi as revealed by electron microscopy. *Biol. Rev.* **40**, 52.

Hess, W. (1969). Ultrastructure of onion roots infected with *Pyrenochaeta terrestris*. *Amer. J. Bot.* **56**, 832.

Hewett, E. J. (1966). A physiological approach to the study of forest tree nutrition. *Forestry* **39**, Suppl., 49–59.

Hillis, W. E., Ishikura, N., Foster, R. C., and Marks, G. C. (1968). Role of extractives in the formation of ectotrophic mycorrhizae. *Phytochemistry* **7**, 409.

Hoffmann, G. (1967). The course of root and shoot growth in young *Quercus robur* plants in open and in shade. *Arch. Forstw.* **16**, 745 (translation).

Hofsten, A. (1969). The ultrastructure of mycorrhiza. 1. Ectotrophic and ectendotrophic mycorrhiza of *Pinus sylvestris*. *Sv. Bot. Tidskr.* **63**, 455.

Horton, J. V., and Keen, N. T. (1966). Sugar repression of endopolygalacturonases and cellulase synthesis during pathogenesis by *Pryenochaeta terrestris* as a resistance mechanism in onion root. *Phytopathology* **56**, 908.

How, J. E. (1943). The mycorrhizal relations of larch. III. Mycorrhiza formation in nature. *Ann. Bot.* (*London*) [N.S.] **6**, 103.

Jackson, L. W. R. (1947). Method of differential staining of mycorrhizal roots. *Science* **105**, 291.

Jahn, E. (1934). Die peritrophe Mykorrhiza. *Ber. Deut. Bot. Ges.* **52**, 463.

Kelly, A. B. (1941). "The Variations in Form of Mycorrhizal Short Roots of *Pinus virginiana* Mill. associated with Certain Soil Series," p. 10. Landenberg Labo., Landenberg, Pennsylvania.

Kessler, K. J. (1966). Growth and development of mycorrhizae on sugar maple (*Acer saccharum* Marsh). *Can. J. Bot.* **44**, 1413.

Kinugawa, K. (1965). Effect of daylength and temperature on growth and formation of mycorrhizal short root of the seedling of *Pinus densiflora*. *Bot. Mag.* **78**, 366.

Kozlowski, T. T. (1971). "Growth and Development of Trees," Vol. 2. Cambial Growth, Root Growth, and Reproductive Growth. Academic Press, New York.

Laiho, O. (1967). Field experiments with ectendotrophic Scotch pine seedlings. *Proc., Int. Union Forest. Res. Organ., 14th, 1967* Sect. 24, p. 149.

Levisohn, I. (1954). Occurrence of ectotrophic and endotrophic mycorrhizas in forest trees. *New Phytol.* **53**, 284.

Lewis, D. H., and Harley, J. L. (1965a). Carbohydrate physiology of mycorrhizal roots of beech. I. Identity of endogenous sugars and utilization of exogenous sugars. *New Phytol.* **64**, 224.

Lewis, D. H., and Harley, J. L. (1965b). Carbohydrate physiology of mycorrhizal roots of beech. II. Utilization of exogenous sugars by uninfected mycorrhizal roots. New Phytol. 64, 238.

Lewis, D. H., and Harley, J. L. (1965c). Carbohydrate physiology of mycorrhizal roots of beech. III. Movement of sugars between host and fungus. New Phytol. 64, 257.

Lihnell, D. (1942). Cenococcum graniforme als mykorrhizabildner von Waldbäumen. Symb. Bot. Upsal. 5, 1.

Linnemann, G. (1969). Erfahrungen bei Synthese-Versuchen, insbesondere mit Pseudotsuga menziesii, (Mirb.) Franco. Zentralbl. Bakteriol., Parasitenk. Infektionskr., Abt. 2 123, 453.

Loub, W. (1963). The development of mycorrhizae under the influence of forest fertilizing. Zentralbl. Gesamte Forstw. 80, 185 (translation).

Lyford, W. H., and Wilson, B. F. (1964). Development of the root system of Acer rubrum L. Harvard Forest Pap. 10, 1–17.

Lyr, H. (1961). Die Wirkungsweise toxischer Kernholz-Inhaltsstoffe (Thuja plicine und Pinosylvine) auf den Stoffwechsel von Mikroorganismon. Flora (Jena) 150, 227.

Lyr, H., and Hoffmann, G. (1967). Growth rates and growth periodicity of tree roots. Int. Rev. Forest Res. 2, 81.

MacDougal, D. R., and Dufrenoy, J. (1944). Mycorrhizal symbiosis in Aplectrum corallorhiza and Pinus. Yearb., Amer. Phil. Soc. p. 170.

MacDougal, D. R., and Dufrenoy, J. (1946). Criteria of nutritive relations of fungi and seed plants in mycorrhizae. Plant Physiol. 19, 440.

McKeen, W. E., Smith, R., and Mitchell, N. (1966). The haustoria of Erysiphe chicoracearum and the host parasite interface on Helianthus annuus. Can. J. Bot. 44, 1299.

McKeen, W. E., Smith, R., and Bhattacharya, P. K. (1969). Alterations of the host wall surrounding the infection peg of powdery mildew fungi. Can. J. Bot. 47, 701.

Mandels, M., and Reese, E. T. (1963). Inhibition of cellulases and glucosidases. In "Advances in Enzymatic Hydrolysis of Cotton and Related Material" (E. T. Reese, ed.), pp. 115–58. Pergamon, Oxford.

Manocha, M. S., and Shaw, M. (1967). Electron microscopy of uredospores of Melampsora lini on rust infested flax. Can. J. Bot. 45, 1575.

Marks, G. C. (1965). Pathological histology of root rot associated with late damping-off in Pinus lambertiana. Aust. Forest. 29, 238.

Marks, G. C., and Foster, R. C. (1967). Succession of mycorrhizal associations on individual roots of radiata pine. Aust. Forest. 31, 193.

Marks, G. C., Ditchburne, N., and Foster, R. C. (1967). A technique for making quantitative estimates of mycorrhiza populations in P. radiata forests. Proc., Int. Union Forest. Res. Organ., 14th, 1967 Sect. 24, p. 67.

Marx, D. H. (1969). The influence of ectotrophic mycorrhizal fungi on the resistance of pine roots to pathogenic infections. II. Production identification and biological activity of antibiotics produced by Leucopaxillus cerealis var piceina. Phytopathology 59, 411.

Marx, D. H., Bryan, W. C., and Davy, C. B. (1970). Influence of temperature on aseptic synthesis of ectomycorrhizae by Thelephora terrestris and Pisolithus tinctorius on loblolly pine. Forest Sci. 16, 424.

Melin, E. (1917). Studier över de norrländskä myrmarkernas vegetation med särskild hansyn till deras skogs-vegetation efter torrläggning. *Akad. Avhandl. Uppsala*, p. 426.

Melin, E. (1923). Experimentelle Untersuchungen über die Konstitution und Ökologie der Mykorrhizen von *Pinus sylvestris* und *Picea abies. Mykol. Unters.* **2**, 73.

Melin, E., and Nilsson, H. (1957). Transport of labelled photosynthate to the fungal associate of Pine mycorrhiza. *Sv. Bot. Tidskr.* **51**, 166.

Meyer, F. H. (1964). The role of the fungus *Cenococcum graniforme* Sow Ferd. et Winge in the formation of mor. "Soil Micromorphology," pp. 23–31. Elsevier, Amsterdam. (reprint)

Meyer, F. H. (1962). Die Buchen und Fichten Mykorrhiza in verschiedenen Boden-typen ihre Beeinflussung durch Mineraldünger sowie für die Mykorrhizabildung wichtige Faktoren. *Mitt. Bundesforschungsanst. Holzwirt.* **54**, p. 1.

Mihovic, A. I. (1952). (A study of the mycorrhizal relations of oak in the forest-steppe zone of the Ukraine, East of the Dnieper.) *Les. Khoz.* **5**, 35.

Mikola, P. (1967). The effect of mycorrhizal inoculation on the growth and root respiration of Scotch Pine seedlings. *Proc., Int. Union Forest. Res. Organ., 14th, 1967* Sect. 24, p. 100.

Mikola, P., and Laiho, O. (1962). Mycorrhizal relations in the raw humus layer of northern spruce forests. *Commun. Inst. Forest. Fenn.* **55**, 1.

Mitchell, H. L., Finn, R. F., and Rosendahl, R. O. (1937). The relation between mycorrhizae and growth and nutrient absorption of coniferous seedlings in nursery beds. *Black Rock Forest Pap.* **1**, 58–73.

Mollenhauer, H. H., Whaley, W. G., and Leech, J. H. (1961). A function of the Golgi apparatus in outer root cap cells. *J. Ultrastruct. Res.* **5**, 1575.

Moore, E. J., and McAlear, J. H. (1962). Fine structure of mycota. Observations on the septa of ascomycetes and basidiomycetes. *Amer. J. Bot.* **49**, 86.

Morrison, T. M. (1956). The mycorrhiza of silver beech. *N. Z. J. Forest.* **7**, 47.

Mullin, R. E. (1963). Growth of white spruce in the nursery. *Forest Sci.* **9**, 68.

Noelle, W. (1910). Studien zur vergleichenden Anatomie und Morphologie der Koniferenwurzeln mit Rücksicht auf die Systematik. *Bot. Ztg.* **68**, 169.

Northcote, D. H., and Pickett-Heaps, J. D. (1966). A function of the Golgi apparatus in polysaccharide synthesis and transport in root cap cells of wheat. *Biochem. J.* **98**, 159.

Oswald, C. T., and Ferchau, H. A. (1966). Bacterial associations of coniferous mycorrhizae. *Plant Soil* **28**, 187.

Peyton, G. A., and Bowen, C. C. (1963). The host parasite interface of *Peronospora manshurion* on *Glycine max. Amer. J. Bot.* **50**, 787.

Redmond, D. R. (1955). Studies in forest pathology. XV. Rootlets, mycorrhiza and soil temperatures in relation to Birch die-back. *Can. J. Bot.* **33**, 595.

Reid, C. P. P. (1968). Nutrient transfer by mycorrhizae. *Diss. Abstr. B* **29**, 429 (abstr.).

Richards, B. N., and Voigt, G. K. (1964). Role of mycorrhiza in nitrogen fixation. *Nature (London)* **201**, 310.

Richards, B. N., and Wilson, G. L. (1963). Nutrient supply and mycorrhiza development in Caribbean pine. *Forest Sci.* **9**, 405.

Richardson, S. D. (1953). Studies on root growth of *Acer saccharinum*. II. Factors affecting root growth where photosynthesis is curtailed. *Proc. Kon. Ned. Akad. Wetensch., Ser. C* **56**, 366.

Richardson, S. D. (1956). Studies on root growth of *Acer saccharinum*. V. The effect of long term limitation of photosynthesis on root growth in first year seedlings. *Proc. Kon. Ned. Akad. Wetensch., Ser. C* 59, 694.

Robertson, N. F. (1954). Studies on the mycorrhiza of *Pinus sylvestris.* 1. Pattern of development of mycorrhizal roots and its significance for experimental studies. *New Phytol.* 53, 253.

Saljaev, R. K. (1958). (The anatomical structure of root tips of adult *Pinus sylvestris* and the development of mycorrhizae on them.) *Bot. Zl.* 43, 869.

Sappa, F. (1940). Ricerche biologiche sul *Tuber magnatum* Pico. La germinatizone delle spore e caratteri della micorriza. *Nuovo Gi. Bot., Ital.* 47, 155.

Scannerini, S. (1968). Sull ultrastruttura delle ectomicorrize. II. Ultrastruttura di una micorriza di Ascomycete *"Tuber albidum"* × *Pinus strobus. Allionia* 14, 77.

Schweers, W., and Meyer, F. H. (1970). Einfluss der Mykorrhiza auf den Transport von Assimilaten in die Wurzel. *Ber. Deut. Bot. Ges.* 83, 109.

Shaw, M., and Manocha, M. S. (1965). The physiology of host-parasite relations. XV. Fine structure in rust infected wheat leaves. *Can. J. Bot.* 43, 1285.

Slankis, V. (1958). The role of auxin and other exudates in mycorrhizal symbiosis of forest trees. *In* "Physiology of Forest Trees" (K. V. Thimann, ed.), pp. 427–443. Ronald Press, New York.

Slankis, V. (1961). On the factors determining the establishment of ectotrophic mycorrhiza of forest trees. *Recent Advan. Bot.* p. 1738.

Slankis, V. (1963). Der gegenwärtege Stand unseres Wissens von der Bildung der ektotrophen Mykorrhiza bei Waldbaumen. *Int. Mykorrhiza Symp., 1960* pp. 175–183.

Slankis, V. (1967). Renewed growth of ectotrophic mycorrhizae as an indicator of an unstable symbiotic association. *Proc., Int. Union Forest Res. Organ., 14th, 1967* Sect. 24, p. 84.

Squire, R. O. (1972). Physiological problems in the early development of *Pinus radiata.* Unpublished information.

Stephenson, G. (1959). Nitrogen fixation in non-nodulated seed plants. *Ann. Bot.* (*London*) [N.S.] 23, 622.

Theodorou, C., and Bowen, G. D. (1971). Influence of temperature on the mycorrhizal associations of *Pinus radiata* D. Don. *Aust. J. Bot.* 17, 59.

Trappe, J. M. (1962). Fungus associates of ectotrophic mycorrhizae. *Bot. Rev.* 28, 538.

Trappe, J. M. (1967). Principles of classifying ectotrophic mycorrhizae for identification of fungal symbionts. *Proc., Int. Union Forest Res. Organ., 14th, 1967* Sect. 24, p. 46.

Treshow, M. (1970). "Environment and Plant Response." McGraw-Hill, New York.

Unger, F. (1840). Beitrage zur kenntnis der parasitischen pflanzen. Erster oder anatomischphysiologischen Theil. *Ann. Wien. Mus. Naturgeschi.* 2, 13.

Van Dyke, C. G., and Hooker, A. L. (1969). Ultrastructure of host and parasite in interactions of *Zea mays* with *Puccinia sorghi. Phytopathology* 59, 1934.

Wichner, S., and Libbert, E. (1968). Interactions between plants and epiphytic bacteria regarding their auxin metabolism. *Physiol. Plant.* 21, 500.

Wilcox, H. E. (1968). Morphological studies of the root of red pine, *Pinus resinosa.* I. Growth characteristics and patterns of branching. *Amer. J. Bot.* 55, 247.

Wood, R. K. S. (1967). "Physiological Plant Pathology." Blackwell, Oxford.

Young, H. E. (1940). Mycorrhizae and the growth of *Pinus* and *Araucaria. Bull. Queensl. Forest Serv.* 13, 108.

Zak, B. (1964). Role of mycorrhizae in root disease. *Annu. Rev. Phytopathol.* 2, 377.

Zak, B. (1969). Characterization and classification of mycorrhizae of Douglas Fir. 1. *Pseudotsuga* and *Poria terrestris* (blue and orange strains). *Can. J. Bot.* **47**, 1833.

Zak, B., and Marx, D. H. (1964). Isolation of mycorrhizal fungi from roots of individual slash pine. *Forest Sci.* **10**, 214.

Zottl, H. (1964). The effect of fertilization on the distribution of fine roots in spruce stands. *Mitt. Staats. forstverw. Bayerns* No. 34, p. 333. (translation).

CHAPTER 2

Classification of Ectomycorrhizae

B. ZAK

I. The Need for Classification

Over a decade ago Trappe (1962) listed 10 orders, 30 families, 81 genera, and over 525 species of fungi believed to be ectomycorrhizal with over 280 species of trees. Included were over 1500 different tree plus fungus combinations based on reports from throughout the world. A list today would undoubtedly be much larger. The actual number of different ectomycorrhizae that exist in nature must be very great indeed. If we conservatively estimate that *Pseudotsuga menziesii* [Mirb.] Franco in the Pacific Northwest (United States) bears over 100 different ectomycorrhizae on its roots and multiply this value by the number of ectomycorrhizal tree species in the world, we arrive at a truly enormous figure.

However, most of the tree plus fungus combinations listed by Trappe are tenuous assumptions based on the consistent association of sporophores of fungi with given tree species, and some are based on mycorrhizae synthesized in pure culture. Less than 10 represent known mycorrhizae in the forest. An additional 10 to 15 have been identified since Trappe's listing. Thus, of the huge number of different ectomycorrhizae in nature, only a handful have actually been named and described; the rest are unknown.

As a consequence, ectomycorrhizal fungi employed in laboratory and field experimentation are usually those proven to be mycorrhizal by pure culture synthesis, or, as Bowen (1965) has indicated, those producing fruiting bodies, rather than fungi known to be associated with natural mycorrhizae. Their mycorrhizae may not even exist in nature, or they may occur only infrequently and be of little significance in the forest. Our concept of the ectomycorrhiza is based, then, on only a very small fraction of the large number of mycorrhizae in nature. To obtain a fuller understanding of this important tree–fungus relationship, we must widen our studies and include more of nature's diverse forms. Before this can be done, however, we need to identify, describe, and classify the mycorrhizae.

A. Physiology Research

An increasing number of physiological studies of ectomycorrhizal fungi are being performed, as summarized by Trappe (1962) and Harley (1969). A recent investigation by Lundeberg (1970) examined a number of fungi for production of extracellular enzymes and for their ability to utilize various nitrogen sources. Another study by Laiho (1970) compared metabolic processes of *Paxillus involutus* (Batsch) Fr. with those of other mycorrhizal fungi. A common feature in all these studies is the physiological diversity displayed by the different fungi. Many functions of the fungus affect the morphology and physiology of the mycorrhiza and thus indirectly influence the physiology of the tree itself. Consequently some ectomycorrhizae may benefit the tree more than others, as demonstrated by Young (1940), Moser (1956), Marx and Zak (1965), and Bowen (1970). It is reasonable to expect that we shall find an even greater diversity of physiological processes among the as yet undiscovered ectomycorrhizae in nature. Possibly we may even discover some mycorrhizae whose fungal partners actually fix atmospheric nitrogen, as some workers have speculated.

The morphology of the fungal component of ectomycorrhizae is highly variable, varying according to species of fungus involved. Some ectomycorrhizae have smooth mantles devoid of attachments, while others are heavily enveloped by the mycelium and rhizomorphs of their fungal symbiont. These differences are probably reflected in the physiologies of the mycorrhiza and tree. It is plausible that, because of their larger effective absorbing surface, mycorrhizae with attached mycelium and rhizomorphs radiating into the soil are better absorbers of nutrients than mycorrhizae lacking them. Thus, for example, it might be expected that *Pseudotsuga menziesii* + *Corticium bicolor*,[1] which has dense wefts of attached mycelium and rhizomorphs radiating into surrounding decayed wood or humus for as much as 1 m², may absorb nutrients better than *P. menziesii* + *Lactarius sanguifluus*,[2] which has a smooth mantle with only an occasional attached rhizomorph. Many such examples may be observed among the as yet nameless ectomycorrhizae in nature.

B. PATHOLOGY RESEARCH

Recently, Marx and Davey (1969a,b) demonstrated that, besides aiding tree nutrition, ectomycorrhizae also probably serve to protect root tissue from attack by pathogens. Three protective mechanisms postulated by Zak (1964) are (i) that the fungal mantle provides a mechanical barrier, (ii) that the fungus may secrete an antibiotic, and (iii) that the fungus may attract a protective rhizosphere population. These mechanisms may be expected to vary widely among the many different ectomycorrhizae in nature. Mantle thickness and density are known to be variable (Fontana and Centrella, 1967). Some fungi secrete antibiotics effective against some pathogens but not against others (Marx, 1969), and it was shown that different ectomycorrhizae attract different populations of fungi and bacteria to their rhizospheres (Neal *et al.*, 1964), although actual protection by the rhizosphere has not yet been demonstrated.

Furthermore, mycorrhizae formed by some fungi may be more resistant to root aphid and nematode attack than those of other fungi (Zak, 1965, 1967). For example, colonies of the aphid, *Rhizomaria piceae* Hartig, were commonly observed infesting several different mycorrhizae along the same *Pseudotsuga menziesii* root, but not the tuberculate mycorrhiza formed by *Rhizopogon vinicolor* A. H. Smith (Zak, 1971a). Apparently, the thick, dense rind of the tubercle prevents the aphid

[1] *Corticium bicolor* Peck.
[2] *Lactarius sanguifluus* (Paulet ex) Fr.

from inserting its stylus into cortical and vascular tissues. Recently
a new function of the ectomycorrhiza and its associated fungal symbiont
was proposed: protection of absorbing roots from soil phytotoxins (Zak,
1971c).

C. Ecology Research

An accurate and workable classification system is especially needed
for ecological study of ectomycorrhizae. It is surprising that we are unable to catalog the different mycorrhizae in even a small and uniform
forest stand, or even in a tree nursery which usually contains only few
forms. We know almost nothing of the frequency and distribution of
specific ectomycorrhizae, or how they vary according to season or
changes in the environment.

Except for a few studies, such as those of Mikola (1948) and Trappe
(1964) on *Cenococcum graniforme* Sow. Ferd. & Winge mycorrhizae,
that of Mikola (1962) on *Corticium bicolor* (*C. sulphureum* Fr. ?)
mycorrhizae, and the recent excellent monograph on *Paxillus involutus*
mycorrhizae by Laiho (1970), most ecological works have been broad
and general, ignoring or not recognizing specific mycorrhizae. Or, at
best, mycorrhizae have been distinguished simply by mantle color, and
in some works they were designated according to Melin's (1927) brief
classification, or more commonly in recent years to the tentative system
devised by Dominik (1956).

D. Application of Research Results

A few often spectacular results have been achieved by introducing
ectomycorrhizal fungi into exotic plantings in various parts of the world,
as, for example, in Puerto Rico (Briscoe, 1959). Mostly, however, our
extensive knowledge of mycotrophy has found little practical application
in the forest or nursery. Again, a major reason has been the lack of a
meaningful and practical classification system based on identity of the
individual mycorrhiza. Assistance to the forester and nursery manager
has consisted largely of vague expressions on the presence and abundance
of mycorrhizae and their supposed condition. Rarely has a specific
mycorrhiza been mentioned.

II. Past and Present Classification Attempts

Although mycorrhizae had been classified earlier, beginning with
Frank's (1885) distinction of endomycorrhizae and ectomycorrhizae,

Melin (1927) provided the first plan for separating ectomycorrhizae. Originally devised for pines, his plan was later modified and extended to trees of other genera. He divided ectomycorrhizae into four basic types and two subtypes based on a few gross morphological characters. However, with increasing awareness of the large number and great variety of ectomycorrhizae in nature, it soon became apparent that a much more detailed and specific classification system was needed.

A major advance toward solving this problem was made by Dominik (1956). He devised a classification scheme, together with an identification key, based on various morphological features of ectomycorrhizae. Tree species were not specified; the system was designed to apply to all ectomycorrhizal trees. Mycorrhizae were first divided into 12 subtypes (designated A–L) according to macroscopic and microscopic characters of the mantle and according to gross structure of the mycorrhiza. Then each subtype was separated into 1–8 "genera" (designed a–h) by color and other mantle features. Thus, for example, the common mycorrhizae formed by *Cenococcum graniforme* were assigned to "genus" Ga (Dominik, 1961).

Dominik's system is noteworthy in that it utilizes detailed macroscopic and microscopic morphological features in defining categories. It relies especially on color, structure, and surface character of the mantle. The latest version (Dominik, 1969) includes a more detailed treatment of the different kinds of setae and cystidia covering the mantle surface. As an example, subtype D specifies monopodial or ramiform mycorrhizae with a regularly woven, dense, and feltlike mantle covered with setae. "Genus" Da further specifies a mantle which appears colorless when viewed microscopically and with setae that are hyaline and not longer than ten times their basal width. However, many other useful features, such as details of attached rhizomorphs and hyphae, are not employed.

A major limitation of Dominik's classification is disregard of the specific tree–fungus or distinct mycorrhiza. Except for two or three, each of his "genera" represents several ectomycorrhizae which, other than the characters which define their particular "genus," may have little in common with each other, especially in their relationship to tree growth. It is an artificial grouping in which one "genus" may include various taxonomically unrelated fungi, or one fungus may be represented in more than one "genus." It completely ignores the natural classification system for tree and fungus.

However, Dominik (1961) conceded that his system was not the final answer, but rather a temporary approach to a difficult problem. He anticipated that as more information became available, his generic categories would be divided into smaller units, eventually leading to individual mycorrhizae. And, like Peyronel (1922), he acknowledged that the ultimate classification must be based upon the identity of the fungal partner of the ectomycorrhiza.

Trappe (1967a) has also emphasized the need for identifying fungal partners of ectomycorrhizae and listed various principles to achieve this goal. His mycorrhizal key to identify fungal symbionts would be based primarily on stable hyphal characteristics, such as the presence and morphology of clamp connections, and structure of septal pores. What he regards as less constant mycorrhizal features, such as color and form, would be used only after stable characteristics were exhausted.

In contrast to Melin's and Dominik's broad and general classifications, others, very recently, have described and classified natural ectomycorrhizae of a single tree species or a single genus. Some descriptions are carefully and accurately detailed and are accompanied by excellent illustrations showing important features. A few even provide identities of fungal symbionts. Most descriptions, however, are much too brief, incomplete, and inadequately illustrated for mycorrhizae to be recognized by others.

An early attempt to characterize distinct ectomycorrhizae was made by Woodroof (1933) for cultivated pecan (*Carya* sp.) in Georgia. She described seven mycorrhizae, which she designated as forms A–G, and named the fungal partners of three of these. Photographs and well-executed line drawings were used to illustrate important macroscopic and microscopic details. Mantles of three mycorrhizae were depicted bearing distinctive setal coverings. Rhizomorphs attached to two mycorrhizae were described and their microscopic features illustrated.

In Australia, seven *Pinus radiata* D. Don ectomycorrhizae collected from a 40-year-old stand were described by Marks (1965). His characterizations included branching habit, color and structure of mantle, and microscopic details of the tannin layer, Hartig net, cortex, and endodermis. A photograph depicted the various mycorrhizae, and photomicrographs illustrated pertinent features in cross sections of mycorrhizae. Six of the mycorrhizae were designated by subtype and "genus" according to Dominik (1956).

Eight distinctive *Eucalyptus* ectomycorrhizae from forests of New South Wales, Australia, were definitively described by Chilvers (1968). Each mycorrhiza was fully characterized, and important details were clearly illustrated by excellent line drawings. Two features regarded as especially useful for distinguishing the mycorrhizae were plan views of lactophenol-cleared mantles and structure of attached rhizomorphs. Also included were details of hyphae comprising mantles and Hartig nets and details of those radiating into the soil. Two fungal symbionts were identified.

During the past 10 or 12 years, Italian workers have described ectomycorrhizae of various coniferous and deciduous trees. Five ectomycorrhizae of *Pinus strobus* L. nursery seedlings were characterized by Fassi and de Vecchi (1962). Included were photographs of mycorrhizae and of fungal symbiont sporocarps. Line drawings showed important details of mantle structure and attached mycelia. Identification of fungi of four of the mycorrhizae was based on proximity of sporocarps and comparison of sporocarp and mycorrhizal mycelia.

Similarly, the *Pinus strobus* ectomycorrhiza formed with *Endogone lactiflua* Berk & Br. was described in detail by Fassi (1965), and one formed by *Tuber albidum* Pico was reported by Scannerini and Palenzona (1967).

Another study of nursery-grown pine seedlings, by Rambelli (1967), described macroscopic and microscopic features of ten *Pinus radiata* ectomycorrhizae. Important characters were illustrated by photographs. Fungal symbionts were not named; mycorrhizae were designated according to color, form, or both, i.e., "chestnut-brown form," "coralloid-brown form," etc. Recently Rambelli (1970) similarily described ten additional *P. radiata* ectomycorrhizae from seedlings planted on various sites.

Ectomycorrhizae of several species of *Populus* and *Salix*, growing in northern Italy, were carefully described by Fontana (1961, 1962, 1963). Well-executed line drawings depicted mycorrhizae, including microscopic details of mantles and attached mycelia. Each mycorrhiza was designated as a numbered "form" and classified according to Dominik's system. Fungal partners of several mycorrhizae were identified.

Similarily, Ceruti and Bussetti (1962) described five *Tilia* sp. ectomycorrhizae and named their fungi, while Luppi and Gautero (1967) detailed ectomycorrhizae of three *Quercus* species and named fungal symbionts of several from the Piedmont region in Italy. Fontana and Centrella (1967) described ectomycorrhizae of *Castanea*, *Carpinus*, *Fagus*, *Pinus*, and *Quercus* formed by species of hypogeous fungi. Fungal partners of eight of the mycorrhizae were identified. And from the African Congo, Peyronel and Fassi (1957) and Fassi and Fontana (1961, 1962) described ectomycorrhizae of several leguminous tree species. Distinctive features were illustrated with line drawings, but symbiotic fungi were not named.

Other attempts to classify ectomycorrhizae of deciduous trees that may be mentioned are those of Jeník (1957) for *Quercus* and of Sen (1961) for *Tilia europaea* L. In the latter study, mycorrhizae were placed into five categories according to gross form, and these were further divided into subtypes largely on the basis of mantle character.

Zak (1969a, 1971a), beginning a series of studies characterizing and classifying *Pseudotsuga menziesii* mycorrhizae, described two mycorrhizae each formed by a distinct strain of *Poria terrestris* (DC. ex Fries) Sacc. and a tuberculate form by *Rhizopogon vinicolor*. Each mycorrhiza was fully described and illustrated, and cultural characteristics of the fungal symbionts listed. Mycorrhizae were designated by tree species plus fungus species and strain.

Zak (1971b) recently proposed a plan for the characterization and identification of ectomycorrhizae of *Pseudotsuga menziesii* which can also be applied to other species. The proposal emphasizes recognition of each distinct mycorrhiza, and naming it to confer on it an identity. The ideal name would include tree and fungus species. However, if the latter were unknown, an arbitrary designation would be applied. In any event, identity of the mycorrhiza would be supported by rigid and definitive characterization. The author's identification key, in contrast with that suggested by Trappe (1967a), would be based more on easily determined gross characters and less on difficult to see microscopic features. The present work further explores this proposal.

III. Characterizing the Ectomycorrhiza

Unlike the endomycorrhiza, the ectomycorrhiza is usually a distinctive structure, differing sharply from the nonmycorrhizal root. It is distinguished by color, form, texture, and various unusual microscopic features. Its unique character is contributed in part by the root and in larger part by the symbiotic fungus. The role of each and of the surrounding environment in determining the morphology of the ectomycorrhiza will now be examined, and features useful for characterizing the ectomycorrhiza and constructing an identification key will be discussed.

A. Role of the Tree

External configuration and internal cell arrangement of the ectomycorrhiza are features inherent in the tree but modified or expressed by interaction with the fungus. Branching pattern exhibits little or no variation between different tree species of the same genus, but striking differences can be observed between different genera. Thus, *Populus* and *Salix* have monopodial and ramiform ectomycorrhizae (Fontana, 1961, 1962); *Castanea, Fagus,* and *Quercus,* in addition to monopodial and ramiform forms, also have irregular pinnate forms (Fontana and

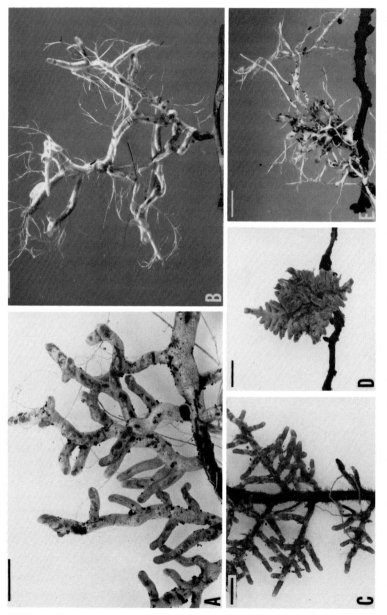

Fɪɢ. 1. Distinctive ramiform and pinnate *Pseudotsuga menziesii* ectomycorrhizae. (A) Unidentified yellow mycorrhiza with reticulate surface and thread-like rhizomorphs. (B) Silvery-white mycorrhiza with lavender tint and with branching rhizomorphs formed by *Cortinarius* sp. (C) Brown with silvery pinkish white patches, open-pinnate mycorrhiza with sparse rhizomorphs formed by *Thelephora terrestris*. (D) Unidentified orange, closed-pinnate mycorrhiza. (E) Unidentified white, open-pinnate mycorrhiza with flat and weft-like rhizomorphs. Scale lines equal 2 mm.

Centrella, 1967); *Abies, Larix, Picea, Pseudotsuga,* and *Tsuga* have monopodial, ramiform and regular, irregular and tightly pyramidally pinnate ectomycorrhizae (Fig. 1A–E); and *Pinus,* besides monopodial mycorrhizae, alone has the characteristic bifurcate form (Fig. 2A–D) (Hatch and Doak, 1933).

Some trees also form ectomycorrhizal nodules or tubercules with certain fungi (Fig. 3A,B). A typical nodule consists of a tightly packed mass of individual ectomycorrhizal elements encased in a loose to dense fungal sheath. Those found on the roots of several pine species are believed by Melin (1927) to be formed by *Boletus* sp. A bright yellow ectomycorrhizal nodule discovered by J. M. Trappe on roots of *Tsuga mertensiana* (Bong.) Carr. in Oregon has *Rhizopogon cokerii* A. H. Smith as its fungal partner. The black ectomycorrhizal tubercle of *Pseudotsuga menziesii,* first reported by Dominik (1963) in Poland and later by Trappe (1965) from the Pacific Northwest, is formed by *Rhizopogon vinicolor* (Zak, 1971a).

Except for the lack of an epidermis and root hairs, occasional hypertrophy, and always spatial distortion of at least the outer cortical cells, the inherent internal structure of the root is unchanged by mycorrhization. Such inherent features as size and number of tiers of cortical cells, diameter of stele, and radial arrangement of primary xylem elements help define the ectomycorrhiza.

B. ROLE OF THE FUNGUS

It is the symbiotic fungus which directly and indirectly transforms the root into a characteristic ectomycorrhiza. It sheaths the root with a distinctive fungal mantle to which mycelium and rhizomorphs may or may not be attached. Its hyphae penetrate into the root and enmesh cortical cells within a hyphal network. Interacting with the root, it may induce enlargement of cortical cells and thus further increase element diameter. It may also stimulate a distinctive growth pattern by the root, such as the forking of *Pinus* mycorrhizae attributed by Slankis (1951) to fungal auxins, the pinnate structure of many other species, and the nodular or tuberculate form of mycorrhizae of some trees.

Different isolates of some ectomycorrhizal fungi grown on nutrient agar media display much variation in color and texture of vegetative growth. Such cultural differences in morphology, however, do not appear to be reflected in the mycorrhiza. For example, mats of several isolates of *Rhizopogon vinicolor* grown on potato-dextrose agar medium vary widely in diameter growth, color, and texture. Yet, *Pseudotsuga menziesii* + *R. vinicolor* tubercles from different sources appear identical in form, color, and texture. However, some fungal species which em-

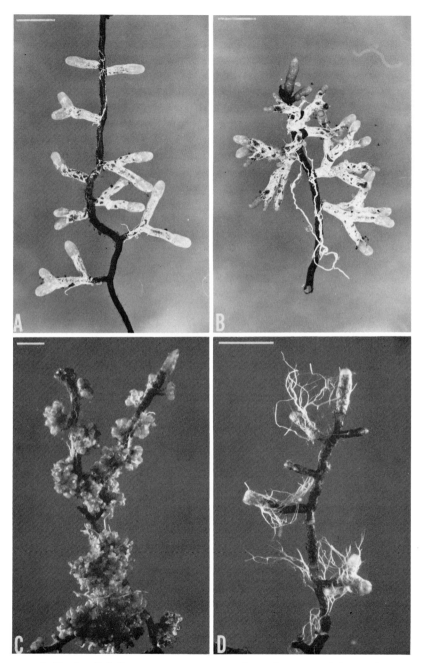

FIG. 2. Unidentified *Pinus contorta* ectomycorrhizae from coastal dunes of Oregon.
A and B. Young and older white, large-forked mycorrhizae with basally attached
rhizomorphs. C. Golden-yellow, almost sessile, small-forked form producing tight
coralloid structures; rhizomorphs connect to loose mycelium surrounding mantles.
D. Young mycorrhizae of white, wooly form; rhizomorphs attach directly to mantle
surface. Scale lines equal 2 mm.

FIG. 3. Tuberculate ectomycorrhizae. A. *Pseudotsuga menziesii* + *Rhizopogon vinicolor*. B. *Tsuga mertensiana* + *R. cokerii*. Scale lines equal 1 cm.

brace several physiologically and morphologically distinct forms may produce as many distinguishably different mycorrhizae. An example is *Poria terrestris* whose four different strains each produce a distinct mycorrhiza with *Pseudotsuga menziesii* (Zak, 1969b). Such discrepancy between fungus and mycorrhiza may justify the establishment of new species.

On the other hand, since most fungal species are differentiated by characters of the spore-bearing structures rather than by character of vegetative mycelium or its physiology, many closely related species would likely produce indistinguishable mycorrhizae with a given tree species. An example may be *Lactarius deliciosus* (L. ex Fr.) S. F. Gray and *L. sanguifluus* mycorrhizae of *Pseudotsuga menziesii* and other conifers. Other examples recently noted by the author in western Oregon are closely similar *Pseudotsuga menziesii* mycorrhizae formed by *Cortinarius zakii* J. F. Ammirati sp. prov. and *C. aureifolius* var. *hesperius* J. F. Ammirati var. prov.[3] Undoubtedly the taxonomic positions of some

[3] Both fungi identified by J. F. Ammarati, University of Michigan. Formerly referred to (Zak, 1971b) as *Cortinarius croceifolius* Peck. and *C. aureifolius* Peck., respectively. Dr. Ammarati has kindly granted permission to use these provisional new names.

fungi will be modified as further identification, description, and cataloging of natural mycorrhizae add to the knowledge of species differentiation.

Domination of the ectomycorrhizal morphology by the fungal symbiont, at least the external morphology, is seen in mycorrhizae formed on widely diverse tree species by a single fungus. The best known example is *Cenococcum graniforme*. The characteristic coal-black, monopodial mycorrhizae with stiff, black, radiating hyphae are found on a large variety of trees and shrubs and even herbs (Trappe, 1964). Despite the great diversity of higher plant tissues, the fungus is largely unchanged and readily identifiable.

A more limited example was recently noted by the author in dune forest along the Oregon coast. Within a few square meters, *Cortinarius aureifolius* var. *hesperius* was found mycorrhizal with *Pinus contorta* Dougl., *Pseudotsuga menziesii*, *Arctostaphylos uva-ursi* (L.) Spreng., and *Polygonum paronychia* Cham. and Schlecht. Mycorrhizae of the first two species are ectomycorrhizal, while those of the latter two are ectendomycorrhizal. Except for size and form, the four mycorrhizae have closely similar external morphologies including attached mycelium and rhizomorphs. Many more similar examples will probably be found as more mycorrhizae and their fungal partners are recognized.

Yet, as Trappe (1962) suggested, some ectomycorrhizal fungi may themselves be morphologically modified when associated with different trees. Although he referred mainly to sporophore morphology, it is even more likely that fungal tissues intimately associated with root tissues in the mycorrhiza may be changed. Thus, a fungus may form ectomycorrhizae with some host plants and endomycorrhizae or ectendomycorrhizae with others, as in the previously cited example with *Cortinarius aureifolius* var. *hesperius*. It is possible, too, that different host plants may induce formation of structurally different mantles by the same fungus.

C. Role of the Environment

Although environmental factors, through their effects on both tree and fungus, may affect morphology of the ectomycorrhiza, they probably are responsible for only minor changes. It is unlikely that the same tree–fungus mycorrhiza will differ significantly from habitat to habitat, making identification difficult. The effect of pH of the growing medium may be cited as an example. Many fungi grown at different pH levels on artificial media display various morphological changes, especially color of mycelium. However, pH-related differences in morphology of natural mycorrhizae have not been reported nor has the author observed such differences. In second-growth *Pseudotsuga menziesii* forests in western Oregon several ectomycorrhizae are found in both mineral soil, pH 5.5–5.8, and decayed wood, pH 3.2–3.8, of stumps, logs, and debris. Both the soil and decayed wood forms of each mycorrhiza appear morphologically identical. Even colors of mantles are the same.

D. USEFUL CHARACTERS

Many macroscopic and microscopic characters can be used to accurately define the ectomycorrhiza. Most are contributed by the fungus; few come from the tree. Some features are quite stable; others are more or less variable. Some have a wider range of expression than others; a few, such as odor and taste, are only occasional criteria. Some can be easily determined, but others require tedious examination of histological preparations. Obviously, the more distinctive a character, the more valuable it may be for describing the mycorrhiza and for use in an identification key. To be meaningful, it is essential that the description of the ectomycorrhiza be as complete as possible and include crisp illustrations showing important details. Macroscopic and microscopic characters which may be employed to describe the ectomycorrhiza are next listed and discussed; they are ranked in each group according to their usefulness, based on stability, range of expression, and ease of determination.

1. *Macroscopic Characters* (× 5–50)

a. Color. This most obvious character, which first strikes the eye, is judged by some as too variable and unstable. Boullard (1965), for example, criticizes its use by Dominik (1956) in his classification of ectomycorrhizae. Marks (1965) and Trappe (1967a) believe that for many mycorrhizae color varies too much between developmental stages, and Dominik (1961), while regarding color as one of the important features of the ectomycorrhiza, cautions its interpretation. Often, he warns, the color seen may not be that of the fungus, which may be colorless, but that of the variable underlying tannin layer.

While color may be poorly defined and confusing in some ectomycorrhizae, in most it is clear and stable and a valuable diagnostic feature. Colors of ectomycorrhizae are many, including practically the entire spectrum. They may be uniform and remain unchanged throughout the life of the ectomycorrhiza, as in the black mycorrhizae formed by *Cenococcum graniforme,* or the mycorrhiza may be one color when young and gradually change to another with age, as *Pseudotsuga menziesii + Lactarius sanguifluus.* When young, it is a dull orange, and, as it ages, it becomes dark verdigris. Other mycorrhizae may be uniformly colored at first but with age acquire discrete stains of another color. Actually, color of the mycorrhiza varies no more than that of other biological materials. In fact, the author has observed several mycorrhizae whose colors are more stable and uniform than colors of overlying sporocarps of their respective fungal symbionts. Usually, as indicated by Zak (1971b), all normal color variations can be observed in a single collection, unless the mycorrhizae are the first of the season or are all old and deteriorated.

b. Attached Mycelium. The presence or absence and the character of attached mycelium are important features of the ectomycorrhiza. Some mycorrhizae appear to lack any radiating hyphae or mycelium (Fig. 4 A–D), while others may be completely enveloped in dense mycelium (Fig. 5A–D). Gradiations from one extreme to the other may be observed among different mycorrhizae in nature. The attached mycelium may be stiff and hair-like, delicately gossamer, fleecy, cottony, or weft-like. It may vary in appearance from very fine to very coarse. Interspersed among the hyphae may be sclerotia or sclerotia-like bodies. When the mycelium of *Corticium bicolor* mycorrhizae is teased apart in water, clouds of fine, bright-yellow particles are released.

c. Attached Rhizomorphs. Some ectomycorrhizae, as those of *Cenococcum graniforme,* lack rhizomorphs; others have only a few, such as *Pseudotsuga menziesii + Lactarius sanguifluus* (Fig. 4C), while others have many rhizomorphs, such as *P. menziesii + Poria terrestris* (blue-staining).

The attached mycelium is usually concolorous with the fungal mantle, but rhizomorphs may be somewhat differently colored and may even be spottily stained, especially when older. They may be long or short, thin or thick, sharply defined and threadlike, or loose and weftlike. In cross section they may be flat, oval, or round. The surface may appear smooth, cottony, velvety, or granular.

Some rhizomorphs are attached to loose mycelia surrounding the mantle; others are connected directly to the mantle surface; and others, observed by the author on *Pseudotsuga menziesii + Cortinarius* sp. mycorrhizae (Fig. 1B), are fastened, as if glued, along their lengths to the mantle surface.

d. Surface Character of Mantle. The mantle surface may be uniform in color, or it may be patchy in color with light and dark hues. Some ectomycorrhizae have mantles with a metallic sheen. Surface texture may be smooth, cottony, wooly, velvety, reticulate, granular, or crusty.

FIG. 4. Ectomycorrhizae with smooth mantles. A. Unidentified white, pinnate *Pseudotsuga menziesii* mycorrhiza. B. Unidentified brown, coralloid *Pinus virginiana* Mill. mycorrhiza. C. Orange-green pinnate *P. menziesii* mycorrhiza formed with *Lactarius sanguifluus*. D. Unidentified olive, pinnate *P. menziesii* mycorrhiza. Scale lines equal 2 mm.

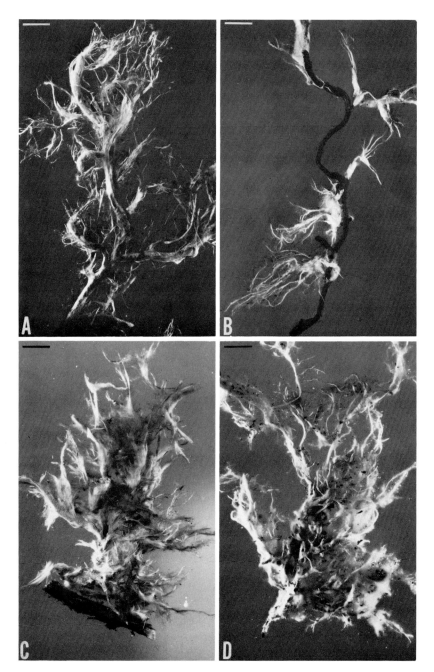

Fig. 5. Four *Pseudotsuga menziesii* ectomycorrhizae with dense covering of mycelium and rhizomorphs. A. Pinkish-red mycorrhiza of *Cortinarius phoeniceus* var. *occidentalis*. B. White *Cantharellus cibarius* Fr. form. C. Bright orange *Corticium bicolor* mycorrhiza. D. White mycorrhiza probably formed by *Amanita vaginata* (Bull. ex Fr.) Vitt. Scale lines equal 2 mm.

e. Form. Monopodial and ramiform (Fig. 1A, B) ectomycorrhizae are most common. Trees of some genera, such as *Abies, Larix, Picea, Pseudotsuga,* and *Tsuga,* also have characteristic pyramidally pinnate structures (Fig. 1C–E and Fig. 4A,C–D). Those formed by some fungi are weakly and irregularly developed, while those formed by other fungi are strongly and regularly pinnate, often into compound fans. Nodular or tuberculate mycorrhizae, formed by various fungi, may be found on roots of some trees. Trees of *Pinus,* alone, in addition to monopodial forms, have characteristic bifurcate ectomycorrhizae which vary in shape and size according to the species of the fungal partner. They may be sessile or, more commonly, stipitate; they may be simple or multiple-forked, sometimes into coralloid structures (Fig. 2B,C and Fig. 4B).

f. Chemical Reagent Color Reaction. Various chemical reagents applied to mycorrhizal mantles and attached fungal tissues produce distinctive color reactions helpful in distinguishing ectomycorrhizae (Peyronel, 1934; Trappe, 1967a; Fassi *et al.,* 1969; Zak, 1969a, 1971a,b). Useful reagents commonly employed in fungal taxonomy, as described by Singer (1962), include 15% KOH, concentrated NH_4OH, concentrated H_2SO_4, 10% $FeSO_4$, and Melzer's solution (iodine-chloral hydrate).

It is important that only fresh mycorrhizae be used; older ones may give an erratic or even a different reaction. Obviously, the more reactions that can be ascribed to a mycorrhiza, the better. Each additional reaction adds to the characterization of the mycorrhiza and makes identification by use of a key more positive. The *Pseudotsuga menziesii* + *Poria terrestris* (blue-staining) mycorrhiza, described by Zak (1969a), exhibited five distinct reactions. Other mycorrhizae may display more or fewer reactions. A search for more useable reagents should be highly rewarding.

g. Ultraviolet Light Fluorescence. May (1961) found that tissues and ethyl alcohol extracts of various hymenomyceteous sporophores fluoresced in various colors under long-wave (3660 Å) ultraviolet (UV) light. Several of the fungi were mycorrhizal species. Trappe (1967a) also observed UV light fluorescence by many mycorrhizal fungi and emphasized its diagnostic value. Fluorescence in long-wave UV light of various *Pseudotsuga menziesii* ectomycorrhizae was reported by Zak (1971b). He also (Zak, 1969a, 1971b) included UV light fluorescence in descriptions of *Pseudotsuga menziesii* mycorrhizae and used this feature to aid identification of their fungal symbionts.

While many ectomycorrhizae fluoresce in UV light, the author has found relatively few which fluoresce strongly. One such mycorrhiza,

Pseudotsuga menziesii + *Cortinarius phoeniceus* var. *occidentalis*[4] (Fig. 6A,B), fluoresces a very bright golden yellow in long-wave UV light. Even water in which the mycorrhiza has been rinsed fluoresces strongly. Both the color and intensity of the fluorescence may vary with age of the mycorrhiza, but, as with color in natural light, all gradations can usually be observed in a single collection.

h. Individual Elements. Form may be straight or tortuous. For example, elements of *Pseudotsuga menziesii* + *Poria terrestris* (blue-staining) (Fig. 7A) are straight, while those of the closely related *P. menziesii* + *P. terrestris* (rose-staining) (Zak, 1971b) are tortuous (Fig. 7C). Different ectomycorrhizae of a given tree species display little or no variation in element diameter, but element diameter uniformity may vary somewhat according to fungal symbiont.

i. Habitat. A description of the site and its higher vegetation should be included in characterization of the ectomycorrhiza. Occasionally this information will help identification of the mycorrhiza. Mycorrhizae of

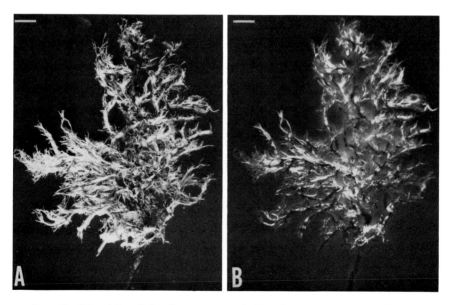

Fig. 6. Ultraviolet light fluorescence of *Pseudotsuga menziesii* + *Cortinarius phoeniceus* var. *occidentalis*. A. In normal light. B. In 3660-Å UV light. Scale lines equal 2 mm.

[4] *Cortinarius phoeniceus* var. *occidentalis* Smith. Identified by J. F. Ammarati. Formerly referred to (Zak, 1971b) as *Cortinarius semisanguineus* (Fr.) Gill.

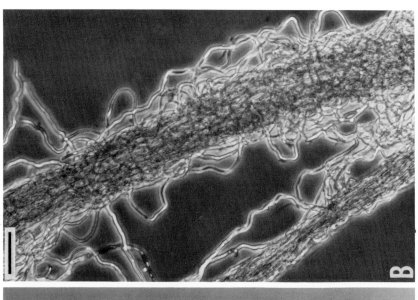

Fig. 7. Two distinct *Pseudotsuga menziesii* ectomycorrhizae formed by different strains of *Poria terrestris*. A. Straight-element pinnate mycorrhiza formed by *P. terrestris* (blue-staining). B. Attached rhizomorph.

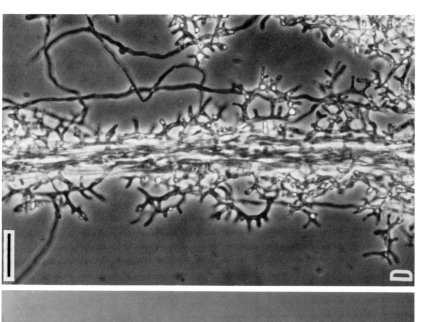

FIG. 7 (Continued). C. Tortuous-element pinnate mycorrhiza formed by *P. terrestris* (rose-staining). D. Attached rhizomorph. Monopodial black mycorrhiza in (C) is by *Cenococcum graniforme*. A and C scale lines equal 2 mm; B and D are with phase contrast light, scale lines equal 20 μm.

Cenococcum graniforme, for example, have never been reported from tree nurseries but are often dominant on dry sites (Trappe, 1962). A record of the growing medium, whether mineral soil, humus, litter, or decayed wood, will be useful. In the Pacific Northwest *Corticium bicolor* mycorrhizae are generally found in decayed wood partially buried in the soil, or in humus or decayed litter.

j. Taste and Odor. Only rarely are these characters sufficiently distinctive to aid identification of the ectomycorrhiza. An example is the *Pseudotsuga menziesii* mycorrhiza reported by Zak (1967). Both the mycorrhiza, especially when crushed, and its isolated fungal symbiont grown on nutrient agar emit a characteristic sharply acrid odor. In fact, the mycorrhiza's presence in a mass of soil and roots is readily detected by this unusual odor.

2. Microscopic Characters (× 100–1000)

a. Attached Mycelium. Simple slide preparations of attached mycelium in water, lactophenol, 5% KOH, Melzer's solution, or Hoyer's medium (Anderson, 1954) can reveal much useful information. Such features as types of hyphae present; their diameter and uniformity; hyphal form, whether straight, undulating, or dendritic; presence or absence of wall deposits and their character; presence or absence of clamp connections and their frequency and character; and types of hyphal branching will help to characterize the mycorrhiza.

b. Attached Rhizomorphs. As Woodroof (1933) demonstrated with pecan (*Carya* sp.), Chilvers (1968) with *Eucalyptus,* and Zak (1969a, 1971a) with *Pseudotsuga* mycorrhizae (Fig. 7B,D), anatomical features of attached rhizomorphs can significantly add to mycorrhizal description. Surface hyphae are often strikingly distinctive and can usually be viewed using simple slide preparations.

c. Effect of Chemical Reagents on Wall Deposits. By mounting a teased weft of mycelium in water on a slide and placing a drop of reagent, to be drawn under along the edge of the cover slip, one can observe its effect on wall deposits. Some deposits remain unaffected; others dissolve leaving walls free. An example is mycelium attached to *Pseudotsuga menziesii* ectomycorrhizae formed by *Corticium bicolor.* Untreated hyphae are coated with yellow spicules or with yellow amorphous deposits, and some with shard-like crystals. When 15% KOH is introduced under the cover slip, spiculate and amorphous incrustations disappear almost immediately. After a few seconds, large clusters of long, yellow, needle-like crystals appear along hyphae and free in the liquid. The natural, shard-like crystals are dissolved only slowly.

d. Mantle Surface. Mantles of many ectomycorrhizae have highly distinctive surface hyphae, cystidia, and setae, as illustrated by Dominik (1956, 1969), Fassi and Fontana (1961), and Chilvers (1968). Especially striking examples are stellate setae on mantle surfaces of *Salix caesia* Vill. mycorrhizae, as described and illustrated by Fontana (1962).

e. Mantle Structure. Mantle tissue of the ectomycorrhiza may vary from a simple prosenchyma to a compacted pseudoparenchyma or synenchyma. Chilvers (1968), describing *Eucalyptus* mycorrhizae, further divided each tissue designation into felt and net prosenchyma and irregular and regular synenchyma. Structure of the mantle may be observed from stained and mounted cross sections and from whole mycorrhizae cleared in lactophenol (Chilvers, 1968) or from mantle scrapings (Schramm, 1966; Zak, 1969a) mounted in 5% KOH, in saturated chloral hydrate, or, for permanency, in Hoyer's mounting medium. Although thickness of mantle is usually included in descriptions, its variability limits its usefulness for identification purposes.

f. Ultraviolet Light Fluorescence. Although never yet used in describing the ectomycorrhiza, Trappe (1967a) suggests that fluorescence microscopy may be more useful even than macrofluorescence. It may be especially productive with ectomycorrhizae whose hyphae bear incrustations which may fluoresce in distinctive colors. Fluorescence of hyphae before and after treatment with 15% KOH and other reagents may provide additional useful characters.

g. Hartig Net. As with mantle thickness, the Hartig net is not sufficiently distinctive and is usually too variable to be a useful character. Occasionally depth of penetration into the cortex may be a usable criterion.

IV. Identification of Fungal Symbiont

A. PURE CULTURE SYNTHESIS

Several different methods (Fig. 8A–E) have been used to identify fungal partners of natural ectomycorrhizae with varying degrees of success. One of these is the pure culture mycorrhiza synthesis technique originated by Melin (1921) and modified by Hacskaylo (1953) and Marx and Zak (1965). Aseptically germinated seedlings are grown together with the test fungus in a flask in sand, vermiculite, or vermiculite mixed with peat moss, and moistened with nutrient solution. Mycorrhizae formed after several months (Fig. 8D) are described with the hope of finding their natural counterparts in the forest or tree nursery.

Fig. 8. Identification of fungal symbionts of ectomycorrhizae. A. Nursery-grown
Pseudotsuga menziesii seedlings bound together by *Thelephora terrestris* sporocarps.

A similar pure culture technique was developed by Pachlewska (1968), using a nutrient agar medium in large test tubes. The medium found best for mycorrhizal formation was a "starvation" medium consisting of only agar, thiamine, and water. Well-formed *Pinus sylvestris* L. mycorrhizae were produced with *Amanita muscaria* (L. ex Fr.) Pers. ex Hooker, *Boletus edulis* Bull. ex Fr., *Cenococcum graniforme*, *Lactarius deliciosus*, *Suillus luteus* (L. ex Fr.) S. F. Gray, and *S. variegatus* (Swartz ex Fr.) O. Kuntze. From *Rhizopogon luteolus* Fr. and Nordh. mycorrhizae, synthesized by this method, Pachlewski and Pachlewska (1968) were able to identify the natural mycorrhiza in a dune *P. sylvestris* stand on the Baltic coast.

Although extremely useful for many studies, pure culture mycorrhizal synthesis techniques are not, in the author's opinion, suitable alone for identifying fungal symbionts of natural mycorrhizae. Formed in a wholly artificial environment, the synthesized mycorrhiza is rarely the same morphologically as its natural counterpart. It may differ significantly in color, form, size, and other characteristics. An example, although extreme, is *Pseudotsuga menziesii* + *Rhizopogon vinicolor*. Tuberculate in nature, it develops only as pyramidally pinnate fans in pure culture synthesis (Zak, 1971a), that are similar to *Pseudotsuga menziesii* mycorrhizae formed by *R. colossus* A. H. Smith and *R. parksii* A. H. Smith.[5] The exceptions are those few mycorrhizae whose fungal partners are sharply distinctive and without much variability such as *Cenococcum graniforme* and *Corticium bicolor*. This method, too, is limited to those fungi which can be grown on artificial media, and even many of these fail to form mycorrhizae under conditions of artificial synthesis.

B. Isolation of Fungus

A second method used to identify the fungal partner of the ectomycorrhiza is isolation of the fungus from the mycorrhiza (Fig. 8E)

[5] Zak (1970).

Scale line equals 3 cm. Fungal partner of mycorrhizae is readily determined by comparing rhizomorphs and mycelium attached to sporocarps with those attached to mycorrhizae. Often rhizomorphs can be traced from sporocarp to mycorrhizae. B. Hypogeous sporocarp of *Hysterangium separabile* Zeller attached by rhizomorph to *P. menziesii* mycorrhizae. C. *P. menziesii* ectomycorrhizae formed by *Hebeloma crustuliniforme*. The characteristic surrounding mycelium containing small, white sclerotia is also found at the bases of overlying sporocarps of *H. crustuliniforme*. D. Regular open-pinnate *P. menziesii* mycorrhiza formed by pure culture synthesis with *T. terrestris*; compare with natural mycorrhizae of Fig. 1C. E. Fungal symbiont emerging onto nutrient agar medium from H_2O_2 surface-sterilized piece of *P. menziesii* mycorrhiza. B–E scale lines equal 2 mm.

and its comparison with known cultures developed from sporocarps, as demonstrated recently by Lamb and Richards (1970). The method is not difficult provided that the fungus will grow on laboratory media. Fresh, initially clean mycorrhizae, requiring little or no cleaning, give best results. Those with bits of debris attached to or embedded in their mantles are rarely treated successfully. Surface sterilants used include solutions of $Ca(OCl)_2$, $Na(OCl)$, $HgCl_2$, and H_2O_2. The author has had good success treating 2- to 5-mm-long mycorrhiza pieces and 5- to 10-mm-long rhizomorph pieces in 30% H_2O_2 for 5 to 20 sec, followed immediately by a 2- to 5-min rinse in sterile water (Zak, 1969a, 1971a,b). Media found best were potato-dextrose agar, as formulated by Lacy and Bridgmon (1962), and Melin-Norkrans medium, as modified by Marx (1969).

Rather than compare the isolated fungus with cultures from a general stock culture collection, it is far better to compare it with cultures developed from likely sporocarps found in the immediate vicinity of the place where the mycorrhizae were collected. Identifying criteria may include rate of mat growth, color, texture, presence or absence of rhizomorphs and/or hyphal strands, and their character, whether raised or appressed, UV light fluorescence, and color reactions produced by various chemical reagents. Use of more than one medium will increase the chance of success. Various microscopic features should be compared. Hyphal fusion may be attempted between the mycorrhizal isolate and known fungi (Buller, 1931). It was used successfully by Lamb and Richards (1970).

Although apparently straightforward, this method could be difficult to apply successfully. Isolates from mycorrhizae are not matched easily with known cultures, even when the latter are from sporophores collected near the mycorrhizae. One reason for this is the high variability of many ectomycorrhizal fungi. Another would be an inadequate collection of known fungi for comparison, and, as with the pure culture mycorrhizal synthesis technique, this method is limited to those fungi which will grow on laboratory media.

C. Tracing Rhizomorphs and Hyphae to Sporocarp

Fungi of some ectomycorrhizae may be identified by actually tracing connecting rhizomorphs from mycorrhiza to sporocarp (Fig. 8A,B). In this manner Woodroof (1933) determined *Xerocomus chrysenteron* (Bull. ex St. Am.) Quél. (*Boletus communis*) as the fungal partner of a pecan mycorrhiza. The method has also been used by Schramm (1966), Zak and Bryan (Zak and Marx, 1964), Chilvers (1968), Zak (1969a), and others to link identified sporocarps to mycorrhizae.

It is also possible to establish a connection between mycorrhiza and sporocarp via mycelium if the mycelium is distinctive and abundant, and if sporocarp and mycorrhiza are close to each other. Such coupling may often be easily established with sporocarps of hypogeous fungi nestled among mycorrhizae (Fig. 8B), as demonstrated by Fassi (1965) for *Pinus strobus* and *Endogone lactiflua*. Other examples are those reported by Peyronel (1922), Melin (1923), Dominik (1961), Fontana (1963), Schramm (1966), Fontana and Centrella (1967), and Laiho (1970). The author has commonly observed in the Pacific Northwest bright, yellow-orange *Pseudotsuga menziesii* mycorrhizae (Fig. 5C) enveloped in concolorous mycelium connected to resupinate sporocarps of *Corticium bicolor.*

Some have regarded rhizomorph or mycelial connections between sporocarp and mycorrhiza as inadequate evidence to establish identity of the fungal partner. It is believed that in some cases the rhizomorphs or mycelium may be merely attached superficially to and not actually joined to and part of the fungal mantle. This, however, can be easily ascertained by microscopic examination. Use of a hand lens (× 10) will usually confirm that the attached rhizomorphs or mycelium belong to the same fungus forming the mantle, because of the color, texture, and other gross characters of the respective tissues. Furthermore, several collections of the mycorrhiza should be examined for the regular presence of the same attached rhizomorphs or mycelium.

D. LINKING SPOROCARP TO UNDERLYING MYCORRHIZAE

A fourth method for identifying the fungal symbiont of the ectomycorrhiza is to compare rigidly mycelium and rhizomorphs at the base of the sporocarp with fungal tissues attached to underlying mycorrhizae (Fig. 8A,C). A cursory form of the method was used by Peyronel (1922) and others to establish tree–fungus associations. Briefly, it consists of locating a likely sporophore, or preferably several in a close group, and gathering these and the underlying mycorrhizae for laboratory analysis. A tentative selection of the mycorrhiza in the field can be made by comparing gross features with a hand lens.

In the laboratory, tissues of sporocarp and mycorrhiza are compared meticulously using characters previously listed for mycelia and rhizomorphs attached to mycorrhizae. Fungi of a few ectomycorrhizae may possibly be identified with confidence by matching one or two unusual and striking characters, as for the *Pseudotsuga menziesii* mycorrhiza formed by *Hebeloma crustuliniforme* (Bull. ex St. Am.) Quél. formerly referred to by Zak (1971b) as *Inocybe xanthomellas* Boursier & Kühner (Fig. 8C). However, most ectomycorrhizae will require comparison of at least three distinctive features to make an accurate identification of the fungal symbiont.

The fourth method is regarded as best for mycorrhizal identification because it provides an accurate and reliable identification of the fungal symbiont of the natural ectomycorrhiza, it can be readily applied, and it does not require that the fungus grow on artificial media. Its only disadvantage is that it cannot be applied everywhere and anytime since it is dependent on sporophore occurrence.

However, no method should be regarded as exclusive. In practice two or more of those described should be employed when feasible, as Laiho (1970) demonstrated with *Paxillus involutus* mycorrhizae. Another example is the identification of two strains of *Poria terrestris* as symbionts of two distinct *Pseudotsuga menziesii* mycorrhizae by Zak (1969a): (i) Sporocarps coupled to mycorrhizae via rhizomorphs were observed; (ii) underside mycelium of resupinate sporocarps was identical to that surrounding mycorrhizae; (iii) rhizomorphs attached to sporocarps were found to be identical to those attached to mycorrhizae; (iv) fungi isolated in pure culture from sporocarps were judged identical to those isolated from mycorrhizae; and (v) *Pseudotsuga menziesii* mycorrhizae were synthesized in pure culture with one of the *P. terrestris* strains.

V. Naming the Ectomycorrhiza

Each distinct tree–fungus ectomycorrhiza, in addition to being described fully, should, the author believes, be identified by a designation to confer an identity on it. A preliminary name can be simply an arbitrary designation by tree species, such as Douglas fir Type I, or by color of mantle or some other easily apparent gross feature as, for example, Douglas fir Tuberculate. If and when the fungus is isolated from the mycorrhiza in pure culture, an arbitrary designation of the fungus may be included, e.g., Douglas fir + M-23. And, finally, when the fungus is identified, the designation becomes *Pseudotsuga menziesii* + *Rhizopogon vinicolor*. Combination mycorrhizae, representing two or more indistinguishable mycorrhizae formed by as many closely related fungi, may be named, in addition to tree species, according to a type species of the fungi embraced.

VI. A Practical Key for Identification of Ectomycorrhizae

To be useful, a key for identification of ectomycorrhizae must be based on characters which are relatively easy to determine but which are stable enough to ensure accuracy. Thus, rather than starting with microscopic characters of the mantle according to Dominik (1969) or stable hyphal characters as Trappe (1967a) suggested, the proposed key would begin with tree species followed by those characters which can be easily discerned at low magnification (× 5–50). These would be followed by more difficult to determine stable microscopic features, and finally by difficult to determine and less stable or less useful characters. Thus, features of the ectomycorrhiza previously listed and described may be ranked accordingly:

1. Tree species.
2. Color of mycorrhiza (× 10).
3. Attached mycelium; presence or absence and gross (× 10–50) character.
4. Attached rhizomorphs; presence or absence and gross (× 10–50) character.
5. Surface texture of mantle (× 10–50).
6. Form or structure (× 5).
7. Ultraviolet light fluorescence (× 10).
8. Chemical reagent color reaction (× 10).
9. Attached mycelium microscopic (× 100–1000) features.
10. Attached rhizomorphs microscopic (× 100–1000) features.
11. Form of individual elements (× 10).
12. Ultraviolet light fluorescence of hyphae (× 100–400).
13. Chemical reagent effect on hyphae (× 100–1000).
14. Character of mantle surface hyphae (× 100–1000).
15. Mantle structure (× 100–1000).
16. Individual element diameter and uniformity in diameter (× 10–50).
17. Character of the Hartig net (× 100–1000).
18. Habitat.
19. Taste and odor.

First keys for identification of ectomycorrhizae and their symbiotic fungi will probably be constructed for individual tree species and for specific habitats, as Trappe (1967a) has suggested. Later, as more evidence is accumulated indicating that mycorrhiza morphology for a given fungus varies little, if any, between tree species in the same genus, tree species keys may be combined into tree genus keys. The individual tree species identity of each mycorrhiza, however, should always be retained.

VII. Discussion

Beginning with Melin's (1927) very abbreviated scheme, attempts have been made to classify ectomycorrhizae to help study these unique structures and better understand their many interrelationships with tree, fungus, and environment. The classification system devised by Dominik

(1956, 1969) is currently in use by various investigators for lack of a better method. It has been used in a number of ecological studies, and its generic designations have been incorporated into descriptions of pure culture synthesized and natural mycorrhizae. While simple in design, it is all encompassing, including all ectomycorrhizal trees and all ectomycorrhizal fungi. Its two main categories, subtype and "genus," are based on distinct macroscopic and microscopic characters which Dominik has clearly illustrated, especially in his 1969 version. Using the straightforward and easy to use key, it is not difficult to arrive at the correct "genus" for most ectomycorrhizae.

Dominik's system has, however, a serious flaw which makes its continued use questionable. Its fault is not that it omits the identity of the fungal symbiont, as such, but rather that it ignores the distinct tree-fungus mycorrhiza and establishes instead artificial morphological groupings, viz., "genera." While two or three of the 75 "genera" listed (Dominik, 1969) represent distinct mycorrhizae, each formed by a specific fungus, such as "genus" Ga with *Cenococcum graniforme* as fungal symbiont, the remainder embrace one or many—one does not know how many—different mycorrhizae. Mycorrhizae belonging to a "genus" have only a few arbitrarily chosen morphological features in common; otherwise they may be totally unrelated and be formed by completely different fungi representing different species and even genera.

The weak base upon which Dominik's "genera" are established is illustrated by the effect of aging on mycorrhizal classification. As noted earlier, many ectomycorrhizae begin with one color and gradually change to another as they become older. According to Dominik's system subtypes A, B, C, F, and H are separated into "genera" solely on the basis of mantle color. For example, mantles of "genera" Ha, Hb, and Hc are, respectively, "colorless or white-grayish," "cream-colored," and "yellowish to yellow." Thus, because a mycorrhiza may be first one color when young and another when older, it may have more than one generic designation. Or as Trappe (1967b) noted with *Pseudotsuga menziesii* + *Rhizopogon colossus* in pure culture, the mycorrhiza began in "genus" Ca, changed to Ce after a few weeks, and ended in "genus" Cf. Possibly other criteria in Dominik's system, such as character of mantle surface hyphae, may also change sufficiently by aging to transfer a mycorrhiza from one "genus" to another.

The question may then be asked, what significance is there in assigning ectomycorrhizae to such artificial and purposeless categories? What, for example, is meant by a mycorrhiza classed Ae? According to Dominik's latest key (1969), these mycorrhizae may be monopodial or ramiform, with a mantle which is light ochre, single-layered, feltlike, and often with mycelial strands on its surface. Obviously, this very broad and general characterization may apply to many different ectomycorrhizae. Is an Ae mycorrhiza of *Pinus strobus* in Europe, for example, the same as an Ae mycorrhiza of *P. strobus* in North America? Are the fungal symbionts the same? If not, and they very likely may not be, the mycorrhizal designation Ae has little meaning. It relates only to an artificial grouping of morphological characters which define its "genus." It bears no real relationship to other ectomycorrhizae or to other ectomycorrhizal fungi.

Several ecological listings of ectomycorrhizae have been prepared based on Dominik's classification scheme. Except for those "genera" which actually represent specific fungal symbionts [i.e., Ad = *Corticium bicolor;* Cd = *Suillus luteus;* and Ga = *Cenococcum graniforme* (Dominik, 1959, 1961, 1963)], cataloging different "genera" and their frequency is of dubious value. There can be but little meaning in, for example, finding "genus" Fg in one habitat and not in another, or finding that "genus" Ab is more abundant than "genus" Be. For the most part, the listings are little more than collections of symbols which express no true relationships within the forest.

In contrast to Dominik's plan, the classification system here proposed has, as its basis, the identifiably distinct ectomycorrhiza. Rather than being a new system, it advocates strict adherence, insofar as possible, to the natural classification system already in use for tree and fungus. Most mycorrhizae will each represent a single tree species and a single fungus species. Some, however, may be "combination" mycorrhizae representing two or more indistinguishable mycorrhizae formed by as many closely related fungi. Newly discovered mycorrhizae will be fully described and, ideally, their fungal symbionts named. Combination mycorrhizae will be named, in addition to tree species, according to a type species of the fungi embraced. Mycorrhizae whose fungi are yet unidentified will be named by tree species and, tentatively, by an arbitrary designation.

A complete and accurate description of each ectomycorrhiza is essential for its identification and classification. It should include photographs and/or drawings clearly illustrating important details. The characterization should be based on several collections of the ectomycorrhiza, and variation in morphology, that may be expected, should be noted. Thus, changes in color and other features of the mantle resulting from aging of the mycorrhiza would be included in the description. Sporocarps used to identify the fungal symbiont should, as Trappe (1967a) has emphasized, be deposited in a herbarium for future reference.

We may expect classification of ectomycorrhizae to benefit taxonomical and ecological studies of their fungal symbionts. Examination of underlying mycorrhizae, for example, may aid identification of sporocarps. Formation of morphologically identical mycorrhizae by some species may be justification for combining these species, while formation of distinct mycorrhizae by different strains of a fungus may suggest the establishment of new species. Also, identification of ectomycorrhizae will allow development of a more complete distribution record for many fungi. This will be especially true for those fungi which fruit infrequently or rarely or have not yet been reported from some regions. A good example is *Rhizopogon vinicolor* associated with the tuberculate mycorrhiza of *Pseudotsuga menziesii* in the Pacific Northwest. Although sporocarps have not yet been reported from Europe, Dominik (1963) and Dominik and Majchrowicz (1967) apparently have described the same tuberculate mycorrhiza of *Pseudotsuga menziesii* in Poland.

Identification of even the more common ectomycorrhizae of only important timber species is a formidable but not impossible task. A start has already been made with the numerous definitive descriptions of ectomycorrhizae that have recently appeared in scientific journals. More will follow, and eventually identification keys based on tree species will be constructed. Because the fungal symbiont appears to dominate ectomycorrhizal morphology, it probably will be feasible to combine tree species keys into tree genus keys. Possibly even some tree genera, such as *Abies, Larix, Picea, Pseudotsuga,* and *Tsuga,* may be placed in one key for identifying the fungal symbiont.

Many new discoveries await us among the great diversity of ectomycorrhizae and their fungi in nature. The key to this knowledge is a meaningful and practical classification system based on the identifiably distinct ectomycorrhiza, and one which reflects the natural classification of both the tree and fungus partners. It will permit more definitive examination of ectomycorrhizae in their natural state. The different types present, their density and frequency, and how they vary from habitat to habitat will be determined. Such a classification will make it possible to test laboratory findings and later apply them in the forest and nursery in order to grow more vigorous and healthier trees. Much painstaking and concerted effort will be needed to achieve this goal; there is no short cut.

References

Anderson, L. E. (1954). Hoyer's solution as a rapid permanent mounting medium for bryophytes. *Bryologist* **57**, 242.

Boullard, B. (1965). Considérations sur la systématique des mycorrhizes ectotrophes. *Bul. Soc. Bot. Fr.* [5] **112**, 272.

Bowen, G. D. (1965). Mycorrhiza inoculation in forestry practise. *Aust. Forest.* **29**, 231.

Bowen, G. D. (1970). Mycorrhizal responses of radiata pine in experiments with different fungi. *Aust. Forest.* **34**, 183.

Briscoe, C. B. (1959). Early results of mycorrhizal inoculation of pine in Puerto Rico. *Carib. Forest.* **20**, 73.

Buller, A. H. R. (1931). "Researches on Fungi," Vol. V. Longmans, Green, New York.

Ceruti, A., and Bussetti, L. (1962). On the mycorrhizal symbiosis of *Boletus subtomentosus, Russula grisea, Balsamia platyspora* and *Hysterangium clathroides* with limes (*Tilia* sp.). *Allionia* **8**, 55.

Chilvers, G. A. (1968). Some distinctive types of eucalypt mycorrhiza. *Aust. J. Bot.* **16**, 49.

Dominik, T. (1956). Tentative proposal for a new classification scheme of ectotrophic mycorrhizae established on morphological and anatomical characteristics. *Roczn. Nauk Les.* **14**, 223. [English transl., U. S. Dept. of Commerce OTS 60-21383 (1962).]

Dominik, T. (1959). Development dynamics of mycorrhizae formed by *Pinus silvestris* and *Boletus luteus* in arable soils. *Pr. Szczecinskiego Tow. Nauk.* **1**, 1. [English transl., U. S. Dept. of Commerce TT 65-50332 (1965).]

Dominik, T. (1961). Studies on mycorrhizae. *Folia Forest. Pol., Ser. A* No. 5, p. 3. [English transl., U. S. Dept. of Commerce TT 65-50333 (1966).]

Dominik, T. (1963). Occurrence of Douglas-fir (*Pseudotsuga taxifolia* Britton) in various Polish stands. *Inst. Bad. Les.* **258**, 29. [English transl., U. S. Dept. of Commerce TT 65-50353 (1966).]

Dominik, T. (1969). Key to ectotrophic mycorrhizae. *Folia Forest. Pol. Ser. A* No. 15, p. 309.

Dominik, T., and Majchrowicz, I. (1967). Studies on the tuberculate mycorrhizae of Douglas-fir (*Pseudotsuga taxifolia* Britton). *Ekol. Pol., Ser. A* No. 15, p. 75.

Fassi, B. (1965). Ectotrophic mycorrhizae produced by Endogone *lactiflua* Berk. on *Pinus strobus* L. *Allionia* **11**, 7.

Fassi, B., and Fontana, A. (1961). The ectotrophic mycorrhizas of *Julbernardia seretii*, Caesalpiniacea of Congo. *Allionia* **7**, 131.

Fassi, B., and Fontana, A. (1962). Ectotrophic mycorrhizae of *Brachystegia Laurentii* and some other Caesalpiniaceae of the Congo. *Allionia* **8**, 121.

Fassi, B., and de Vecchi, E. (1962). Researches in ectotrophic mycorrhizae of *Pinus strobus* in nurseries. I. Description of some of the most common forms in Piedmont. *Allionia* **8**, 133.

Fassi, B., Fontana, A., and Trappe, J. M. (1969). Ectomycorrhizae formed by Endogone *lactiflua* with species of *Pinus* and *Pseudotsuga*. *Mycologia* **61**, 412.

Fontana, A. (1961). First contribution to the study of poplar mycorrhizas in Piedmont. *Allionia* **7**, 87.

Fontana, A. (1962). Researches on the mycorrhizae of the genus *Salix*. *Allionia* **8**, 67.

Fontana, A. (1963). Mycorrhizal symbiosis of *Hebeloma hiemale* Bres. with a willow and a poplar. *Allionia* **9**, 113.

Fontana, A., and Centrella, E. (1967). Ectomycorrhizae produced by hypogeous fungi. *Allionia* **13**, 149.

Frank, A. B. (1885). Über die auf Wurzelsymbiose beruhende Ernährung gewisser Bäume durch unterirdische Pilze. *Ber. Deut. Bot. Ges.* **3**, 128.

Hacskaylo, E. (1953). Pure culture synthesis of pine mycorrhizae in Terra-lite. *Mycologia* **45**, 971.

Harley, J. L. (1969). "The Biology of Mycorrhiza." Leonard Hill, London.

Hatch, A. B., and Doak, K. D. (1933). Mycorrhiza and other features of the root systems of *Pinus*. *J. Arnold Arboretum, Harvard Univ.* **14**, 85.

Jeník, J. (1957). Kořenový systém dubu letního a zimního. *Rozpr. Cesk. Akad. Ved, Rada Mat. Prirod. Ved* **64**, 1.

Lacy, M. L., and Bridgmon, G. H. (1962). Potato-dextrose agar prepared from dehydrated mashed potatoes. *Phytopathology* **52**, 173.

Laiho, O. (1970). *Paxillus involutus* as a mycorrhizal symbiont of forest trees. *Acta Forest. Fenn.* **106**, 1.

Lamb, R. J., and Richards, B. N. (1970). Some mycorrhizal fungi of *Pinus radiata* and *P. elliottii* var. *elliottii* in Australia. *Trans. Brit. Mycol. Soc.* **54**, 371.

Lundeberg, G. (1970). Utilisation of various nitrogen sources, in particular bound soil nitrogen, by mycorrhizal fungi. *Stud. Forest. Suec.* **79**, 1.

Luppi, A. M., and Gautero, C. (1967). Researches on the mycorrhizae of *Quercus robur, Q. petraea*, and *Q. pubescens* in Piedmont. *Allionia* **13**, 129.

Marks, G. C. (1965). The classification and distribution of the mycorrhizas of *Pinus radiata*. *Aust. Forest.* **29**, 238.

Marx, D. H. (1969). The influence of ectotrophic mycorrhizal fungi on the resistance of pine roots to pathogenic infections. I. Antagonism of mycorrhizal fungi to root pathogenic fungi and soil bacteria. *Phytopathology* **59**, 153.

Marx, D. H., and Davey, C. B. (1969a). The influence of ectotrophic mycorrhizal fungi on the resistance of pine roots to pathogenic infections. III. Resistance of aseptically formed mycorrhizae to infection by *Phytophthora cinnamomi*. *Phytopathology* **59**, 549.

Marx, D. H., and Davey, C. B. (1969b). The influence of ectotrophic mycorrhizal fungi on the resistance of pine roots to pathogenic infections. IV. Resistance of naturally occurring mycorrhizae to infections by *Phytophthora cinnamomi*. *Phytopathology* **59**, 559.

Marx, D. H., and Zak, B. (1965). Effect of pH on mycorrhizal formation of slash pine in aseptic culture. *Forest Sci.* **11**, 66.

May, C. (1961). An exploratory survey for fluorescent substances in Hymenomycetes collected in the vicinity of Washington, D. C. *Plant Dis. Rep.* **45**, 777.

Melin, E. (1921). Über die Mykorrhizenpilze von *Pinus silvestris* L., und *Picea abies* (L.) Karst. *Sv. Bot. Tidskr.* **15**, 192.

Melin, E. (1923). Experimentelle Untersuchungen über die Konstitution und Ökologie der Mykorrhizen von *Pinus silvestris* L. und *Picea abies* (L.) Karst. *Mykol. Unters. Ber.* **2**, 73.

Melin, E. (1927). Studier över barrtradsplantans utveckling i råhumus. II. Mykorrhizans utbildning hos tallplantan i olika råhumusformer. [German summary]. *Medd. Skogsforsoks anstalt, Stockholm* **23**, 433.

Mikola, P. (1948). On the physiology and ecology of *Cenococcum graniforme* especially as a mycorrhizal fungus of birch. *Commun. Inst. Forest. Fenn.* **36**, 1.

Mikola, P. (1962). The bright yellow mycorrhiza of raw humus. *Proc., Int. Union Forest Res. Organ., 13th, 1961* No. 24-4.

Moser, M. (1956). Die Bedeutung der Mykorrhiza für Aufforstungen in Hochlagen. *Forstwiss. Centralbl.* **75**, 8.

Neal, J. L., Jr., Bollen, W. B., and Zak, B. (1964). Rhizosphere microflora associated with mycorrhizae of Douglas-fir. *Can. J. Microbiol.* **10**, 259.

Pachlewska, J. (1968). Studies on mycorrhizal synthesis of pine (*Pinus silvestris* L.) in pure cultures on agar. *Inst. Bad. Les.* 345, 3.

Pachlewski, R., and Pachlewska, J. (1968). *Rhizopogon luteolus* Fr. in a synthesis with pine (*Pinus silvestris* L.) in pure culture in agar. *Inst. Bad. Les.* 346, 77.

Peyronel, B. (1922). Nuovi casi di rapporti micorizici tra Basidiomiceti e fanerogame arboree. *Bol. Soc. Bot. Ital.* 4, 50.

Peyronel, B. (1934). Il sapore e alcune reazioni microchimiche delle micorrize ectotrofiche prodotte da Russule e Lattarii. *Nuovo G. Bot. Ital.* [N. S.] 41, 744.

Peyronel, B., and Fassi, B. (1957). Micorrize ectotrofiche in una Cesalpiniacea del Congo Belga. [In Italian, French summary.] *Atti Acad. Sci. Torino* 91, 1.

Rambelli, A. (1967). Atlas of some mycorrhizal forms observed on *Pinus radiata* in Italy. [Italian-English.] *Cent. Sper. Agr. Forest.* 9, Suppl., 1.

Rambelli, A. (1970). Second atlas of some mycorrhizal forms observed on *Pinus radiata* in Italy. [Italian-English.] *Cent. Sper. Agr. Forest.* 10, Suppl., 1.

Scannerini, S., and Palenzona, M. (1967). Researches on ectomycorrhizae of *Pinus strobus* in nurseries III. Mycorrhizae of *Tuber albidum* Pico. *Allionia* 13, 187.

Schramm, J. R. (1966). Plant colonization studies on black wastes from anthracite mining in Pennsylvania. *Trans. Amer. Phil. Soc.* [N. S.] 56, 1.

Sen, D. N. (1961). Root ecology of *Tilia europea* L. I. On the morphology of mycorrhizal roots. *Preslia* 33, 341.

Singer, R. (1962). "The Agaricales in Modern Taxonomy." Cramer, Weinheim.

Slankis, V. (1951). Über den Einfluss von β-Indolylessigsäure und andere Wirksstoffen auf das Wachstum von Kiefernwurzeln. *Symb. Bot. Upsal.* 11, 1.

Trappe, J. M. (1962). Fungus associates of ectotrophic mycorrhizae. *Bot. Rev.* 28, 538.

Trappe, J. M. (1964). Mycorrhizal hosts and distribution of *Cenococcum graniforme*. *Lloydia* 27, 100.

Trappe, J. M. (1965). Tuberculate mycorrhizae of Douglas-fir. *Forest Sci.* 11, 27.

Trappe, J. M. (1967a). Principles of classifying ectotrophic mycorrhizae for identification of fungal symbionts. *Proc., Int. Union Forest Res. Organ., 14th, 1967* Sect. 24, p. 46.

Trappe, J. M. (1967b). Pure culture synthesis of Douglas-fir mycorrhizae with species of *Hebeloma, Suillus, Rhizopogon, and Astraeus. Forest Sci.* 13, 121.

Woodroof, N. (1933). Pecan mycorrhizas. *Ga. Agr. Exp. Sta., Bull.* 178.

Young, H. E. (1940). Mycorrhiza and growth of *Pinus* and *Araucaria. J. Aust. Inst. Agr. Sci.* 6, 21.

Zak, B. (1964). Role of mycorrhizae in root disease. *Annu. Rev. Phytopathol.* 2, 377.

Zak, B. (1965). Aphids feeding on mycorrhizae of Douglas-fir. *Forest Sci.* 11, 410.

Zak, B. (1967). A nematode (*Meloidodera* sp.) on Douglas-fir mycorrhizae. *Plant Dis. Rep.* 51, 264.

Zak, B. (1969a). Characterization and classification of mycorrhizae of Douglas-fir. I. *Pseudotsuga menziesii* + *Poria terrestris* (blue- and orange-staining strains). *Can. J. Bot.* 47, 1833.

Zak, B. (1969b). Four *Poria terrestris* (DC ex Fries) Sacc. strains mycorrhizal with roots of Douglas-fir. *Abstr. 11th Int. Bot. Congr., 1969* p. 247.

Zak, B. (1970). Unpublished data.

Zak, B. (1971a). Characterization and classification of mycorrhizae of Douglas-fir. II. *Pseudotsuga menziesii* + *Rhizopogon vinicolor. Can. J. Bot.* 49, 1079.

Zak, B. (1971b). Characterization and identification of Douglas-fir mycorrhizae. *In*

"Mycorrhizae" (E. Hacskaylo, ed.), USDA Misc. Publ. No. 1189, pp. 38–53. US Govt. Printing Office, Washington, D. C.

Zak, B. (1971c). Detoxication of autoclaved soil by a mycorrhizal fungus. U. S. *Forest Serv., Res. Pap. PNW*-**159**, 1.

Zak, B., and Marx, D. H. (1964). Isolation of mycorrhizal fungi from roots of individual slash pines. *Forest Sci.* **10**, 214.

CHAPTER 3

Distribution of Ectomycorrhizae
in Native and Man-Made Forests

F. H. MEYER

I. Ectomycorrhizae in Native Forests

Frank, a pioneer in the research on mycorrhizae, coined this term in 1885. Later in 1887 he distinguished two types of mycorrhizae and called these ectotrophic and endotrophic, according to the presence of fungal mycelium outside or within the cells of the root cortex. Recently, however, the terminology has been changed by Peyronel *et al.* (1969) to ecto- and endomycorrhizae. Frank stated that in ectomycorrhizae there must be a close symbiotic relationship between the tree and the mycorrhizal fungus. In his theory of ectomycorrhizae he proposed that water and mineral nutrients were absorbed by, and translocated through, the fungus into the tree.

Later, there were differences of opinion regarding the function of ectomycorrhizae until Melin (1917) published observations which supported the ideas of Frank. Melin examined pine and spruce seedlings

79

which had been planted on freshly drained peat bogs and were found to lack ectomycorrhizal fungi. The seedlings at first remained stunted and became yellow in color. Later, only those trees that grew normally had roots that had been infected with mycorrhizal fungi, which colonized the bog after it had been drained. As shown by Melin (1917), Hatch (1936, 1937), Melin and Nilsson (1950, 1952, 1955, 1958), Melin *et al.* (1958), and Harley and Brierley (1955), ectomycorrhizae play a significant role in the mineral nutrition of trees. Nevertheless, tree species bearing ectomycorrhizae under normal woodland conditions can be cultivated aseptically in an Erlenmeyer flask without any fungal associate, as long as a solution of suitable nutrient is present (Lundeberg, 1960; Melin, 1939; Schweers and Meyer, 1970). Therefore, the trees are not absolutely dependent on a fungal associate. However, in natural substrates they become stunted or even die if ectomycorrhizal fungi are absent. Therefore, important theoretical and practical interest is connected with the questions of which tree species are dependent on ectomycorrhizal fungi, in which plant formations do these trees occur, and under what soil conditions the mycorrhizae are formed.

A. Occurrence in Systematic Categories of Woody Plants

As shown by Janse (1897), Stahl (1900); Gallaud (1905), Asai (1935), and Maeda (1954), mycorrhizae are widely distributed among the phanerogams. But most of these mycorrhizae belong to the type of endomycorrhizae and only about 3% of the phanerogams exhibit ectomycorrhizae. A compilation of the known ectomycorrhizal trees can be found in Table I. Some of the genera mentioned have ecto- as well as endomycorrhizae, for instance, *Juniperus, Cupressus, Salix, Malus, Pyrus, Tilia, Eucalyptus,* and *Arbutus.* As shown in Table I, ectomycorrhizae are most common in the families Pinaceae, Salicaceae, Betulaceae, and Fagaceae; in other families, they only exist in some genera. Their presence in distinct systematic categories might have various reasons:

(a) Certain chemical compounds are involved in the mechanism regulating the formation of the mycorrhizae such as orchinol (Gäumann *et al.,* 1960). These substances apparently are synthesized in some taxa and not in others, or they might be formed in definite amounts. In this respect there is a unexplored field for investigation.

(b) The woody plants with ectomycorrhizae succeed only in regions or in soils where the symbiotic fungus is present or where it is capable of growing. Thus, for instance, in Poland, Dominik (1948) found that *Pyrus communis* produced ectomycorrhizae only within a forest (forest soil with a layer of decomposing litter) and not outside in gardens. In

TABLE I

GENERA OF PHANEROGAMS WITH ECTOMYCORRHIZA[a]

Gymnospermae	Fagaceae (*Continued*)
Pinaceae	*Pasania*
Abies	*Quercus*
Cathaya	*Trigonobalanus*
Cedrus	Urticales
Keteleeria	Ulmaceae
Larix	*Ulmus*
Picea	Guttiferales
Pinus	Dipterocarpaceae
Pseudolarix	(no special genus mentioned)
Pseudotsuga	Rosales
Tsuga	Rosaceae
Cupressaceae	*Crataegus*
Cupressus	*Malus*
Juniperus	*Pyrus*
Angiospermae	*Sorbus*
Juglandales	Leguminosae (Caesalpinioideae)
Juglandaceae	*Afzelia*
Carya	*Anthonotha*
Juglans	*Brachystegia*
Salicales	*Gilbertiodendron*
Salicaceae	*Julbernardia*
Populus	*Monopetalanthus*
Salix	*Paramacrolobium*
Fagales	Sapindales
Betulaceae	Sapindaceae
Alnus	*Allophylus*
Betula	Aceraceae
Carpinus	*Acer*
Corylus	Malvales
Ostrya	Tiliaceae
Ostryopsis	*Tilia*
Fagaceae	Myrtiflorae
Castanea	Myrtaceae
Castanopsis	*Eucalyptus*
Fagus	Ericales
Lithocarpus	Ericaceae
Nothofagus	*Arbutus*

[a] Compiled from the following references: Rivett (1924), Levisohn (1958), Lobanow (1960), Singer and Morello (1960), Fassi and Fontana (1961, 1962), Trappe (1962), Fassi (1963), Chilvers and Pryor (1965), Bakshi (1966), Singh (1966), Moser (1967), and Redhead (1968).

this respect we can distinguish between tree species which form ectomycorrhizae under all environmental conditions and others that are to a certain degree "elastic" (Levisohn, 1954; Lobanow, 1960), suggesting that they have ectomycorrhizae on some sites and not on others.

Under natural conditions, trees of the first group are incapable of developing without ectomycorrhizae, although, as already mentioned, we can cultivate these aseptically in Erlenmeyer flasks. This group of trees could be named obligate ectomycorrhizal. The reason for their debilitation in the absence of ectomycorrhiza might be due to the following. The truly ectomycorrhizal fungus acts primarily as a parasite and takes advantage of the tree. On infection the tree develops defensive mechanisms. Only when a state of equilibrium has been reached between the pathogenic capacity of the fungus and the defensive mechanisms of the host is the balance maintained and can both partners live together "eusymbiotically" for a longer period, provided that the association is advantageous to both partners.

In the absence of truly mycorrhizal fungi, other fungi may penetrate into the root. The tree responds to this invasion by forming toxic compounds such as the tannins which fill the outer cells of the root cortex. Furthermore, the cell walls of the root cortex become suberized and, as in true ectomycorrhizae, the formation of root hairs is suppressed (cf. Figs. 1 and 2). Under such conditions the normal function of the root as an organ for absorption might be hindered or greatly diminished and consequently the trees decline. Björkman (1942), Bergemann (1955), and Meyer (1962) consider that the absence of the fungal sheath or a thin, weakly developed mantle and intracellular infection of the cortex in trees that normally have ectomycorrhizae are signs of a pathogenic relationship between fungus and tree. Poorly growing trees often exhibit a high degree of such anatomical features. But in other cases involving a distinct fungus, no disease accompanies intracellular infection and simultaneous formation of a fungal sheath (Mikola, 1965; Meyer, 1968). Obligately ectomycorrhizal trees are, for example, members of the genera *Abies, Larix, Picea, Pinus, Carpinus, Fagus,* and *Quercus.* Plantations of these trees may fail to grow well in the absence of truly ectomycorrhizal fungi.

Typical, facultative ectomycorrhizal genera are *Cupressus, Juniperus, Salix, Betula, Corylus, Alnus, Ulmus, Pyrus, Acer,* and *Eucalyptus.* These trees are capable of thriving in the absence of true ectomycorrhizal fungi. Most of them are the first invaders of wasteland and act as pioneers in forest succession. While plantations of the obligate ectomycorrhizal genus *Pinus* in subtropical regions may succumb because of the absence of ectomycorrhizal fungi (Mikola, 1969), plantations of the facultative ectomycorrhizal *Eucalyptus* and *Cupressus* have been successful. The facultative ectomycorrhizal trees improve the site in many respects for the trees that follow and under their influence true ectomycorrhizal fungi may colonize the area.

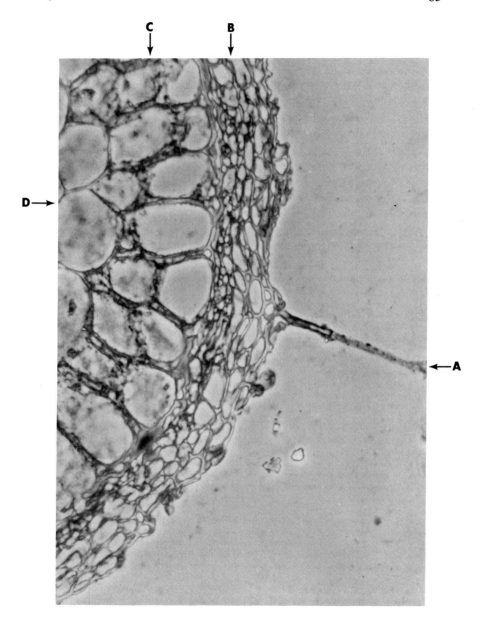

FIG. 1. Ectomycorrhiza of *Betula pendula.* ×527. A. Hyphae of fungal partner radiating into soil. B. Fungal sheath. C. Hartig net that facilitates the exchange of substances between tree and fungus. D. Normal cortex cells.

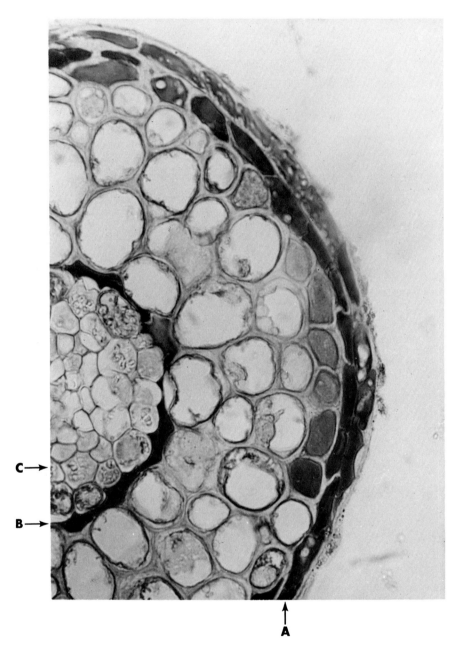

FIG. 2. "Pseudomycorrhiza" of *Picea abies*. ×666. A. Outer layer of root cortex filled with tannins. B. Endodermis. C. Cells of stele filled with starch grains.

B. Geographical Distribution

1. Occurrence in Different Forest Types

Most genera of woody plants mentioned in Section I,A, are important forest components of cool and temperate regions. Ectomycorrhizal tree species play a dominant role there and often forestry gave an impetus for research on ectomycorrhizae.

In boreal coniferous forests members of the Pinaceae are abundant (*Picea abies, P. obovata, P. glauca, P. mariana, Pinus sylvestris, P. sibirica, P. pumila, Abies balsamea, A. sibirica, Larix gmelinii,* and *L. laricina*). Broad-leaved trees of this zone (*Betula, Populus*) are also ectomycorrhizal. Coniferous forests of similar appearance extend far south of the coniferous forest belt. But southward of the boreal zone they mainly develop at the subalpine stage and often form the timberline. Subalpine forests in Europe consist of *Pinus cembra, P. mugo, P. sylvestris, Larix decidua, Picea abies,* and *Abies alba.* In the Mediterranean they contain *Cedrus atlantica, C. libani,* and *Juniperus excelsa;* in northern Anatolia and in the Caucasus they contain *Abies bornmülleriana, A. nordmanniana,* and *Picea orientalis.* In the Himalayas we find *Abies pindrow, A. webbiana, Cedrus deodara, Pinus griffithii, Larix griffithiana, Picea morinda, Juniperus* species, *Quercus semicarpifolia,* and *Betula utilis* (Schweinfurth, 1957). Widely distributed species in the subalpine forests of western North America are *Picea engelmannii, Abies lasiocarpa, Pinus contorta,* and *Pinus flexilis.* From Japan through Eurasia to eastern North America the temperate deciduous forests form a belt adjacent to boreal coniferous forests. The trees of the temperate deciduous forest exhibit endo- as well as ectomycorrhizae. Important ectomycorrhizal genera that form large forests are *Quercus, Fagus, Salix, Populus, Castanea, Betula, Carpinus, Alnus, Tilia, Carya,* and *Ostrya.* It must be emphasized that members of the genus *Quercus* extend to the timberline against treeless dry areas (steppe, prairie), e.g., *Quercus robur* (Russian Steppe); *Quercus pedunculiflora, Q. pubescens* var. *anatolica,* and *Q. haas* (Anatolian Steppe); *Q. stellata, Q. marilandica, Q. macrocarpa, Q. alba, Q. velutina,* and *Q. ellipsoidalis* (North American Prairie). In the northern grassland–forest boundary aspens are present, such as *Populus tremuloides* in North America and *P. tremula* in Eurasia.

The evergreen sclerophyllous forest which is climatically characterized by mild and rainy winters as well as hot and dry summers is spread over various parts of the earth (coastal regions of the Mediterranean, parts of California, southwest and southeast of Australia, southwest of the Cape Province, and in central Chile between La Serna and Talca). Important ectomycorrhizal trees in the evergreen sclerophyllous forest of the Mediterranean are *Quercus* species (*Q. ilex, Q. suber,* and *Q. coccifera*) and *Pinus* species (*P. pinea, P. halepensis, P. brutia,* and *P. pinaster*).

In California we find members of the genus *Pinus, Cupressus, Quercus, Castanopsis,* and *Lithocarpus. Pinus radiata,* the Monterey pine, which lives in a narrow strip along the Pacific coast of southern California became a famous tree for afforestation of wasteland in Mediterranean climates. Eucalypts are the dominant trees in the evergreen sclerophyllous forests of Australia. Over 400 species have been recorded. No ectomycorrhizal trees were described from the indigenous sclerophyllous forests of the Cape Province and of central Chile.

In the temperate rain forests of the Southern Hemisphere, which occur in south Chile, New Zealand, Tasmania, and southwest Australia, that extend up to the subalpine timberline, the ectomycorrhizal genus *Nothofagus* plays an important role (*N. dombeyi, N. obliqua, N. procera, N. pumilio, N. cunninghamii,* and *N. menziesii*). All the *Nothofagus* species of New Zealand and South America are ectomycorrhizal (Morrison, 1956; Singer, 1964). In tropical forests, especially rain forests, the proportion of ectomycorrhizal trees is, as known to date, low. In comparison to the several tens of thousands of woody plants within the tropics, the few members of the *Caesalpinioideae* and *Dipterocarpaceae* (Table I) with ectomycorrhizae are negligible. Among the conifers, *Pinus merkusii,* the most tropical of all *Pinus* species, must be mentioned. It is a useful pioneer tree in north Sumatra, Burma, and Thailand; it grows in tropical lowlands as well as in the uplands, and it often has been used successfully for afforestation of burned areas.

Woody plants with endomycorrhizae also appear dominant on the mountains of the tropical belt, although some ectomycorrhizal genera of the *Fagaceae,* namely, *Castanopsis, Quercus, Lithocarpus* and *Pasania,* grow in Southeast Asia.

In the Andean mountains, *Alnus jorulensis* forms large forests (Hueck, 1966) and partly reaches the timberline (Moser, 1967). In the higher regions of Central America and of South America (Colombia) *Quercus* species were also found (Bader, 1960).

2. The Ectotroph

As shown in the previous sections, endomycorrhizal trees are abundant in tropical forests of the lowlands, while the ectomycorrhizal ones occur there only sporadically. The ectomycorrhizal trees are often pioneers on wastelands.

In contrast, in the subalpine timberline of temperate regions, especially in the colder boreal forests, ectomycorrhizal trees dominate

(Fig. 3). According to Singer and Morello (1960), a climax forest with predominantly ectomycorrhizal trees is distributed mainly in areas with a larger amplitude in the annual periodicity of temperature. Moser (1967) emphasized that in the temperate zones the timberline is formed almost entirely by ectomycorrhizal trees. Under conditions of shorter growing seasons, as at the boreal and subalpine timberline, and at the timberline against the steppe and prairie, the superiority of ectomycorrhizal trees may be due to, among other reasons, their enlarged absorbing surface (Lobanow, 1960). The mycorrhizal fungi spreading into the soil promote the absorption capacity for nutrients and water and thus enable the ectomycorrhizal trees to complete their annual growth cycle and harden off within a relatively shorter time. Consequently they suffer less from the early frosts. While in the region of the timberline adjacent to steppe and prairie improved water uptake may be the decisive factor, in the region of subalpine and boreal timberline it might be the uptake of nutrients. According to Mikola (1969), the advantage of ecto-mycorrhizal trees in the region of the boreal timberline probably is also connected with the supply of organic nitrogen through the fungal part-ner. In cool climates the mineralization of organic nitrogen is hampered and often the humus form mor (raw humus) is present. Under such conditions the nitrogen requirement of trees might be better fulfilled with the aid of symbiotic mycelia. Melin and Nilsson (1953) gave evi-

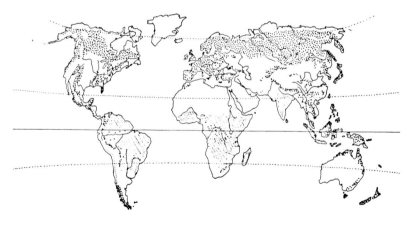

▨ predominant ectomycorrhizal forest

▨ predominant nonectomycorrhizal forest

Fig. 3. Distribution of ectomycorrhizal and nonectomycorrhizal forests (after Moser, 1967).

dence that labeled nitrogen from glutaminic acid could be transferred to pine seedlings through the hyphae of a strain of *Boletus variegatus*. When comparing the conditions for tree growth in different parts of the world and at various sites, ectomycorrhizal trees appear to be better adapted to unfavorable environmental conditions and, on such sites, seem to have greater vigor than other trees.

As already mentioned in Section I,A, the essential substrates required by many of the trees mentioned in Table I can only be supplied by their fungal partner. They do not survive in the absence of a functional symbiotic partner. Symbiosis is obligatory for these trees, as is often the case with the symbiotic partners of lichens. On the other hand, a large number of the mycorrhizal fungi are dependent on trees. The obligate mycorrhizal fungi develop their fruiting bodies only after a symbiotic relationship has been established with their hosts (Singer, 1963).

Since there are close connections and reciprocal relationships between ectomycorrhizal trees and their fungal partners, Singer and Morello (1960) compared this type of symbiosis with lichen symbiosis and coined the term "ectotroph" for the complex tree/fungus organism. With respect to its biology, Singer and Morello place the ectotroph alongside the lichens, the major difference being that distinctly new morphological units arise in lichens, but this does not apply to the upper parts of trees, only the root morphology being changed. Generally, a tree may form mycorrhizal partnerships with several fungi (Wojciechowska, 1960) unlike lichens. For these reasons it seems more appropriate to refer to an "ectomycorrhizal tree" instead of an ectotroph.

C. Occurrence in Soils of Native Forests

In native forests there are many fungi which form ectomycorrhizae and are best adapted to the conditions prevailing on that particular site. In the competition between different tree species and fungi, an equilibrium has been reached which enables the best adapted species to form ectomycorrhizal forests of a relatively great stability.

However, especially in the temperate zone, man has often influenced natural forests to varying degrees, for instance, by selectively felling or planting some trees species and creating even-aged stands. Obviously, such forests are not truly natural, but, in many respects, the conditions come close to being natural. By such forest management the natural state is modified to a certain degree as are the mycorrhizal conditions. However, on a given woodland site, if the tree composition remains similar to the original, the mycorrhizal conditions might remain more or less unaltered. We must consider that many mycorrhizal fungi are capable of living in symbiosis not only with one tree species, but also with several others (Trappe, 1962). On the other hand, a given tree species can form mycorrhizae with a series of fungi, so that in many cases tree and fungal species are interchangeable. There are also examples of fungal species forming a mycorrhiza with only a particular host tree, for example, *Suillus grevillei* and *Suillus tridentinus* with *Larix*. In the management of only slightly modified forests, ectomycorrhizae require no more consideration than that given to those in native forests, and much of the data given below are derived from forests influenced by man. Greater problems arise when ectomycorrhizal trees are cultivated outside of the natural range of ectomycorrhizal forests. Although mycorrhizae are widespread in ectomycorrhizal forests influenced by man, differences in their percentage distribution have been observed in soils.

1. Depth of Soil

In general, mycorrhizae are not equally distributed throughout the soil profile. In the upper humus layers more root tips are converted into mycorrhizae than in deeper layers (Fig. 4). Apparently the conditions for mycorrhizal formation are better in humus than in the lower mineral B-horizon. And in the B-horizon the proportion of mycorrhizae declines still further with increasing soil depth. Mycorrhizae were found extending to considerable depths as shown by the following examples: 1.25 m with *Populus tremula* and *Picea abies* (Siren and Bergman, 1951); 1.50 m with *Pinus sylvestris* (Lobanow, 1960); 1.90 and 2.60 m with *Pinus sylvestris* and *Fagus sylvatica* (Werlich and Lyr, 1957); 3.00 m with *Quercus* sp. (Grudsinskaja, 1955). Lyr (1963) discussed the

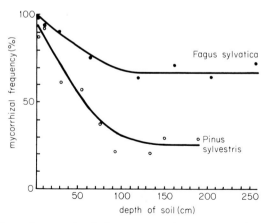

Fig. 4. Decline of mycorrhizal frequency with increasing depth of soil (after Lyr, 1963).

reasons for the decrease of mycorrhizal frequency (percentage of root tips converted into mycorrhizae) with increased depth of soil. Several factors might be involved: (a) Decrease in oxygen content. This factor, according to Lyr, may play a subordinate role. (b) Increase in CO_2 concentration. According to Penningsfeld (1950), below 1 m the CO_2 concentration can increase to 3–4%. (c) Changes in microflora, in the composition of organic soil components, or in nutrition status of soil and roots.

Lyr (1963) suspected that, in addition to changes in the concentration of oxygen and carbon dioxide, the deeper root systems may be more resistant to infection by symbiotic fungi. He suggested that the improved nutrient supply provided by the superficial mycorrhizae could increase the resistance of the deeper roots to infection. Besides these, another important factor might be associated with the decrease in mycorrhizal frequency. The rate of respiration of mycorrhizae is higher than that of nonmycorrhizal roots (Mikola, 1967; Schweers and Meyer, 1970), and mycorrhizae, therefore, require improved replacement of the oxygen consumed. The supply of additional oxygen to the subsoil depends on the rate at which it diffuses through the soil above. The rate of oxygen diffusion through the soil is correlated with bulk density and porosity of soil (Taylor, 1950; Bertrand and Kohnke, 1957). A close relationship between the water-free pore volume and the rate of oxygen diffusion was recorded, and the rate of oxygen diffusion increased with porosity.

Blume (1968) investigated the rate of oxygen diffusion in different soils (brownearth, parabrownearth, pseudogley, and gley) and always found the highest rate in the upper humus horizon and a lower one in the subsoil. High water content diminished the rate of O_2 diffusion, and in dryer periods it was higher. In this connection it is of interest to note that the deepest mycorrhizae (3 m) were found on *Quercus* in the steppe region. Beside the water regimes the rate of diffusion of oxygen into the subsoil might be decisive in this case.

2. Soil Type and Humus Properties

a. Mycorrhizal Frequency. Differences in mycorrhizal frequency in ectomycorrhizal forests arise from differences in soil properties and types of tree associations. Various theories concerning mycorrhizal frequency and its relation to soil properties have been proposed. Stahl (1900) considered mycorrhizae as a special manifestation of soils with poor nutrient content. There competition for mineral salts between trees and soil microorganisms is especially intense, and the mycorrhizal fungi might help the tree to procure the requisite mineral nutrients.

According to Melin (1925), development of the mycorrhiza is influenced by the activity of the fungus infecting the roots. The activity of the fungus, however, is related to a variety of soil properties. A pH value of about 5 is optimal for many mycorrhizal fungi. In Sweden, Melin found that best conditions for mycorrhizal formation were in a microbially active mor, with fewer mycorrhizae formed in inactive mor and in mull. Melin suggested that, in an evolutionary context, mycorrhizal formation would have initially taken place in forest soils with mull, and symbiosis would have resulted from continuous association. This enabled the trees to advance into climates where the decomposition of organic soil material was slow and soils contained greater amounts of raw humus. Under these conditions mycorrhizae aid the tree in competition for minerals, especially nitrogen. Hatch (1937) studied the relationships between mineral content of soil and formation of mycorrhizae. A deficiency of soil nitrogen, phosphorus, potassium, or calcium stimulated development of mycorrhizae. The internal concentration of these elements in the roots supposedly determined whether a mycorrhiza did or did not form.

Björkman (1942) found the amount of nitrogen and phosphorus in the soil to be decisive factors for establishment of mycorrhizae, while potassium deficiency did not influence mycorrhizal frequency. Likewise, addition of calcium or potassium did not influence mycorrhizae as long as the pH of the substrate was maintained at a constant level. Björkman stated that nitrogen and phosphorus did not act directly, but rather influenced the carbohydrate metabolism of the roots. A surplus of N and P stimulates protein synthesis in the plant and thereby lowers the amount of available, soluble carbohydrates which are required by the mycorrhizal fungi in the root. This again hampers the development of mycorrhizae. Thus, according to Björkman, formation of mycorrhizae is determined by the status of the tree, "through the surplus of energy nourishment present in the roots."

In German forests, Meyer (1962, 1966) found *Fagus* and *Picea* mycorrhizae fairly well developed in mull soils, whereas in raw humus (mor) the frequency of mycorrhizae was lower. The activity of the fungus is important among the factors regulating mycorrhizal formation. This was demonstrated by an experiment in which the roots of a single *Fagus sylvatica* plant spread simultaneously into mull and mor (Fig. 5). Both humus forms differed by their microbial activity and by the amount of nutrients (Table II).

Although mull has a higher nitrogen and potassium content than mor, the frequency of mycorrhizae was not lowered as suggested by Björkman; rather it was higher in mull than in mor. The same *Fagus* plant produced roots simultaneously in both the humus forms. Therefore, deviations in

TABLE II

Nutrient Content, Respiration of Mull and Mor, and Mycorrhizal
Frequency of *Fagus sylvatica* Grown in These Substrates

	P content[a] of soil (in % of dry matter)	Nitrification (ppm mineralized N after 6 weeks)	Soil respiration (μl O₂/gm dry organic material × hr)	Mycorrhizal frequency
1. Roots of the same plant only in one substrate				
Solely in mull	0.258	251	80	88
Solely in mor	0.011	79	19	51
2. Roots of the same plant simultaneously in two substrates				
On mullside	0.258	251	80	86
On morside	0.011	79	19	60

[a] Disintegration by perchloric acid.

the properties of both substrates must have affected mycorrhizal formation. Among these effective properties the microbial activity (including that of mycorrhizal fungi) could be of importance. Additional factors stimulate the activity of mycorrhizal fungi and mycorrhizal frequency:

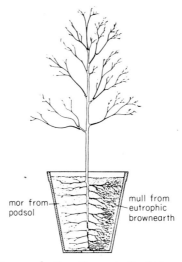

mor from podsol   mull from eutrophic brownearth

Fig. 5. Seedling of *Fagus sylvatica*, the root of which spread simultaneously into podsol and eutrophic brownearth (after Meyer, 1966).

(a) An adequate supply of oxygen. In pot experiments, the well-aerated roots at the periphery always exhibit a higher mycorrhizal frequency than the inner roots. (b) Most mycorrhizal fungi require a weakly acid substrate. Yet all mycorrhizal fungi do not develop best in an acid soil. Richards (1961) found that mycorrhizal fungi survived in neutral or weakly alkaline soils as long as nitrification remained low. Lobanow (1960) observed that abundant mycorrhizae formed on *Quercus* growing on alkaline soils. (c) Sufficient water supply. Most mycorrhizal fungi are affected to a greater extent than other soil fungi by lack of water. However, some, such as the dark-colored *Cenococcum graniforme,* withstand water deficiency comparatively better than others (Worley and Hacskaylo, 1959). The abundance of *Cenococcum graniforme* in podsol is, *inter alia,* a consequence of occasional dryness (Meyer, 1964). Since the decomposition of dead *Cenococcum* hyphae is slow (Meyer, 1970), the humus of a podsol often contains considerable amounts of dead hyphae.

b. Absolute Number of Mycorrhizae. As shown in the foregoing sections, mycorrhizal frequency is high in microbiologically active soil. It is especially high in weakly acid mull. In mor with a slower rate of organic matter decomposition, mycorrhizal frequency decreases. If we express the mycorrhizal state in forest soils, not by the percentage of root tips converted into mycorrhizae, but by the absolute number of mycorrhizae, quite a different result is obtained. As shown in Fig. 6 the number of root tips of *Fagus sylvatica* varies considerably in different soil types. The number of root tips increases from the eutrophic brown-earth with about 500 root tips per 100 ml of soil to the podsol with 45,600 in the OF_3-layer in the A-horizon. The better the site and the biological conditions, the fewer the root tips that have to be formed by *Fagus sylvatica* to obtain the necessary amounts of minerals and water, and the more evenly the root tips are distributed within the profile.

Kern *et al.* (1961) stated there is a direct relationship between the number of tender roots and the nutrient supply in the main rooting zone. From the results of Kern *et al.* (1961) and from Fig. 6 we can conclude that in nutrient-poor soils trees have to expend a higher proportion of their assimilates than in rich soil, building up a root system that can furnish it with an adequate nutrient supply. In podsols the number of root tips is especially high, but mycorrhizal frequency is relatively low. A certain percentage of root tips in podsol are not converted into typical ectomycorrhizae. These so-called pseudomycorrhizae have a shorter life span than true mycorrhizae (Laiho and Mikola, 1964) and, because of their anatomy (Figs. 1 and 2), they are less efficient in absorbing water

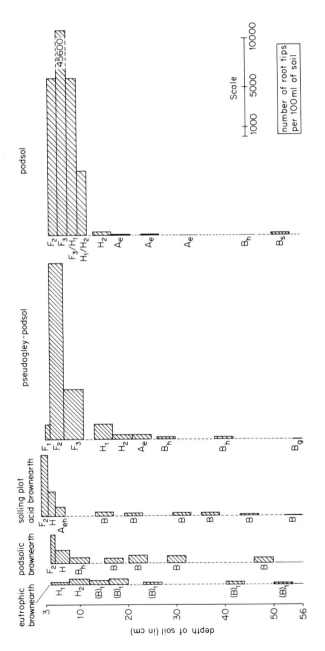

Fɪɢ. 6. Number of root tips of *Fagus sylvatica* in different soil types (after Meyer and Göttsche, 1971).

and nutrients (Meyer, 1967). Consequently there is no direct relationship between the number of root tips and the absorbing capacity of a tree. In mull the root system of *Fagus sylvatica* had a mycorrhizal frequency of 88 and the number of root tips was about 500, whereas in mor the number of root tips was up to 45,600 and mycorrhizal frequency was only 51. This means that in mor, irrespective of low mycorrhizal frequency, the absolute number of ectomycorrhizae far exceeds the number in mull.

Mycorrhizal frequency is directly related to the conditions of mycorrhizal formation, these being favorable in mull and unfavorable in mor, and the number of root tips corresponds inversely to the nutritional state of the soil, good in mull, bad in mor. These findings support the theory of Melin (stated earlier), namely, that the initial constitution of ectomycorrhizae occurred in forests with mull as the humus form, and the well-balanced symbiosis developed in this environment enabled the trees to advance towards locations where site factors were suboptimal. Under these conditions, true ectomycorrhizae are more advantageous to trees than "pseudomycorrhizae." Within the truly ectomycorrhizal trees the genera *Pinus* and *Quercus* are the most numerous. Members of these genera can grow successfully under the most unfavorable woodland conditions.

In this connection it is interesting to note that, except for the tropical *Podocarpus*, within the phylogenetically old group of gymnosperms the ectomycorrhizal genus *Pinus* is still developing (Look, 1950). It contains more than 90 species, whereas the nonectomycorrhizal members of gymnosperms, such as the Taxodiaceae, are relicts of a formerly larger group and include several monotypic genera.

Thus, it seems that not only do ectomycorrhizae enable trees to penetrate into new and less favorable environments, but also provide an impetus for further phylogenetic development.

II. Ectomycorrhizae in Man-Made Forests

A. Forest Sites

As already mentioned in Section I, man has influenced natural forests in various ways so that today nearly all forests differ more or less from their original state. Human interference could have occurred in the following ways:

(a) Without substantially changing the natural composition of the forest. This is the case in boreal and temperate forests, as for instance in *Fagus sylvatica* forests. The tree communities conform closely to the original.

(b) By exploiting commercially important timbers as in the tropical rain forest and in the dry sclerophyll forests of Southern Australia. The composition of tree communities is altered by this procedure, but the original characteristics prevail.

(c) By stopping succession in the subclimax stage (for instance, *Pseudotsuga* forests in Northwest America). In the examples (a)–(c) the natural mycorrhizal conditions are in all probability not influenced significantly.

(d) By sowing or planting tree species which are not natural to a given site, but where they do occur in a greater area in the surrounding associations, as for instance *Picea abies* on broad-leaved forest locations. In general, mycorrhizal problems do not arise by this type of management, because many mycorrhizal fungi have a broad host spectrum and they can live in symbiosis with tree species of different taxonomic position. Thus, many fungi grow with conifers as well as with broad-leaved trees; *Picea abies* planted on a *Fagus sylvatica* site can be infected with the following fungi which normally form symbiotic associations with beech and spruce (Trappe, 1962): *Amanita citrina, A. muscaria, A. pantherina, A. phalloides, Boletus edulis, Cantharellus cibarius, Cenococcum graniforme, Clitopilus prunulus, Cortinarius largus, C. purpurascens, Craterellus cornucopioides, Hydnum repandum, Lactarius scrobiculatus, Paxillus involutus, Phallus impudicus, Ramaria aurea, Russula delica, R. furcata, R. nauseosa, R. ochroleuca, R. olivacea, R. rubra, Scleroderma aurantium, Tricholoma terreum,* and *Xerocomus subtomentosus.* Since tree and fungal species are often interchangeable in the above-mentioned example, mycorrhizae require little or no attention.

(e) By sowing or planting tree species which are alien to the region in question. Then mycorrhizal fungi of local woody species might infect the newly introduced trees. For instance, in south Australia the exotic *Pinus radiata* enters into symbiosis with indigenous mycorrhizal fungi from *Eucalyptus* (Bowen, 1963).

However, problems may arise when obligate ectomycorrhizal tree species are introduced into regions of nonectomycorrhizal trees. As a consequence of the absence of proper symbionts, afforestations or breedings of tree seedlings in nurseries may fail completely. In the last few decades, members of the genus *Pinus* (*P. canariensis, P. caribaea, P. elliottii, P. halepensis, P. khasya, P. merkusii, P. oocarpa, P. patula, P. pinaster, P. radiata,* and *P. taeda*) especially have been used mainly for plantations in subtropical or tropical regions with periodically dry seasons (Streets, 1962). In nurseries these introduced pines repeatedly exhibited yellow needles and symptoms of starvation; in some cases they could not be cultivated (Kessell, 1927; Roeloffs, 1930; Becking, 1950; Dale *et al.*, 1955; Briscoe, 1959). The cessation of growth could not be avoided by shading, irrigation, or manuring. Only the introduction of suitable mycorrhizal fungi reversed stagnation; this was accomplished either by inoculating the soils with humus containing the fungal mycelium or, more effectively, through planting healthy mycorrhizal pines into the diseased nursery beds. In the latter case, the mycelium spread from the mycorrhizae of healthy pines to the roots of diseased ones and, with the commencement of the ectomycorrhizal state, the seedlings became green and began to grow normally. Details of the various methods of introducing mycorrhizal fungi are described by Mikola in Chapter 10 of this volume.

B. PRIMARY OR SECONDARY TREELESS SITES

Man often tries to extend forests or tree plantations to sites which either have never borne trees or were treeless for a long period as in prairies, peat soils, or industrial wastelands. Little is known about the survival of ectomycorrhizal fungi in soils that were free of woody plants for very long periods. Some faculative mycorrhizal fungi such as *Cenococcum graniforme*, *Xerocomus subtomentosus*, or *Laccaria laccata* might survive for several years as free-living saprophytic mycelia after the trees were removed. It must be assumed, however, that spores and possibly mycelia of obligate ectomycorrhizal fungi may reach such soils, but their fate in the absence of trees is unknown. Wilde (1954) observed ectomycorrhizae on seedlings which were raised in fields that had been denuded 60 years previously and stated: "Once forest soil, always forest soil." But the origin of the mycorrhizal fungi was not studied in his experiments. Active mycelium may either have survived in the soil, or have been transferred from surrounding forests by tools, shoes, and airborne spores (cf. Robertson, 1954; Lamb and Richards, 1971). Such open questions deserve further attention.

1. *Locations above the Present Subalpine Timberline*

The forest in the region of the subalpine timberline stands on the boundary of its range, chiefly owing to the short growing season. It therefore is more sensitive to interference by man. Forests on such sites can be easily destroyed; the subalpine timberline has thus been lowered by about 100 m or more in the case of the European Alps (Ellenberg, 1963, 1966). However, subalpine forests can be very important for lowland areas as a barrier against avalanches. Therefore, in Austria enormous efforts have been made to restore subalpine forests, and in this connection the problem of ectomycorrhizae has been investigated thoroughly by Moser (1956, 1958, 1963, 1967) and by Göbl (1965). The most important mycorrhizal fungus for *Pinus cembra, Suillus plorans,* is absent in prospective afforestation areas above the present timberline. Hence methods have been developed to inoculate *Pinus cembra* in nurseries.

2. Prairie and Steppe

In prairie and steppe soils, obligate ectomycorrhizal fungi are absent. This is a consequence of the absence of ectomycorrhizal trees (Hatch, 1936; Mikola, 1953; Wilde, 1954; Lobanow, 1960). In prairies or in steppe regions, landscape management often requires shelterbelts to reduce the adverse influence of drying winds on the cultivated plants. It has been shown that ectomycorrhizal trees (as for instance, *Pinus resinosa* and *P. sylvestris*) can be raised for shelterbelts only in the presence of the proper symbiotic partner (Wilde, 1954). As an inoculum, soil which contained mycelia of mycorrhizal fungi has been used often. The addition of organic substances, together with the mycorrhizal inoculum, was most beneficial (Lobanow, 1960). In an alkaline chernozem, inoculation of *Larix* with mycorrhizal fungi was found to be successful only with an additional acidification and with an application of such trace elements as B, Mn, Mo, and Zn (Zerling, 1958).

The advantage of ectomycorrhizae for trees in prairie and steppe region lies in improved water uptake and enhanced drought resistance. Seedlings lacking ectomycorrhizae die away in the first growing season (Zerling, 1960).

Lobanow (1960) stated that inoculation of *Quercus* and *Pinus* seedlings was considered unnecessary for the forest–steppe zone. Nevertheless, mycorrhizal formation and seedling growth could be stimulated by improving the conditions for fungal growth, i.e., by addition of organic matter, by phosphorus fertilization, and by watering.

3. Peat Bogs

Melin indicated as early as 1917 the importance of ectomycorrhizae for the afforestation of drained peat bogs. *Pinus sylvestris* and *Picea abies,* when planted into a peat bog, remained stunted as long as the fungal partners were absent. As the fungi spread gradually into the drained peat bogs, the trees recovered. Björkman (1941) showed that moderate fertilization with wood ash stimulated tree growth and mycorrhizal formation on a drained bog.

In this connection it would be of interest to discuss the natural occurrence of trees on peat bogs. There are two different types of peat bogs in Europe: the oceanic treeless bogs and continental ones, the latter containing trees such as *Pinus sylvestris* var. *turfosa*, *Betula pubescens*, and some *Pinus mugo* (Ellenberg, 1963). The reason for the differences in the vegetation of these two types of peat bogs might be due to aeration of the upper layers of the bog. While in oceanic areas the peat bog is saturated with water during the whole year, dry periods occur in bogs in the continental region during the growing season. This results in better aeration, which, in turn, is necessary for mycorrhizal viability. Of the woody plants inhabiting peat bogs, at least the obligate, ectomycorrhizal pines are dependent on mycorrhizae. A drained peat bog is a substrate which enables only relatively few ectomycorrhizal fungi to exist. Here we find, for instance, *Cortinarius pholideus*, *Inocybe napipes*, *Lactarius helvus*, and *Russula venosa*.

4. *Calluna Heathland*

In many parts of northwestern Europe large areas are covered with *Calluna* heath. This landscape is only to a small extent of natural origin mainly in sandy coastal regions. In most cases the heath has developed under the influence of man, through extensive sheep grazing. In the last century a great deal of *Calluna* heathland has been afforested mainly with *Pinus sylvestris*. In some cases difficulties that arose during afforestation were closely connected to the mycorrhizal problem.

In heathland soils of southern England, Brian *et al.* (1945) found saprophytic fungi (mainly *Penicillium* species) which secreted substances (gliotoxin, griseofulvin, and others) (Brian, 1949) that were harmful to mycorrhizal fungi. Handley (1963) investigated the reasons for suppression of tree growth on *Calluna* heathland. In the raw humus under vigorously growing *Calluna* he found water-soluble substances which inhibited the development of mycorrhizal hymenomycetes. Aqueous extracts of raw humus collected under old or shaded *Calluna* plants with reduced vitality, were found to be less inhibitory. The inhibiting factor seemed to be associated with the roots of healthy *Calluna* plants. Therefore Handley suggested that the endophyte in the *Calluna* roots excreted antibiotic substances into the vicinity of the roots. As a precondition for the afforestation of *Calluna* heathland he therefore recommended suppression of *Calluna*. In his experiments, Handley found that certain strains of *Leccinum scabrum*, isolated from sporophores under *Betula* in *Calluna* heathland, showed greater resistance to the inhibitory factor. Apparently these strains were specific for those *Betula* and *Pinus* species that are able to live in the *Calluna* heathland, while "heather-sensitive" trees such as *Picea abies* might have less resistant fungal partners.

5. *Industrial Wasteland*

In highly industrialized countries we often find wastelands as a consequence of industrial activity. Such wastelands may arise after deterioration of environmental conditions by industrial by-products (as, for instance, by air pollution) or by heaping waste shale around the openings of underground mines. In Europe industrial wastelands are often situated in or near centers of dense population; they therefore need special attention in the case of afforestation.

If the industrial wasteland was originally forest land, afforestation might be possible under certain circumstances, i.e., after the original conditions for normal plant growth were restored. Unfortunately however, mine tailings comprise a medium, low in organic matter, that never bore trees. Afforestation of such a medium is faced with difficulties originating partly from the special kind of mine deposits. Among the properties adverse to normal tree growth are absence or lack of humus in the rock material and unfavorable pH value of the parent stone either directly or as a result of weathering. For example, the oxidation of pyrites results in formation of sulfuric acid. Thus, for several years plants do not colonize this type of material. In addition, factors that are directly injurious to higher plants, conditions for formation of ecto-mycorrhizae are unsatisfactory because of absence or lack of humus, coarse structure with very low water retention capacity, or compacted material with poor aeration as in the case of spoil banks, and high in-solation. However, once the ectomycorrhizal trees have colonized the wasteland and can actually grow there, mycorrhizae will promptly form. In general, most industrial wastelands are located within the region of ectomycorrhizal forests where they can be reached by the fungal spores. Mycorrhizal fungi are virtually ubiquitous in regions that have long been forested. Diaspores of mycorrhizal fungi can be transferred from forest to wasteland by different ways; even air-borne spores may cause infection. Up to now little is known about germination and initiation of mycorrhizae by airborne spores. Robertson (1954) observed easy propagation of mycorrhizal fungi by these spores. Schramm (1966) and Meyer (1968) demonstrated that mycorrhizal formation proceeds in a relatively short time after coal mine refuse has been spontaneously colonized, for instance, by *Betula lenta, B. populifolia, Populus tremuloides, P. grandi-dentata, Salix humilis,* and *Betula pendula.* Thus, ectomycorrhizal fungi are able to advance into the industrial wasteland via the pioneer trees.

Mine deposits were seeded by Schramm (1966) with *Pinus* and *Quercus* species. He then observed that those plants growing normally and with green color invariably were mycorrhizal, whereas nonmycorrhizal seedlings were pale green to chlorotic and exhibited little or no growth. Schramm applied different nutrients to chlorotic seedlings of *Pinus virginiana*. After nitrate fertilization, the seedlings regained normal color, but this recovery was only temporary and soon the seedlings were again moribund. This can be interpreted to mean that the seedlings growing in substrates low in nitrogen can obtain their nitrogen requirements only with the aid of ectomycorrhizae.

Certainly, the special environmental conditions of the mining wastes must be selective for mycorrhizal fungi. Schramm (1966) observed *Pisolithus tinctorius*, a fungus which is adapted to relatively high temperatures (Marx and Davey, 1969). He also found *Thelephora terrestris*, *Amanita rubescens*, and *Scleroderma aurantium*. On the mining wastes of the Ruhr district, *Scleroderma aurantium* was most frequent and *Pisolithus tinctorius* occurred also (Meyer, 1968). Up to now all studies of ectomycorrhizal fungi from mine deposits allow the conclusion that only a small proportion of the numerous ectomycorrhizal fungi are capable to thrive in this unshaded, highly insolated substrate.

References

Asai, T. (1935). Über das Vorkommen und die Bedeutung der Wurzelsymbiose in den Landpflanzen. *Jap. J. Bot.* **7**, 107.

Bader, F. J. W. (1960). Die Verbreitung borealer und subantarktischer Holzgewächse in den Gebirgen des Tropengürtels. *Nova Acta Leopold.* [N. S.] **23**, No. 148.

Bakshi, B. K. (1966). Mycorrhiza in Eucalyptus in India. *Indian Forest.* **92**, 19.

Becking, J. H. (1950). Der Anbau von *Pinus merkusii* in den Tropen. *Schweiz. Z. Forstw.* **101**, 181.

Bergemann, J. (1955). Die Mykorrhiza-Ausbildung einiger Koniferen-Arten in verschiedenen Böden. *Z. Weltforstwirt.* **18**, 184.

Bertrand, A. R., and Kohnke, H. (1957). Subsoil conditions and their effects on oxygen supply and the growth of corn roots. *Soil Sci. Soc. Amer., Proc.* **21**, 135.

Björkman, E. (1941). Die Ausbildung und Frequenz der Mykorrhiza in mit Asche gedüngten und ungedüngten Teilen von entwässertem Moor. *Medd. Skogsförsöksanst. Stockholm* **32**, 255.

Björkman, E. (1942). Über die Bedingungen der Mykorrhizabildung bei Kiefer und Fichte. *Symb. Bot. Upsal.* **6**, 1.

Blume, H. P. (1968). "Stauwasserböden," Arb. Univ. Hohenheim, Vol. 42. Ulmer, Stuttgart.

Bowen, G. D. (1963). Cited in Mikola (1969).

Brian, P. W. (1949). Studies on the biological activity of griseofulvin. *Ann. Bot.* (*London*) [N. S.] **13**, 59.

Brian, P. W., Hemming, H. G., and McGowan, J. C. (1945). Origin of a toxicity to mycorrhiza in Wareham heath. *Nature* (*London*) **155**, 637.

Briscoe, C. B. (1959). Early results of mycorrhizal inoculation in pine in Puerto Rico. *Carrib. Forest.* **20**, 73.

Chilvers, G. A., and Pryor, L. D. (1965). The structure of Eucalypt mycorrhizas. *Aust. J. Bot.* **13**, 245.

Dale, J., McComb, A. L., and Loomis, W. F. (1955). Chlorosis, mycorrhizae and the growth of pines on a high lime soil. *Forest Sci.* **1**, 148.

Dominik, T. (1948). Mycorrhizae on wild orchard trees growing in forests. [In Polish, French summary.] *Acta Soc. Bot. Pol.* **19**, 169.

Ellenberg, H. (1963). "Vegetation Mitteleuropas mit den Alpen." Ulmer, Stuttgart.

Ellenberg, H. (1966). Leben und Kampf an den Baumgrenzen der Erde. *Naturwiss. Rundsch.*, p. 133.

Fassi, B. (1963). Die Verteilung der ektotrophen Mykorrhizen in der Streu und in der oberen Bodenschicht der Gilbertiodendron dewevrei-(Caesalpiniaceae)-Wälder im Kongo. *Int. Mykorrhiza Symp., 1960* pp. 297–302.

Fassi, B., and Fontana, A. (1961). Le micorrhize ectotrofiche di Julbernardia seretii, Caesalpiniaceae del Congo. *Allionia* **7**, 131.

Fassi, B., and Fontana, A. (1962). Micorrize ectotrofiche di Brachystegia laurentii e di alcune altre Caesalpiniaceae minori del Congo. *Allionia* **8**, 121.

Frank, A. B. (1885). Über die auf Wurzelsymbiose beruhende Ernährung gewisser Bäume durch Pilze. *Ber. Deut. Bot. Ges.* **3**, 128.

Frank, A. B. (1887). Über neue Mycorrhiza-Formen. *Ber. Deut. Bot. Ges.* **5**, 395.

Gallaud, J. (1905). Etudes sur les Mykorrhizes endotrophes. *Rev. Gen. Bot.* **17**, 5.

Gäumann, E., Nüesch, J., and Rimpau, R. H. (1960). Weitere Untersuchungen über die chemischen Abwehrreaktionen der Orchideen. *Phytopathol. Z.* **38**, 274.

Göbl, F. (1965). Die Zirbenmykorrhiza im subalpinen Aufforstungsgebiet. *Zentralbl. gesamte Forstw.* **82**, 89.

Grudsinskaja, I. A. (1955). Beiträge zur Mykotrophie der Holzpflanzen in Steppenbedingungen. Cited in Lobanow (1960).

Handley, W. R. C. (1963). Mycorrhizal associations and *Calluna* heathland afforestation. *Forest. Commission Bull.* No. 36.

Harley, J. L., and Brierley, J. K. (1955). The uptake of phosphate by excised mycorrhizal roots of the beech. VII. Active transport of P^{32} from fungus to host during uptake of phosphate from solution. *New Phytol.* **54**, 297.

Hatch, A. B. (1936). The role of mycorrhizae in afforestation. *J. Forest.* **34**, 22.

Hatch, A. B. (1937). The physical basis of mycotrophy in the genus *Pinus*. *Black Rock Forest Bull.* **6**, 1.

Hueck, K. (1966). "Die Wälder Südamerikas." Fischer, Stuttgart.

Janse, J. M. (1897). Les endophytes radicaux de quelques plantes javanaises. *Ann. Jard. Bot. Buitenzorg* **14**, 202.

Kern, K. G., Moll, W., and Braun, H. J. (1961). Wurzeluntersuchungen in Rein- und Mischbeständen des Hochschwarzwaldes. *Allg. Forst- Jagdztg.* **132**, 241.

Kessel, S. L. (1927). Soil organisms. The dependence of certain pine species on a biological soil factor. *Emp. Forest J.* **6**, 70.

Laiho, O., and Mikola, P. (1964). Studies of the effect of some eradicants on mycorrhizal development in forest nurseries. *Acta Forest. Fenn.* **77**, 1.

Lamb, R. J., and Richards, B. N. (1971). Survival potential of sexual and asexual spores of ectomycorrhizal fungi. *Proc., Int. Union Forest Res. Organ., Gainesville, Florida, 1971 15th,* Sect. 24.

Levisohn, I. (1954). Occurrence of ectotrophic and endotrophic mycorrhizas in forest trees. *Forestry* **27**, 145.

Levisohn, I. (1958). Mycorrhizal infection in Eucalyptus. *Emp. Forest Rev.* **37**, 237.

Lobanow, N. W. (1960). "Mykotrophie der Holzpflanzen." Verlag Wiss., Berlin.

Look, E. E. M. (1950). The pines of Mexico and British Honduras. *Union S. Afri., Dep. Forest. Bull.* **35**, 1.

Lundeberg, G. (1960). The relationship between pine seedlings (*Pinus silvestris* L.) and soil fungi. Some experiments with a new method for sterilized water cultures. *Sv. Bot. Tidskr.* **54**, 346.

Lyr, H. (1963). Über die Abnahme der Mycorrhiza- und Knöllchenfrequenz mit zunehmender Bodentiefe. *Int. Mykorrhiza Symp., 1960* pp. 303–313.

Maeda, M. (1954). The meaning of mycorrhiza in regard to systematic botany. *Kumamoto J. Sci., Ser. B* **3**, 57.

Marx, D. H., and Davey, C. B. (1969). The influence of ectotrophic mycorrhizal fungi on the resistance of pine roots to pathogenic infections. III. *Phytopathology* **59**, 549.

Melin, E. (1917). Studier över de norrländska myrmarkernas vegetation med särskildhänsyn till deras Skogsvegetation efter torrlägging. *Akad. Afh. Uppsala* **7**, 1.

Melin, E. (1925). "Untersuchungen über die Bedeutung der Baummykorrhiza." Fischer, Jena.

Melin, E. (1939). Methoden der experimentellen Untersuchung mykotropher Pflanzen. *In* "Handbuch der biologischen Arbeitsmethoden" (E. Abderhalden, ed.), Sect. II, pp. 1015–1108. Urban & Schwarzenberg, Berlin.

Melin, E., and Nilsson, H. (1950). Transfer of radioactive phosphorus to pine seedlings by means of mycorrhizal hyphae. *Physiol. Plant* **3**, 88.

Melin, E., and Nilsson, H. (1952). Transfer of labelled nitrogen from an ammonium source to pine seedlings through mycorrhizal mycelium. *Sv. Bot. Tidskr.* **46**, 281.

Melin, E., and Nilsson, H. (1953). Transfer of labelled nitrogen from glutaminic-acid to pine seedlings through the mycelium of *Boletus variegatus* (SW.) Fr. *Nature (London)* **171**, 134.

Melin, E., and Nilsson, H. (1955). Ca^{45} used as an indicator of transport of cations to pine seedlings by means of mycorrhizal mycelia. *Sv. Bot. Tidskr.* **49**, 119.

Melin, E., and Nilsson, H. (1958). Translocation of nutritive elements through mycorrhizal mycelia to pine seedlings. *Bot. Notis.* **111**, 251.

Melin, E., Nilsson, H., and Hacskaylo, E. (1958). Translocation of cations to seedlings of *Pinus virginiana* through mycorrhizal mycelium. *Bot. Gaz. (Chicago)* **119**, 243.

Meyer, F. H. (1962). Die Buchen- und Fichtenmykorrhiza in verschiedenen Bodentypen, ihre Beeinflussung durch Mineraldüngung sowie für die Mykorrhizabildung wichtige Faktoren. *Mitt. Bundesforschungsanst. Forst- Holzwirt.* **54**, 1.

Meyer, F. H. (1964). The role of the fungus Cenococcum graniforme (Sow.) Ferd. et Winge in the formation of mor. *In* "Soil Microbiology" (E. A. Jongerius, ed.), pp. 23–31. Elsevier, Amsterdam.

Meyer, F. H. (1966). Mycorrhiza and other plant symbioses. *In* "Symbiosis" (S. M. Henry, ed.), Vol. 1, pp. 171–255. Academic Press, New York.

Meyer, F. H. (1967). Feinwurzelverteilung bei Waldbäumen in Abhängigkeit vom Substrat. *Forstarchiv* **38**, 286.

Meyer, F. H. (1968). Mykorrhiza. *In* "Haldenbegrünung im Ruhrgebiet" (K. Mellinghoff, ed.), Schriftenr. Siedlungsverb. Ruhrkohlenbezirk 22, pp. 118–123.

Meyer, F. H. (1970). Abbau von Pilzmycel im Boden. *Z. Pflanzenern. Bodenkd.* **127**, 193.

Meyer, F. H., and Göttsche, D. (1971). Distribution of root tips and tender roots of beech. In "Ecological Studies. Analysis and Synthesis" (H. Ellenberg, ed.), Vol. 2, pp. 48–52. Springer-Verlag, Berlin, and New York.

Mikola, P. (1953). An experiment on the invasion of mycorrhizal fungi into prairie soil. Karstenia 2, 33.

Mikola, P. (1965). Studies on the ectendotrophic mycorrhiza of pine. Acta Forest. Fenn. 79, 5.

Mikola, P. (1967). The effect of mycorrhizal inoculation on the growth and root respiration of Scotch pine seedlings. Proc., Int. Union Forest Res. Organ., 14th, 1967 Vol. V, p. 100.

Mikola, P. (1969). Afforestation of treeless areas. Unasylva 23, 35–48.

Morrison, T. M. (1956). Mycorrhiza of silver birch. N. Z. J. Forest. 7, 47.

Moser, M. (1956). Die Bedeutung der Mykorrhiza für Aufforstungen in Hochlagen. Forstwiss. Centralbl. 75, 8.

Moser, M. (1958). Der Einfluß tiefer Temperaturen auf das Wachstum und die Lebenstätigkeit höherer Pilze mit spezieller Berücksichtigung von Mykorrhiza-pilzen. Sydowia 12, 386.

Moser, M. (1963). Die Bedeutung der Mykorrhiza bei Aufforstungen unter besonderer Berücksichtigung von Hochlagen. Int. Mykorrhiza Symp., 1960 pp. 407–422.

Moser, M. (1967). Die ectotrophe Ernährungsweise an der Waldgrenze. Mitt. Forstl. Bundesversuchsanst. Wien 75, 357.

Penningsfeld, F. (1950). Über Atmungsuntersuchungen an einigen Bodenprofilen. Z. Pflanzenern., Düng., Bodenkd. 50, 135.

Peyronel, B., Fassi, B., Fontana, A., and Trappe, J. M. (1969). Terminology of Mycorrhizae. Mycologia 61, 410.

Redhead, J. F. (1968). Inocybe sp. associated with ectotrophic mycorrhiza on Afzelia bella in Nigeria. Commonw. Forest. Rev. 47, 63.

Richards, B. N. (1961). Soil pH and mycorrhiza development in Pinus. Nature (London) 190, 105.

Rivett, M. (1924). The root tubercles in Arbutus unedo. Ann. Bot. (London) 38, 661.

Robertson, N. F. (1954). Studies of the mycorrhiza of Pinus silvestris. I. The pattern of development of mycorrhizal roots and its significance for experimental studies. New. Phytol. 53, 253.

Roeloffs, J. W. (1930). Over kunstmatige Verjonging van Pinus Merkusii Jungk. et de Vr. en Pinus Khasya Royle. Tectona 23, 874.

Schramm, J. R. (1966). Plant colonisation on black wastes from anthracite mining in Pennsylvania. Trans. Amer. Phil. Soc. 56, 1, 1.

Schweers, W., and Meyer, F. H. (1970). Einfluß der Mykorrhiza auf den Transport von Assimilaten in die Wurzel. Ber. Deut. Bot. Ges. 83, 109.

Schweinfurth, U. (1957). Die horizontale und vertikale Verbreitung der Vegetation im Himalaya. Bonner Geogr. Abh. 20, 1.

Singer, R. (1963). Der Ectotroph, seine Definition, geographische Verbreitung und Bedeutung in der Forstökologie. Int. Mykorrhiza Symp., 1960 pp. 223–231.

Singer, R. (1964). Areal und Ökologie des Ektotrophs in Südamerika. Z. Pilzk. 30, 8.

Singer, R., and Morello, J. H. (1960). Ectotrophic forest tree mycorrhiza and forest communities. Ecology 41, 549.

Singh, K. G. (1966). Ectotrophic mycorrhiza in equatorial rain forests. Malay. Forest. 29, 13.

Sirén, G., and Bergman, F. (1951). Mycorrhizal fungi and our forest trees. [In Finnish, English summary.] *Skogsbruket* **21**, 39.

Stahl, E. (1900). Der Sinn der Mykorrhiza-Bildung. *Jahrb. Wiss. Bot.* **34**, 539.

Streets, R. J. (1962). "Exotic Forest Trees in the British Commonwealth." Oxford Univ. Press (Clarendon), London and New York.

Taylor, S. A. (1950). Oxygen diffusion in porous media as a measure of soil aeration. *Soil Sci. Soc. Amer., Proc.* **14**, 55.

Trappe, J. M., (1962). Fungus associates of ectotrophic mycorrhizae. *Bot. Rev.* **28**, 538.

Werlich, I., and Lyr, H. (1957). Über die Mykorrhizaausbildung von Kiefer (*Pinus silvestris* L.) and Buche (*Fagus silvatica* L.) auf verschiedenen Standorten. *Arch. Forstw.* **6**, 1.

Wilde, S. A. (1954). Mycorrhizal fungi: Their distribution and effect on tree growth. *Soil Sci.* **78**, 23.

Wojciechowska, H. (1960). Studien über den Mykotrophismus der Fichte (*Picea excelsa* Lk.). [In Polish, German summary.] *Folia For. Pol., Ser. A* **2**, 123.

Worley, J. F., and Hacskaylo, E. (1959). The effect of available soil moisture on the mycorrhizal association of Virginia pine. *Forest Sci.* **5**, 267.

Zerling, G. I. (1958). Die Mykorrhiza der Lärche und ihre Wirkung auf das Wachstum und den Zustand der Sämlinge in Karbonat-Tschernosem-Böden des Transwolga-Gebietes. *Mikrobiologija* **27**, 450.

Zerling, G. I. (1960). Die Mykorrhizabildung bei Lärche unter den Bedingungen des Transwolga-Tschernosems und ihre Stimulierung. *Mikrobiologija* **29**, 401.

CHAPTER 4

Growth of Ectomycorrhizal Fungi around Seeds and Roots

G. D. BOWEN and C. THEODOROU

I. Introduction

Study of the growth of ectomycorrhizal fungi around seeds and roots is highly relevant to proposed inoculation programs both in the technology of application and in management practices likely to lead to ready establishment of the inoculum. It may be necessary to select mycorrhizal fungi not only on their ability to give a desired tree response but also on their ability to colonize the root readily in a range of conditions and persist from season to season. Apart from this, however, growth of the fungus on the root surface has significant roles in disease resistance, in spread of inoculum from tree to tree and along a root (Robertson, 1954; Wilcox, 1968), and in nutrient uptake in some cases, even without infection. Thus Neumann (1959) reported growth stimulation of *Eucalyptus camaldulensis* by the mycorrhizal fungus *Pisolithus tinctorius* in the absence of infection, and Lamb and Richards (1971a) have confirmed this with mycorrhizal fungi on *Pinus radiata* and *P. elliotii* var. *elliottii*. In addition leachates of soil in which mycorrhizal pines are growing can induce forking of short roots (Levisohn, 1960), and this could be due to fungal exudates or to exudates of the mycorrhizal association. Such observations as those above are consistent with findings (Bowen and Rovira, 1969) that noninfecting rhizosphere organisms can affect root morphology, physiology, and ion uptake.

Growth of hyphae into soils away from the root will be examined in Chapter 5 because of its importance in solute uptake from soils. In this Chapter the establishment phases of the fungus on the root will be examined, since this is critical to the success of inoculation programs. Microflora on the seed often have to face extreme, fluctuating environmental conditions (especially in direct seeding in the field) because of their proximity to the soil surface. The seed surface is also a site of great susceptibility to microbial competition since seed diffusates provide a ready nutrient source for many microorganisms. Later, colonization of the emerging root allows movement away from the variable surface conditions, and the somewhat selective environment of the rhizosphere (Rovira, 1965) may reduce competition and antagonism from soil microorganisms to some extent. However it is not until infection occurs in the much more selective and protected environment within the root that some security of tenure is achieved by the fungus.

The specialized behavior of mycorrhizal fungi, e.g., inability to use complex polysaccharides as sources of carbon and energy, their frequent requirements for various growth factors, and inhibition by humus extracts, identifies them as typical rhizosphere fungi, despite the lack of general study of a possible saprophytic behavior in the absence of the host and the ability of some to decompose litter. Many studies have been performed on requirements of mycorrhizal fungi in pure culture (see Harley, 1969) but these will be discussed here only as they relate to growth of fungi in the rhizosphere. Some of the laboratory culture studies on mycorrhizal fungi have only limited relevance to fungus growth in a natural environment, and, similarly, some studies on colonization of roots in simplified defined situations (e.g., Theodorou and Bowen, 1969) have limitations in relating them to much more complex natural situations. However, studies on laboratory media and the simplified systems we describe are useful steps in understanding growth on roots and in suggesting hypotheses, for experimental testing, on colonization in the field.

Such colonization studies are relevant also for noninfecting rhizosphere microorganisms and plant pathogens in their preinfection phase.

II. Survival and Germination of Propagules

Robertson (1954) demonstrated that wind-dispersed basidiospores brought about mycorrhizal formation in previously uninfected soils. By filtering air entering a plant growth room, Marx and Bryan (1969) eliminated casual mycorrhizal infection on *P. echinata,* but outside such rooms 90% infection occurred in trees grown in fumigated soils without inoculation. It is well known that some fungal spores can be distributed over very large distances by wind but no detailed studies have been performed on mycorrhizal fungi; two of the main problems in such studies are the detection of the organisms in small quantities and the unambiguous identification of them. Wind dispersal of spores is the most likely reason for Bowen's observations (1963) that mycorrhizal fungi for *P. radiata,* an introduced species to natural *Eucalyptus* areas, were almost invariably found in soils up to 2 miles from established pine forests but beyond that they were frequently absent. The distance over which spores are carried *in sufficient quantity* to inoculate trees will depend largely on the fungus, the prevailing wind direction at the time of fruiting, and factors such as temperature, humidity, and resistance of the spores to desiccation. There are good grounds for the assumption that mycorrhizal fungi are readily available to seedlings of *indigenous* tree species. However, even close to existing pine forests, Theodorou and Bowen (1970) found naturally occurring mycorrhizal fungi to be erratic in distribution and inoculation to be beneficial to pine growth.

Few studies have been carried out on the persistence of mycorrhizal fungi in the absence of their host plant. Because of the ready wind dispersal of spores (Robertson, 1954) there is some doubt about Wilde's conclusion (1946) that, once introduced to soil, mycorrhizal fungi will survive for considerable periods in the absence of the host. Theodorou and Bowen (1971b) demonstrated the ability of some mycorrhizal fungi to grow in the rhizosphere of nonhost plants and this would aid persistence in the absence of a host.

A. SURVIVAL OF PROPAGULES

A number of studies have shown that while mycelial forms of inoculum can survive successfully after inoculation into sterile soils (e.g., Theodorou, 1967) and in natural soils under ideal conditions, they frequently fail or establish poorly under other field conditions, e.g., where soil dries out or where biological antagonisms occur (Theodorou, 1971). Lamb and Richards (1971b) examined the effect of temperature and desiccation on basidiospores of *Suillus granulatus, Rhizopogon roseolus,* and *Pisolithus tinctorius;* chlamydospores of three unidentified symbionts of *P. elliottii* and *P. radiata;* and oidia of *Xerocomus subtomentosus.* The hyphae of six of these fungi were killed by a 48-hour exposure to temperatures between 28° and 38°C and that of *P. tinctorius* by 45°C. [In other studies Marx *et al.* (1970) found no growth of hyphae of *P. tinctorius* above 40°C.] The chlamydospores were killed by temperatures of 32° to 36°C, the oidia by 47°C, and the basidiospores by 46° to 54°C. Unfortunately, it is not known if the *X. subtomentosus* culture was of the litter-decomposing or of the mycorrhizal type for these can differ substantially in growth properties (Lundeberg, 1970). Chlamydospores did not survive storage at 50% relative humidity for 60 days, whereas basidiospores maintained viability at 30% relative humidity. The resistance of basidiospores (and the susceptibility of mycelia) to relatively high soil temperatures is particularly relevant to seeding in subtropical and tropical climates where summer temperatures of 64.6°C at the surface and 45.3°C at a 5 cm depth have been recorded (Ramdas and Dravid, 1936) and where temperatures in excess of 40°C can frequently occur at 2.5 cm for up to 6 hours (Bowen and Kennedy, 1959, latitude 27°S).

At the other extreme, Moser (1958) recorded variation between different mycorrhizal fungi in resistance of mycelia to temperatures of −11° to −12°C (on agar) and advocated selection of fungi on this basis for alpine reafforestation. Fries (1943) found storing of spores of mycorrhizal fungi at −10°C prolongs their viability, and spores thus provide a survival mechanism in sites even where the mycelium is killed by high or low temperatures.

Resistance of basidiospores to drying has great ecological importance in natural wind-borne dispersal and also has technological advantages in inoculation practices. Theodorou (1971) showed the efficacy of spore inoculation of *P. radiata* seed in the field in both fumigated and nonfumigated soil, and Table I (Theodorou and Bowen, 1973) shows (i) that freeze-dried basidiospores can act as effective inocula 3 months later, (ii) that spores inoculated to seed and dried for 2 days are effective, and (iii) that spores added to soil and dried for 2 months can still lead to good mycorrhizal production.

B. GERMINATION OF PROPAGULES

Fries (1966) has reviewed germination of basidiospores. Germination of basidiospores of mycorrhizal fungi has been a rather neglected field of study because if the spores can be made to germinate at all in

TABLE I

EFFECTS OF DRYING ON BASIDIOSPORE INOCULA FOR *Pinus radiata*[a]

Treatment	Total No. short lateral roots	Percent mycorrhiza[b]
A. Uninoculated seed	219	2
Seed inoculated with freeze-dried basidiospores, 3 months old[c]	262	43
B. Uninoculated soil	336	0
Seed inoculated with basidiospores and sown immediately[c]	348	66
Seed inoculated, dried for 2 days[c] before sowing[c]	404	64
C. Uninoculated soil	146	0
Basidiospores in dry soil for 2 months at 25°C[d]	231	41

[a] In all the experiments four replicate pots each with four plants were grown for 4 months.

[b] Percentage of short roots becoming mycorrhizal.

[c] Basidiospores of *R. luteolus* were applied to seed at approximately 5×10^6 spores/ seed. Experiment A was in nonsterile soil and B and C were in sterile soil.

[d] 3×10^4 spores/pot.

laboratory media, only a very low percentage does so (often less than 0.1%) and germination is very slow. The percentage of germination of basidiospores can vary considerably from one fruiting body to another. Most studies with mycorrhizal fungi have been restricted to either attempting spore germination of hymenomycetes and gasteromycetes in simple solutions, or on synthetic media with growth substances added, although Melin (1959) added excised roots to synthetic media. In the most detailed study of germination of basidiospores of mycorrhizal fungi yet made, Fries (1943) found even within the one genus basidiospores of some species germinated in distilled water only, some needed the addition of malt extract, others were considerably assisted by diffusates from the yeast *Torulopsis sanquinea,* and others could not be germinated at all. As a rule, *Suillus* spores could be germinated by adding unidentified activators from yeast. However, Fries (1966) subsequently reported that *S. luteus* spores would germinate in a simple glucose– mineral salts solution with ammonium tartrate on thoroughly washed agar, but that activators reduced the germination time from 3–4 weeks to 1 week. Fries found no need for a period of "maturation" of the spore before it would germinate. It is obvious from his studies that inorganic composition of laboratory media affects spore germination profoundly.

Lamb and Richards (1971b) have also found an unidentified activator (from *Rhodotorula glutinis*) for germination of basidiospores of *S. granulatus, R. roseolus,* and *P. tinctorius.* Germination of oidia of their *Xerocomus subtomentosus* culture did not require an activator. Oidia of this fungus could be produced in culture in large numbers, and as they have a broad temperature and pH range for germination and are very resistant to desiccation, if such strains are also mycorrhizal they have considerable attractiveness for inoculum production. Germination of basidiospores of *S. granulatus, R. roseolus,* and *P. tinctorius* was optimal at pH 5–5.5. With the first two fungi germination was restricted mainly to pH 4–6.5 but with *P. tinctorius* germination had a somewhat wider range. Although all three fungi had optimum germination at 20°–25°C, germination of *P. tinctorius* occurred between 5° and 40°C, a considerably wider range than the other two fungi, whose germination range was 10°–30°C. Because the chemical composition of the rhizosphere is likely to vary with temperature (Theodorou and Bowen, 1971a) it is difficult to extrapolate from laboratory temperature studies to germination around roots in soil; Lamb and Richard's finding that six times as many spores of *R. roseolus* and *S. granulatus* germinated at 20°C than at 10°C suggests a need for heavier inoculation with spores in many cool-temperature soils in inoculation programs.

A number of organisms can produce stimulators to germination and these include plant roots, yeasts, various soil fungi, and pieces of fruiting bodies. Fries (1943) found that the mycorrhizal fungus *Cenococcum graniforme* could stimulate spore germination of three *Suillus* species in laboratory culture. However, as metabolites produced in laboratory media are almost certainly quite different from those produced in the soil, the ecological importance of such a finding is doubtful. The stimulatory effects of growing roots on basidiospore germination has obvious ecological implications. Basidiospores of *Russula* have proven very intractable to germination on usual laboratory media, but Melin (1959) was able to induce germination of some (but not all) species by addition of excised pine roots to a nutrient medium containing B vitamins and amino acids. Sometimes tomato roots were also effective but in other instances pine roots were needed, possibly suggesting at least quantitative differences if not qualitative differences in stimulator requirements of spores of different fungi. Melin also showed excised roots to stimulate spore germination of species of *Suillus, Amanita, Paxillus, Cortinarius,* and *Lactarius.* Failure to obtain germination by adding excised roots to synthetic media may reflect a toxin production by the roots, a toxin in the medium, or a need for factors from living roots; Marx and Ross (1970) were unable to induce basidiospore germination of *Thelephora terrestris* with root extracts or detached roots of *P. taeda* but the spores germinated to form mycorrhizae with intact roots of *P. taeda* in axenic culture.

The nature of the activator for basidiospore germination is still unknown; several compounds may be involved. The activator yeast is usually separated from the basidiospore on an agar plate, and Fries (1966) suggested therefore, activators must be readily diffusible. The removal of the stimulus by removal of the activator organism suggests involvement of volatile compounds, a very labile compound, or one readily used by spores. Modern analytical tools such as gas chromatography have not yet been employed in these studies.

Fries (1966) pointed out germination stimulation can also be due to removal of toxins in agar media and that asparagine prevents germination of *S. luteus* spores. It is of some interest therefore that asparagine has been recorded in exudates from tree roots (see Table II).

Placing these studies on basidiospore germination in a wider context of rhizosphere biology, plant root stimulation of germination of spores and sclerotia of plant pathogenic fungi is well known (Warcup, 1967). In some such cases specific substances may be involved but in others the response is of a more general nature, for the rhizosphere environment differs from soil in factors such as pH, chemical composition, and oxygen and carbon dioxide concentrations as well as being a zone of intense microbial activity (see Section IV). Some stimulators could be due to the rhizosphere microflora rather than the plant root.

Another soil phenomenon difficult to accommodate in laboratory studies of germination is mycostasis (Dobbs and Hinson, 1953). Such fungistasis is thought to be due to metabolites from other soil organisms but chemical identification of these metabolites has seldom been made (see Garrett, 1970; Lockwood, 1964). Fungistasis can be overcome by addition of readily available energy sources, e.g., glucose, which occur in most root exudates (see Table II), and the triggering of germination by the proximity of a root probably has great conservation value for fungi such as mycorrhizal fungi which seem to be largely restricted to growing on roots. Where specificity of roots in stimulating spore germination occurs (Melin, 1959) greater conservation advantages would occur. Germination of basidiospores of mycorrhizal fungi around plant roots *in soil* is an area much in need of further study.

III. Growth around Seeds

Germinating seeds liberate sugars, amino acids, and other substances into the spermatosphere and these lead to increased microbial activity around the seed (Picci, 1959; Fries and Forsman, 1951). However not all seed diffusates are stimulatory, and Bowen (1961) showed that seed diffusate of the legumes *Centrosema pubescens* and *Trifolium subterraneum* contain an antibiotic toward a wide range of gram-negative and gram-positive soil bacteria. Some organisms were insensitive to the toxin and were stimulated by the seed diffusate. Ferenczy (1956) recorded seed diffusates toxic to gram-positive bacteria from species of *Abies, Picea, Pinus,* and *Pseudotsuga.*

Melin (1925) observed growth stimulatory substances for mycorrhizal fungi coming from germinating pine seeds. Similarly, we have observed that mycelia of *R. luteolus* and *S. granulatus* will grow well on the seed coat of *P. radiata* seeds germinating on water–agar. The extremely limited study of growth of mycorrhizal fungi around seeds would suggest seed diffusates to be stimulatory and not inhibitory to them. It is not known what happens to growth of mycorrhizal fungi around seed in soil with other microorganisms present. If considerable growth of inoculated fungi took place around the seed this would lead to a higher inoculum potential for early colonization of the emerging radicle.

Much spore or mycelial inoculum adhering to the testa is often wasted when the testa is carried above the soil, following germination. The success of spore inoculation of pine seed (Theodorou, 1971) indicates that sufficient spores remain around the emerging root to establish infection, but in seed inoculation some advantage will lie in incorporating inoculum in materials which detach from the testa soon after planting.

IV. Growth around Roots

It is necessary to sustain growth of the fungus in the rhizosphere for at least some weeks after sowing before infection occurs. In nursery soil in Finland, with mean air temperatures around 15°C, Laiho and Mikola (1964) first observed mycorrhizal infection of pine and spruce 6 to 7 weeks after sowing; under probably colder soil conditions in England, Robertson (1954) first observed mycorrhizae on *P. sylvestris* some 18 weeks after sowing in March. The reasons for delay in infection are unknown. The suggestion that sufficient carbohydrate is not available for the fungus until photosynthesis rates become high with the advent of the first true leaves, by itself does not seem tenable, since the fungus grows in the rhizosphere on root exudates well before this.

Marked ectotrophic growth (according to Garrett, 1970) around and along roots precedes infection of pine, spruce (Laiho and Mikola, 1964), beech (J. Warren-Wilson, in Harley, 1969), and *Eucalyptus* (Chilvers and Pryor, 1965). Chilvers and Pryor also considered spread of infection along and between roots of *Eucalyptus* was by surface growth, whereas both internal and external growth are important in spread of the infection with pine and spruce roots (Laiho and Mikola, 1964; Robertson, 1954; Wilcox, 1968). With root-infecting pathogens, ectotrophic growth usually precedes infection of underlying cells, and the ectotrophic growth habit reduces host resistance to infection. A full discussion of ectotrophic growth of pathogens is given by Garrett (1970), who considered a concept of ectotrophic growth overcoming host resistance inappropriate to the harmonious relationships exhibited by an ectomycorrhizal fungus and its host. However, to discard such a concept for ectomycorrhizae is premature in the absence of any detailed study on infection and developmental processes of ectomycorrhizae. It is interesting that in the spectrum of microorganism–plant interactions from noninfecting rhizosphere microorganisms to mycorrhizae to plant pathogens, there are marked similarities in many properties which can often be distinguished (Bowen and Rovira, 1969).

Our interest in ectotrophic growth on roots arose from studies on establishing inocula around roots. However it soon became apparent that most, or all, previous studies on the effects of environment on mycorrhizal production did not distinguish between the two fundamental steps: (i) effects on fungus growth around seedling roots prior to infection and (ii) effects on infection processes and subsequent mycorrhizal development (see Chapter 1). It also became apparent that there was a great lack of experimental approach toward quantitatively evaluating existing concepts and enunciating new ones in the "microecology" of fungi on roots.

This section deals first with a definition of the rhizosphere environment for growth, then with experimental approaches to fungus colonization of roots, and lastly with the dynamics of microbial colonization of roots in soil.

The root in soil is a dynamic, interacting, three-compartment system in series, i.e., the soil, a somewhat discontinuous microbial layer, and the plant root itself. The "rhizosphere" was defined by Hiltner (1904) as the region of soil around a root in which a root influence on microflora occurred. However the extent of this influence will depend on the amount of substances lost from roots, on the rate of diffusion of soluble and volatile materials away from the root in the particular soil, and on the sensitivity of different organisms to diffusates from the root. The term "rhizoplane" is applied to the immediate soil–plant interface, and "rhizosphere" is applied to an ill-defined narrow zone of soil (about 1–3 mm wide) surrounding the root and root hairs. Here we will refer to these collectively as the rhizosphere. It differs substantially from the soil around it because (i) the intact root loses organic substances—"root exudates"—thus stimulating microbial activity greatly. (In soil microbiology literature the term "root exudates" is distinct from other plant physiological connotations which refer to bleeding sap of decapitated roots.) (ii) The roots secrete inorganic ions, including hydrogen ions and bicarbonate ions. (iii) Carbon dioxide and oxygen concentrations change as a result of root and microbial respiration. (iv) Solutes in the soil move to the roots in water or by diffusion; some may be completely depleted because of high uptake by the root and others may accumulate. (v) The root frequently sloughs old material such as root cap and epidermal cells which release their contents (see Chapter 8).

Certain groups of soil organisms are selectively stimulated in the rhizosphere (see Rovira, 1965), in part owing to the above factors. Similarly the immediate environment of fungal hyphae and mycorrhizae is rather different from surrounding soil and adjacent uninfected roots, thus leading to increased microbial activity and microbial specificity in the "mycorrhizosphere" (Katznelson *et al.*, 1962; Neal *et al.*, 1964; Foster and Marks, 1967; Oswald and Ferchau, 1968).

A. ROOT EXUDATES OF TREES

A number of organic compounds capable of being used as energy and growth factor sources for microorganisms have been recorded to come from living plant roots, e.g., sugars, amino acids, vitamins, organic acids, nucleotides, flavonones, enzymes, terpenes, and many unidentified compounds (see Chapter 8).

1. Methods of Study and Sites of Exudation

Most root exudate studies with tree species have involved collection of exudates over several days, using nonradioactive or radioactive techniques. The plants are grown under sterile conditions, usually in solution or sand culture, from which the exudates are collected at intervals of several days, concentrated, and analyzed. In contrast to collections over very short periods (possible with radioactive tracers) this exudate is "net loss" from the root, i.e., the balance between actual loss and reabsorption by the plant, since it is well known that most solutes move both in and out of cells, i.e., influx and efflux, respectively (Briggs *et al.*, 1961). Such exudate collection will not detect labile compounds or volatile compounds not trapped by the growth medium and will often include material lost from senescent and dying cells. The use of radioisotope labeling such as $^{14}CO_2$ used by Subba-Rao *et al.* (1962) with tomato and Slankis *et al.* (1964) with *P. strobus* increases the sensitivity of detection of exudates and sites of exudation considerably. Using 9-month-old *P. strobus* seedlings, Slankis *et al.* (1964) found some 0.8% of $^{14}CO_2$ supplied was recovered in the solution bathing the root during 8 days, and considering the very large effects of growth conditions on exudation, this is of the same order as found with young herbaceous plants (Rovira, 1969a).

With annual plants, seedlings are usually grown with their roots continuously in sterile media (Rovira, 1965). Collection of exudates from mature root systems of trees poses a special problem; W. H. Smith (1970) obtained exudates from mature sugar maple by air layering, then surface sterilizing new roots and aseptically introducing them into sealed sterile test tubes containing nutrient solution. As Smith pointed out, the possibilities of root damage by the sterilizing agents must be minimized, and he surface sterilized roots with a 10–15-second submersion in an antibiotic solution of cycloheximide, streptomycin sulfate, and griseofulvin. Changes in permeability of roots due to sterilizing agents should be easily checked by preliminary experiments using loss of ^{36}Cl as an indicator, as efflux of this is usually passive.

"Pulse" labeling with $^{14}CO_2$ has been employed by MacDougall and Rovira (1965, 1970), Rovira (1969b), and MacDougall (1970) with wheat over much shorter times than those above. Collection of exudates over very short periods, e.g., a few minutes to a few hours, measures exudation relatively free of reabsorption and of loss by senescing cells but will record exudation only from metabolic pools in the root which have become labeled with ^{14}C. Unless all such pools are labeled to the same specific activity, comparison of amounts of different compounds exuded could be in error. A solution to this problem is to grow plants in an atmosphere including $^{14}CO_2$ continuously to ensure uniform labeling as much as possible, or to allow a long period between labeling and collection of exudates.

Exudates are usually collected from solution-grown plants or as eluates of sand culture systems, but occasionally synthetic soil systems have been used. Considering the markedly different physical appearance of roots grown in solution and grown in sand it is not surprising (in retrospect) that Boulter *et al.* (1966) found large differences in exudation from pea roots grown in culture solution and sand culture; up to 700% increase of some amino acids was found with plants grown in quartz sand. Some workers have examined leachates of planted nonsterile soils (e.g., Harmsen and Jager, 1963; Martin, 1971) but such leachates may be partly of microbial origin and partly of root exudates

TABLE II ROOT EXUDATES OF TREE SPECIES AND WHEAT

	Pinus strobus[a]	Pinus lambertiana[b]	Pinus banksiana[b]	Pinus rigida[b]	Pinus radiata[b]	Pinus radiata[c]	Pinus resinosa[d]	Robinia pseudoacacia[b]	Acer saccharum[e]	Eucalyptus pilularis[f]	Wheat[g]
	1	2	3	4	5	6	7	8	9	10	
Amides and amino acids											
α-Alanine		−[h]	−	+[i]	+	+	+	−	+	+	+ +
β-Alanine											
γ-Aminobutyric acid		+	+	+	+	+	+	−		+	+
Arginine						+					+
Asparagine	+	−	−	−	−	+ + +	+ +	+ +	+		+
Aspartic acid	−	+	+	+	+		+	+	+		+
Cystathionine									+	+	+
Glycine	+	+	+	+	+	+ + +	+ +	+	+		+
Glutamic acid	−					+	+		+	+ +	+
Glutamine	+									+	
Homoserine											
Leucine/isoleucine		−	+	+	+	+	+ +	+	+		+
Lysine							+	+			+
Methionine		−	−	−	−			−			+
Phenylalanine		+ + +	+ +	+ + +	+	(+)[j]	+ +	−	+		+
Proline						(+)		+	+		+
Serine		−	−	−	−	+ +	+ +	−	+		+
Threonine			−		−			+	+		+
Tyrosine					−	+ +	+			+	+
Tryptophan											+
Sugars											
Arabinose	+									+ + +	
Fructose		+	−	−	−		+ +	−	+		+
Galactose											+
Glucose	+	+	+	+	+		+	+	+		+
Maltose											

	1	2	3	4	5	6	7	8
Oligosaccharides								
Raffinose								
Rhamnose	++++				−	+	−	+
Ribose		+	+++	+	−	+	−	+
Sucrose				+				
Xylose		+						
Organic acids								
Acetic	+++		+	+	+	+	+	+
Butyric								
Citric	+++	+	+					
cis-Aconitic	+							
Fumaric				−		−	+	−
Glycolic	++				−			
Malic	+ ++				+	−	−	++
Malonic		+		+	−	+	+	
Oxalic	+		+	−	+	−	+	
Oxalacetic			+			−		
Propionic								
Shikimic	+							
Succinic	++	+	+	+	+	+	+	+
Valeric								

[a] Slankis *et al.* (1964).
[b] W. H. Smith (1969).
[c] Bowen (1969).
[d] Agnihotri and Vaartaja (1967).
[e] W. H. Smith (1970).
[f] Cartwright (1967).
[g] Rovira and MacDougall (1967).
[h] Minus sign designates absence.
[i] Plus sign designates presence.
[j] (+) designates present in some unreported studies.

not decomposed or absorbed by the soil and its associated microorganisms.

Pearson and Parkinson (1961) with nonradioactive experiments, MacDougall (1968) with ^{14}C, Bowen (1968) with [36]Cl, and Rovira and Bowen (1970) with ^{32}P allowed exudation onto moist filter paper on either side of a root and thus located sites of loss from the root. In the last three cases, counting of radioactivity along the root and the filter paper enabled loss to be calculated as a fraction of radioactivity in each part of the root. Subsequent one-dimensional chromatography of the filter paper strip enables chemical characterization of the radioactive exudates. Greatest exudation usually occurs from the elongating zone of the root from a few mm to 2–5 cm from the root apex, but appreciable loss can occur all along the root, especially where root damage occurs, such as from emergence of lateral roots, microfauna feeding, and mechanical damage (see Rovira, 1969a). In an extension of the filter paper method, Rovira (1969b) showed peaks of loss of ^{14}C labeled substances near the apex of wheat roots were largely due to poorly diffusible substances which may be the mucigel described by Jenny and Grossenbacher (1963). Readily diffusible exudates occurred in approximately equal amounts along the whole of 20-cm root studied. By sowing spores of nutritional mutants of the fungus *Neurospora crassa* along sunflower roots, Frenzel (1960) showed some amino acids were lost from the root tip and others from the root hair zone.

2. Composition of Tree Root Exudates

Table II summarizes the present records of exudates from forest tree species; wheat data has been included for comparison. As for other plants a wide range of sugars, amino acids, and organic acids have been detected in root exudate from tree species. Although the lists must be considered far from complete—the main compounds investigated are those for which analytical techniques are well developed—they do include the most abundant groups of soluble chemical components of plant cells. In addition, W. H. Smith (1969) reported the vitamin, niacin, in *P. radiata* exudates.

The neglect of volatile compounds in root exudate studies may have seriously underestimated the extent of exudation. Rovira and Davey (1972) referred to wheat studies in which for every unit of carbon exuded as water-soluble material, some 3–5 units were released as non-water-soluble mucilaginous material and root cap cells, and some 8–10 units of materials volatile under acidic conditions were released; much of the volatile material was probably carbon dioxide. Krupa and Fries (1971) and Melin and Krupa (1971) have recently demonstrated the main volatiles of root extracts of *P. sylvestris* to be mono- and sesquiterpenes but no examination of root exudates appears to have been made; substances such as terpenes may well affect the microbial composition of the rhizosphere.

Within the one set of conditions (e.g., W. H. Smith, 1969) exudates may vary both qualitatively and quantitatively between species (Table II, columns 2–6). Smith also showed the amount of exudate in seedlings was related to size (see Table III) but this only partially explains the species differences. Richter *et al.* (1968) and Rovira (1969a) have reported extremely large variation in amounts of exudates (up to a hundred- or thousandfold differences) released by a species depending on the plant growth conditions and the method of collection of exudates. Within the one species under different conditions large differences in exudate composition occur (Table II, columns 5 and 6).

For exudates from forest tree species the only detailed quantitative comparison between amino acids, sugars, and organic acids was made by W. H. Smith (1969, 1970) who showed organic acids were by far the major component of the exudates of both seedlings and mature trees (Tables III and IV). Slankis *et al.* (1964) found the major exudate in their studies was malonic acid. The concentration of particular exudates in the rhizosphere in soil will depend on their rate of movement from the root by diffusion and their uptake by the rhizosphere microflora.

3. *Mechanisms of Exudation*

Mechanisms of exudation of organic compounds have not been studied in detail but generally loss is assumed to occur by passive leakage. Studies of loss mechanisms should be placed in a strict biophysical framework (see Chapter 5) viz., passive loss is loss down a concentration gradient (for uncharged solutes) or an electrochemical potential gradient

TABLE III

EFFECT OF SPECIES ON EXUDATION[a]

Species	Plant dry wt. (mg)	Root dry wt. (mg)	Carbohydrates (μg/seedling)	Amino acids (μg/seedling)	Organic acids (μg/seedling)
P. radiata	21.1	3.6	13.0	149	320.7
P. lambertiana	64.1	9.6	33.8	301	546
P. banksiana	6.4	1.9	0.1	64	270
P. rigida	8.0	1.9	1.0	29	66
Robinia pseudoacacia	14.8	4.2	9.0	44.3	77

[a] Data of W. H. Smith (1969). Exudation over 10 days in complete nutrient solution.

(for charged ions), the rate of loss being a function not only of the gradient but also the permeability of cell membranes to the particular solute. "Active" processes move against the electrochemical potential gradient. For most organic exudates there is almost certainly a lower electrochemical potential on the outside of the root than inside the cells, thus suggesting a purely passive leakage of exudates, but permeability needs to be considered also with regard to *rates* of loss. Cirillo (1961) has described carrier-facilitated diffusion for entry of sugars into yeast cells down a concentration gradient but this possibility does not appear to have been explored with loss of organic substances by higher plants, and it seems unlikely.

Results of Boulter *et al.* (1966) with pea indicate that analyses of exudates do not always precisely reflect the composition of the root, for much depends on where solutes are localized in the cell and on the modes of loss of different solutes and the membrane permeability to each solute. Nevertheless, some relationship between root exudates and cell composition would be expected, and in future more cognizance could well be given to the effects of various treatments on cell composition when studying root exudate composition, e.g., high potassium or high cation media are well known to lead to accumulation of organic acids in plant cells, and although most of this is in the vacuole, we may well expect an increase in organic acids in the root exudates as well. Following Krupa and Fries' (1971) studies on *P. sylvestris* exudates we might reasonably expect some mono- and sesquiterpenes in its root exudate but little ethanol, acetoin, or isobutyric acid.

4. Conditions Affecting Exudation

Rovira (1969a) has indicated factors affecting root exudation by a range of plants. Below we apply these to studies on tree species.

a. Plant Age. W. H. Smith's results (1970) (Table IV) show that although seedling sugar maple (*Acer saccharum*) produced a wider range of sugars and more sugar than did young roots on a 55-year-old tree, sugars and amino acids were minor components of the exudate compared with organic acids, especially in the mature trees. Roots of these produced six times more organic acids/mg dry weight than did seedling roots. Acetic acid was the major organic acid in both instances; in addition, roots on mature trees exuded citric and malonic acids and seedling roots lost oxalacetic acid. Bowen (1964) found the amino acids in root exudates of *P. radiata* seedlings decreased from 2 to 6 weeks but the amino acid composition of the exudates did not change.

b. Nutrition. Nutrition effects on root exudation have not been

TABLE IV

EFFECT OF TREE AGE ON EXUDATION BY *Acer saccharum*[a]

Carbohydrates[b]	Seedling	Mature	Amino acids/amides[b]	Seedling	Mature	Organic acids[b]	Seedling	Mature
Fructose	9.8 ± 1.5[c]	0.5 ± 0.1	Alanine	0.2 ± 0.1	2.4 ± 0.2	Acetic	67.3 ± 5.0	495.7 ± 18.2
Glucose	4.7 ± 0.8		Glutamine	0.3 ± 0.1	0.3 ± 0.1	Citric		46.7 ± 4.5
Rhamnose	0.7 ± 0.1		Glutamic acid		0.2 ± 0.1	Malonic		12.0 ± 2.5
Ribose	1.2 ± 0.1		Glycine	0.2 ± 0.1	0.8 ± 0.1	Oxalacetic	23.3 ± 2.7	
Sucrose	20.6 ± 2.1	5.8 ± 1.4	Homoserine		0.5 ± 0.1			
			Leucine/isoleucine	0.4 ± 0.1	1.4 ± 0.2			
			Methionine	0.2 ± 0.1	0.6 ± 0.1			
			Phenylalanine	0.7 ± 0.1	1.6 ± 0.1			
			Serine	0.2 ± 0.1	1.3 ± 0.2			
			Threonine	0.8 ± 0.1	1.7 ± 0.1			
			Tyrosine		0.8 ± 0.1			
			Valine	0.2 ± 0.1	0.6 ± 0.1			

[a] From W. H. Smith (1970), by permission of *Phytopathology*.
[b] Data in mg × 10⁻⁴ of each carbohydrate, amino acid/amide, and organic acid released per mg of oven dry roots of maple seedlings or mature tree during a 14-day growth period.
[c] Mean and standard error of three replicate determinations using one composite exudate sample from 200 seedling or 20 mature tree roots.

extensively studied. Bowen (1969) found phosphate deficiency in *P. radiata* led to two and a half times more amino acid/amide in exudate than from control plants, and nitrogen-deficient plants exuded only a quarter as much amino acid/amide as control plants, independent of effects on growth (Table V). Cartwright (1967) found that varying the nitrogen and phosphorus levels in culture solutions had no effect on organic acid exudation of *Eucalyptus pilularis* seedlings. Amino acid changes were consistent with those above while the quantity of sugars exuded increased both with decreasing phosphorus supply and decreasing nitrogen supply. Maltose and fructose were detected only at high nitrogen levels (70 and 140 mg N/liter) and galactose only at a low nitrogen level (14 mg N/liter). Nitrogen was supplied as nitrate.

The increase in net amino–amide N loss from *P. radiata* with low phosphorus was examined further by Bowen. Phosphate level had no effect on uptake of ^{14}C labeled amino acids, thus eliminating the possibility that the greater exudation of phosphorus-deficient plants was due

TABLE V

Effect of Nutrition on Amino Acid–Amide Exudation
of *P. radiata* Seedlings[a]

Amide or amino acid	Nutrient solution		
	Complete	Phosphate deficient	Nitrogen deficient
Asparagine	10.9[b]	32.5[b]	3.0[b]
Glutamine	23.6[c]	52.0	2.8
γ-Aminobutyric acid	5.2	13.8	1.0
α-Alanine	1.6	2.8	1.2
Aspartic acid	4.4	9.6	2.0
Glutamic acid	6.0	19.7	2.0
Glycine	7.3	14.0	3.4
Leucine	3.0	5.6	1.8
Serine	4.8	8.0	2.0
Threonine	1.4	2.0	—
Valine	1.8	4.0	0.1
Total amido/ amino nitrogen[d]	104.5	248.5	25.1

[a] From Bowen (1969), by permission of *Plant and Soil*. Amide and amino acid exudates from *Pinus radiata* seedlings 2–4 weeks. Roots were of similar length in all treatments.
[b] Moles × 10^{-9}/plant.
[c] Some arginine was also present but only in small amounts.
[d] Including the two NH_2 groups of asparagine and glutamine.

to a greater reabsorption of lost amino acids by phosphorus-sufficient plants. An effect of phosphate deficiency on increasing permeability of root cells was eliminated by examining the passive loss of ^{36}Cl from pine roots, which was the same for both phosphorus-sufficient and -deficient plants. However phosphorus deficiency gave a doubling of free amino–amide nitrogen in the root and this was the most likely reason for the increased exudation of amino acids and amides with low phosphorus plants.

c. Light. With herbaceous plants, shading has decreased amino acid exudation, but sugars and organic acids have apparently not been studied. The only tree species information is indirect, e.g., Harley and Waid's finding (1955) of different proportions of *Trichoderma* and *Rhizoctonia* on roots of beech, depending on the light regime.

d. Temperature. As with light, effects of temperature on root exudation by *P. radiata* can be deduced by an inordinate reduction of colonization by some strains of *R. luteolus* when reducing soil temperatures from 20° to 15°C compared with growth in laboratory media (Theodorou and Bowen, 1971a, see Section G,4, Table VII).

e. Microbial Effects. Microbial effects on exudation are the most difficult to study. This question is of crucial importance because if microorganisms greatly affect exudation, most of the studies carried out under asepsis, have to be qualified. Furthermore, effects by specific organisms will greatly enhance the competitive ability of that organism in the rhizosphere. Rovira (1969a) derived from Harmsen and Jager's results (1963) that exudation from wheat roots into synthetic soil was increased at least fourfold by microorganisms.

Possible ways in which microorganisms could affect root exudates are (i) by producing substances affecting root permeability. Various antibiotics produced by soil microorganisms in culture can increase permeability (Norman, 1955, 1961), but whether these compounds are produced in sufficient quantity in the rhizosphere to have such effects is yet unknown. (ii) By affecting root metabolism. This may affect the concentration of solutes in the cell or may affect loss mechanisms directly. Metabolic effects of rhizosphere microorganisms have been indicated by Bowen and Rovira (1969) and the phenomenon is well known with plant pathogen-host interactions. Mycorrhizal fungi are known to produce hormones and cytokinins; Highinbotham (1968) found hormones to increase permeability of *Avena* coleoptiles and an increase in passive efflux of potassium. D. Smith *et al.* (1969) noted that the translocation stream of autotrophic higher plants is diverted toward the site of association with fungi, including mycorrhizae. Krupa and Fries (1971) recorded a two- to eightfold increase in volatile compounds in roots of *P. sylvestris* which were mycorrhizal. (iii) Where loss of a solute is by passive diffusion, thought to be the usual case, absorption of exudates by microorganisms would maintain a steep concentration gradient (or electrochemical potential gradient) to drive solute diffusion. The movement of carbohydrates from beech roots to the mycorrhizal sheath, studied so well by D. H. Lewis and Harley (1965a,b,c) is a special and striking case of microorganisms external to cells (i.e., the Hartig net and the fungus mantle) affecting loss of solutes from cells and is directly analogous to a rhizosphere microorganism situation. The mechanism of the loss, as with other symbioses, is unknown but removal of sucrose by the fungus and elaboration into trehalose and polyols maintains a steep concentration gradient for further sucrose loss. D. Smith *et al.* (1969) have indicated a specificity in transfers of carbohydrates in symbiotic systems.

B. Organic Compounds Other Than Root Exudates

Very little information exists on organic substances in the rhizosphere other than root exudates. Reference has been made above to loss of organic substances by senescence of root cap cells or sloughed epidermal cells. Rovira (1956a) found pea roots released 0.5 mg dry weight of cell debris per plant over 21 days. No data are available for forest tree species, although sloughing of epidermal cells from seedling pine roots is very marked.

Small amounts of free organic acids and amino acids probably occur in soil, although analytical techniques usually do not distinguish between such substances in free form and those occurring from death of microorganisms during extraction procedures. Moodie (1965) has suggested that movement of such substances in soil toward roots by convection in water might be important to the nutrition of rhizosphere microorganisms, but the concentrations of free amino acids and organic acids in soil are probably too small for this to be significant. Products of microorganisms growing in the rhizosphere will also contribute to the rhizosphere's chemical composition.

C. Inorganic Composition of the Rhizosphere

Inorganic compounds in the rhizosphere do not usually serve as an energy source and therefore have not attracted study by microbiologists. Measurements of oxygen and carbon dioxide concentrations in the rhizosphere have received little attention, although there is general agreement that oxygen levels will be lowered and carbon dioxide levels raised by such processes as respiration by roots and microorganisms. The metabolic activities of the root itself will not be impaired until the oxygen concentration at the surface of the root is very low (Greenwood, 1969). The isolation of *Clostridium* from roots indicates that small anaerobic pockets do occur on root surfaces.

Soil chemists interested in ion movement to roots have studied the inorganic composition of the rhizosphere. Ions move to roots by convection in water (mass flow) and diffusion, and depending on the absorbing power of the root for that particular ion, it will be completely absorbed or will accumulate around the root. Thus phosphate concentrations in the rhizosphere are usually much lower than in soil solution (D. G. Lewis and Quirk, 1967), while Barber and various associates (Barber, 1962; Barber *et al.*, 1963; Barber and Ozanne, 1970) showed autoradiographically that ^{90}Sr, ^{35}S, and ^{45}Ca usually accumulate around roots. In the studies by Barber and Ozanne (1970), *Lupinus digitatus* absorbed enough calcium to cause depletion around roots. Riley and Barber (1969, 1970) showed an approximate doubling of soluble salts in the rhizosphere soil around soybean (1–4 mm distant from the root) and approximately sixfold increases at the rhizoplane. The rhizoplane soil was 0.3–1.3 pH units higher than surrounding soil. This pH increase was probably due to bicarbonate efflux from the roots during nitrate absorption. Where ammonium ions supplant nitrate, and also when high cation absorption occurs, hydrogen ions tend to be lost from the root thus decreasing pH, and in sand cultures a pH of 3 has been recorded with ammonium nutrition (Nightingale, 1934).

Many inorganic ions as well as bicarbonate, hydroxyl, and hydrogen ions are lost by roots. Using the filter paper method above, Bowen (1968) demonstrated loss of 7.2% of the chloride from the apical 2–3 cm of *P. radiata* seedling roots and 3.3% from basal parts of the roots. Tracer studies of this type measure efflux and need not be interpreted as leading to an accumulation of these ions in the rhizosphere; where influx exceeds efflux there is no net accumulation of the ion in the rhizosphere.

D. Growth on Surfaces

The microenvironment immediately around surfaces can be markedly different from bulk solution (e.g., Mitchell, 1951). Dynamics of growth along and on surfaces have been little studied. Most classical microbiological studies of growth kinetics have been made in stirred solutions.

E. Root Exudates and Growth of Mycorrhizal Fungi

Any treatment of growth of mycorrhizal fungi around roots would be quite incomplete without reference to Melin's extensive studies on root substances and nutrition of mycorrhizal fungi (see Section II,B also). Melin (1963) reviewed his studies on growth of mycorrhizal fungi in culture which showed (i) growth of many mycorrhizal fungi in culture was considerably enhanced by addition of vitamins and amino acids. (ii) Growth of many fungi was enhanced still further by incorporation of unidentified diffusates from cultures of excised roots of pine, sterile exudates from attached roots, and extracts from dead roots obtained at 100°C. The term "M factor" has been given to these unidentified compounds. (iii) Exudates of other than pine species, e.g., tomato could stimulate growth of the fungi, and (iv) high concentrations of M-factor preparations were inhibitory to growth. Melin interpreted these results to suggest the presence of both diffusible and indiffusible stimulatory factors and of an inhibitory factor. It is possible that the stimulation of growth he observed immediately around killed extracted roots was an inert surface effect, i.e., a thigmotropic response which commonly occurs with fungi and other microorganisms. However studies by Melin and others on root colonization of live pine seedlings (e.g., Theodorou and Bowen, 1969) clearly indicate the stimulation of growth of mycorrhizal fungi by root exudates, free of considerations of the thigmotropic response. Melin's studies and other studies of nutrition of mycorrhizal fungi (see Harley, 1969) were made on laboratory media and the search for the identity of the M factor(s) has been pursued without success using replacement techniques.

What do root exudate analyses have to contribute to fungal growth studies in synthetic media?

(i) Amino acids and sugars readily used by mycorrhizal fungi (Harley, 1969; Palmer and Hacskaylo, 1970; Lundeberg, 1970) occur in root exudates of tree seedlings and other plants. [This latter point is consistent with Theodorou and Bowen's findings (1971b) that mycorrhizal fungi can grow in the rhizosphere of nonhost plants, e.g., grasses.] W. H. Smith (1969, 1970) has shown an abundance of organic acids in exudates from tree species, but use of organic acids by ectomycorrhizal fungi has received scant attention; Palmer and Hacskaylo (1970) found some ectomycorrhizal fungi can use citrate as a sole carbon source, but acetic acid (as acetate), the most abundant organic acid in Smith's studies, was little used as a pure carbon source by the six fungal isolates they studied.

(ii) So far no identity of the M factor has been advanced. As considerable variation occurs between mycorrhizal fungi in specific growth requirements it would not be at all surprising to find the M factor (if it exists) is not one compound but several. Root exudate analyses reveal examination of such groups of compounds as amino acids and amides as possible M factors have been by no means exhaustive. For example, media employing casein hydrolysate as the sole amino acid sources include alanine, arginine, aspartic acid, glycine, glutamic acid, leucine, proline, phenylalanine, serine, threonine, valine, and tryptophan, but notable omissions are asparagine, glutamine, and γ-amino butyric acid, all of which are major components in tree seedling exudates (W. H. Smith, 1969; Bowen, 1969). Furthermore, ratios of the various amino acids common to both root exudates and casein hydrolysates are different. Antagonisms between amino acids are also well known (but little understood). Similarly, amino acids in high concentration can be toxic to mycorrhizal fungi: Melin (1963) recorded *Boletus versipellis* and *Lactarius rufus* to be sensitive to all but very low levels of glutamic acid, and Lundeberg (1970) has shown mycorrhizal fungi to be very responsive to asparagine at low levels but toxic at higher levels. Theodorou and Bowen (1968) have found growth of *R. luteolus* to be markedly depressed by 1.15 gm/liter of casein hydrolysate, but that this is relieved by addition of amino acids which are present in exudate of *P. radiata* but absent from casein hydrolysate. In view of these complex interactions and the exhibition of stimulatory and inhibitory effects by the one compound, it is possible that the idea of a specific M factor(s) is illusory; the balance of a number of substances might be the important factor. If specific M factors exist, they may not be easy to find by *ad hoc* replacement techniques alone. To base replacement studies on analyses of exudates is desirable but difficult; analytical techniques are quite laborious and unusual compounds are especially likely to be missed. A further difficulty in relating replacement studies to a real situation lies in demonstrating that the concentration of a particular substance active in laboratory media occurs in the natural situation.

(iii) Mycorrhizal research so far has done little to elucidate the morphogenic phenomenon of mantle formation occurring with many ectomycorrhizae. Melin's studies showed abundant mycorrhizal growth around roots—is this due to an indiffusible M factor? Is mantle development a thigmotropic response and if so, what is the reason for this? Is it due to a nonspecific accumulation of nutrients and energy sources or other environmental conditions at the root–soil interface?[1]

F. Experimental Studies on Ectotrophic Growth along Roots

There are now some data on the effect of soil and biological factors on mycorrhizal production but these rarely distinguish between effects

[1] *Note added in proof.* D. J. Read and W. Armstrong (1972), *New Phytol.* **71,** 49, have found internal oxygen supply to silicone rubber "roots" to be important in sheath formation by *Boletus variegatus.*

on growth in the rhizosphere and on infection processes (and subsequent mycorrhizal development). Extensive data also exist on effects of environmental variables on growth of ectomycorrhizal fungi in laboratory media (see Harley, 1969), but the relationship between such observations and colonization of roots in soil is often not clear. There appear to be no published quantitative data on the rate of ectotrophic growth of mycorrhizal fungi along roots in field soil.

1. Methods

In the last 3 years we have approached colonization of roots in an experimental manner. Our usual analysis of the effects of a particular variable is to examine (i) growth of a range of test mycorrhizal fungi in laboratory media with the nutritional or environmental variable imposed; (ii) growth of the fungus, applied as mycelial inoculum, on the surface of pine seedling roots in soil, usually for 4 weeks (but up to 8), i.e., before infection occurs; and (iii) mycorrhizal production in soil after growth of the test plant for 3–5 months under greenhouse conditions.

Soil conditions may affect the growth of the fungus in the rhizosphere either directly or indirectly via the effects on the plant and root exudation. Colonization studies must also recognize the possibility of surface or thigmotropic effects. For these reasons controls in our colonization studies consist of 1-mm-diameter bundles of 13-μm-diameter sterile glass fibers which are inoculated in soil in precisely the same way as the pine seedling root. Sterile seedlings are planted to sterilized soil in 20 $\times$ 3-cm test tubes. The inoculum is applied as growing mycelium in a 3-mm-diameter disk of Melin-Norkrans agar medium (Melin, 1959) from which residual nutrients have been removed by washing in sterile water. After the colonization period, the soil is gently removed from the root either by washing or allowing it to drop off in water. The roots are then stained with 1% cotton blue in lactophenol, mounted in water, and the length of root colonized is measured and a rating (0–4) given for intensity of colonization. Most of the variation is between seedlings, rather than between tubes, and a coefficient of variation of about 15% was obtained by using 10 tubes each with two seedlings per tube. Full details were given by Theodorou and Bowen (1969).

This simple approach has the value that the importance of single factors can be studied under defined conditions. While it falls far short of the complex dynamic situation obtaining in the field, it has been useful in drawing attention to certain factors presenting unsuspected difficulties in establishing mycorrhizal fungi, e.g., soil temperature factors. The eventual aim is to make the system more complex by controlled introduction of other microorganisms in order to understand better the dynamics of root colonization in soil by mycorrhizal fungi and other microorganisms.

2. *Plant Species*

Theodorou and Bowen (1971b) found that *R. luteolus* could colonize living roots of the grasses *Lolium perenne* and *Phalaris tuberosa,* the tree species *Eucalyptus leucoxylon* and *E. camaldulensis,* and also the clover *Trifolium subterraneum.* They concluded from this that the absence of mycorrhizal fungi for *P. radiata* from grassland areas distant from forests (in contrast to their presence in many eucalypt stands) (Bowen, 1963) was the nonarrival of spores in sufficient quantity to become established in competition with the native microflora, i.e., the absence from grasslands was probably not due to toxin production by the living grass roots. Their studies did suggest however that *decomposing* grass roots could be toxic to the fungus. Mycelial inoculum was used in these colonization studies, and an alternative interpretation of the failure of mycorrhizal fungi to establish in natural grasslands is that some spores may arrive but the grass roots may not stimulate spore germination. The ability of *R. luteolus* to grow in the rhizosphere of nonhost plants may be important in survival and spread of the fungus in the absence of a suitable mycorrhizal-forming host.

Theodorou and Bowen (1971b) also noted growth of *R. luteolus* was discontinuous along the grass roots, rather than compact and continuous as on pine roots. The detailed reasons for this are unknown.

Claims have been made that certain heath plants (or their associated endomycorrhizal fungi) are toxic to ectomycorrhizal fungi which may be introduced with a tree species, e.g., *Calluna* (Handley, 1963). Colonization experiments similar to those above may help to evaluate such hypotheses, as distinct from effects of plant residues in soil, e.g., Handley (1963).

3. *Nutrition*

a. pH and Nitrate. Decrease of mycorrhizal production in alkaline soils could be due to a high pH directly or to associated effects, e.g., high nitrification and consequent nitrate depression of mycorrhizal formation as suggested by Richards (1961). Growth of *R. luteolus* in the rhizosphere of *P. radiata* in sterilized forest soil is given in Table VI. The main points to note are (i) the absence of growth in soil and rhizosphere at pH 8 and very low soil nitrate, thus suggesting that poor mycorrhizal formation at alkaline pH may be due as much to inhibition of growth of the fungus in the rhizosphere as to an effect of nitrate on infection and

TABLE VI

Effects of Soil pH and Nitrate Concentration on the Colonization of Roots of *Pinus radiata* Seedlings and Glass Fibers by *Rhizopogon luteolus*[a]

pH	Nitrate concn. (ppm)	Number colonized		Av. length colonized (mm)		Fungal growth intensity								Root length per plant (mm)
		Fibers	Seedlings	Fibers	Seedlings	Fibers[b]				Seedlings[b]				
						vh	h	m	l	vh	h	m	l	
6.2	4.5	15 (20)[c]	14 (19)	9.8	16.8	0	3	3	9	7	3	2	2	88.8
5.0	12	11 (20)	19 (20)	4.8	26.5	0	2	0	9	17	2	0	0	83.3
5.0	115	14 (20)	16 (18)	8.5	16.0	0	2	6	6	11	3	2	0	90.2
8.0	12	3 (20)	0 (16)	0.8	0.0	0	0	0	3	—	—	—	—	86.2
8.0	115	5 (20)	0 (5)	1.4	0.0	0	0	1	4	—	—	—	—	65.8
				LSD										LSD
				$P = 0.05$	5.83									$P = 0.05$ 13.9
				$P = 0.01$	7.91									$P = 0.01$ 15.8

[a] From Theodorou and Bowen (1969), by permission of *Aust. J. Bot.*
[b] Very heavy (vh), heavy (h), moderate (m), and light (l).
[c] Numbers in parenthesis represent the number surviving out of 20.

mycorrhizal development. (ii) At pH 5, 115-ppm NO_3–N depressed colonization significantly but had little effect (or the reverse effect) on growth on the glass fiber; thus a possible nitrate effect on plant root exudate is suggested. (iii) The importance of root exudation is shown by the much greater colonization of roots than of fibers. Studies of mycorrhizal formation showed ectendomycorrhizae to completely supplant ectomycorrhizae at pH 8, an experimental demonstration of a prediction by Melin (1953) that different mycorrhizal fungi could be associated with trees under acid and under alkaline conditions.

b. Phosphate. Despite the increased exudation of sugars and amino acids with phosphate deficiency (see Section IV,A,4), we have been unable to show any stimulation of colonization of the rhizosphere of phosphate-deficient *P. radiata* by *R. luteolus*. Perhaps this is not surprising as organic acids in exudates are more abundant than amino acids, and Cartwright (1967) could not detect a phosphorus-deficiency effect on organic acid exudates of *E. pilularis*. We suggest increased mycorrhizal production of pines at relatively low phosphate levels (e.g., Björkman, 1942; Purnell, 1958) are attributable to effects on infection phases and not external root colonization by the fungus.

4. Physical Factors

a. Soil Temperature. Studies on the effect of soil temperature on colonization of *P. radiata* roots (Theodorou and Bowen, 1971a) arose from the realization that the optimum temperatures for growth of mycorrhizal fungi in laboratory media, viz., 18°–29°C (Hacskaylo *et al.*, 1965; Harley, 1969) were frequently well above those recorded at sowing time in many cool temperate climates. Studies on growth of *R. luteolus, S. granulatus,* and *S. luteus* in Melin–Norkrans media and the rhizosphere of *P. radiata* with controlled soil temperatures gave the data of Table VII. Statistical ratings of intensity of growth gave similar results to those indicated by the length of root colonized.

With each of the five fungi, the length of root colonized was significantly less at 16°C soil temperature than at 25°C, and, with the exception of *S. granulatus* No. 5, less than at 20°C. In some cases (*S. luteus* No. 1 and *R. luteolus*) there was almost no colonization at 16°C, despite growth of *R. luteolus* at 16°C on laboratory media being half that at 25°C. Predictions of root colonization based on growth in rich laboratory media at 16°C would have been correct with only two of the five fungi studied. This indicates a temperature effect on root metabolism of *P. radiata* (e.g., also Bowen, 1970) and root exudation; since an inordinate reduction occurred with only some fungi, a specific effect rather

TABLE VII

COLONIZATION OF *Pinus radiata* ROOTS AND GROWTH ON MELIN-NORKRANS
AGAR BY STRAINS OF MYCORRHIZAL FUNGI AT DIFFERENT TEMPERATURES[a,b]

Fungus	Soil temperature (°C)	Root length (mm)	Length colonized (mm)		Colony diameter in Melin-Norkrans media (mm)[c]
			Seedlings	Fibers	
Suillus luteus No. 1	16	73.4	6.2	4.5	8.9
	20	77.6	15.2	7.4	70.9
	25	82.9	24.9	6.9	63.8
S. luteus No. 3	16	85.6	18.6	5.1	37.7
	20	91.7	29.4	5.1	60.7
	25	92.9	31.8	9.1	61.3
S. granulatus No. 8	16	75.8	16.0	6.1	7.0
	20	90.7	28.0	6.5	41.0
	25	87.1	26.3	8.1	44.8
S. granulatus No. 5	16	90.3	16.8	5.5	63.2
	20	86.7	17.3	7.0	66.7
	25	87.2	24.4	5.5	57.5
Rhizopogon luteolus No. A[d]	16	41.9	4.1	4.4	40.0
	25	60.1	27.3	3.2	82.0
LSD: *Suillus*			5.1	($P = 0.05$)	
			6.7	($P = 0.01$)	
R. luteolus			7.4	($P = 0.01$)	

[a] From Theodorou and Bowen (1971a).

[b] Growth in soil for 4 weeks.

[c] Growth for 28 days, except for *R. luteolus* which was for 11 days.

[d] This experiment was carried out at a different time from that with the species of *Suillus*.

than a reduction in common energy sources at the lower temperature appears to be involved. With *S. granulatus* No. 8 colonization at 16°C was markedly greater than would have been expected from studies in synthetic media. These results indicate the difficulty in extrapolating to growth in the rhizosphere from studies in laboratory media. Hacskaylo *et al.* (1965) showed growth of mycorrhizal fungi at various temperatures is influenced greatly by the laboratory medium, and perhaps therefore a good relationship should not be expected between growth in synthetic media and the vastly different environment of the rhizosphere.

The very poor colonization by some isolates at soil temperatures of 16°C compared with that at 20°C, i.e., as more realistic soil temperatures to those in southern Australian soils at planting are approached, has revealed a possible need to select fungi capable of colonizing roots

4. Fungal Growth around Seeds and Roots

under field conditions as well as showing efficiency in plant stimulation. Even within species large differences occurred between isolates in root colonization at 16°C (Table VII, S. *luteus* No. 1 and S. *luteus* No. 3).

We also found better and faster mycorrhizal infection at 20° and 25°C than at 15°C but we do not know if this was due only to better colonization of the root alone or to temperature effects on infection process also. Marx *et al.* (1970) reported ectomycorrhizal development and mycelial growth in pure culture at different temperatures were correlated for isolates of *Thelephora terrestris* but not for *Pisolithus tinctorius*.

b. Soil Moisture. Mycorrhizal fungi differ greatly in their growth at low soil moisture. In laboratory media growth of *R. luteolus* ceases in relative humidities approximating those of soil at wilting point, but *C. graniforme* can grow at much lower relative humidities (Bowen, 1964). Colonization of the rhizosphere of *P. radiata* by *R. luteolus* declines markedly above field capacity and below 50% field capacity (Table VIII), i.e., still in the range of relative humidities supporting growth in laboratory media. Possible differences between fungi in this respect could be a factor in the replacement of some types of mycorrhizae by others in dry soils, reported by Worley and Hacskaylo (1959).

5. Microbial Factors

a. Previous Fungal Nutrition. Competition for energy sources in the rhizosphere is likely to decrease the chances of establishment of mycelial forms of inoculum. Using mycelia with high carbohydrate reserves could

TABLE VIII

EFFECT OF SOIL MOISTURE ON COLONIZATION OF ROOTS OF *Pinus radiata* BY *Rhizopogon luteolus*

Moisture level	Length colonized (mm)[a]		Root length[a] (mm)
	Fibers	Seedlings	
25% field capacity	6.6	7.0	101
50% field capacity	10.6	26.0	87
75% field capacity	11.8	22.4	80
Field capacity	10.3	20.0	73
125% field capacity	3.6	0	0
	LSD 4.4 $P = 0.05$		LSD 10 $P = 0.05$
	5.8 $P = 0.01$		13 $P = 0.01$
	7.4 $P = 0.001$		

[a] Mean of 20 plants grown for 4 weeks in sterilized soil. Field capacity was 25% moisture.

enhance growth of the mycorrhizal fungus in the rhizosphere; colonization by hyphae grown in Melin-Norkrans medium with 200 gm/liter glucose was twice that of hyphae grown with 20 gm/liter glucose. Mycorrhizal fungus colonization of the rhizosphere is stimulated by adding as little as 0.2 gm/liter glucose solution to sterile soil (Bowen and Theodorou, 1971); obviously one of the limitations to growth is the energy source in the exudates.

b. Competition and Antagonism. Mycorrhizal fungi are generally considered susceptible to competition from other soil organisms (Harley, 1969) and the successful introduction of mycelial forms of inoculum into fumigated soil (Theodorou, 1967) may be attributed partly to elimination of competitors or antagonists. Evidence exists in laboratory media for antagonism of some fungi to ectomycorrhizal fungi; Brian *et al.* (1945) in investigating a reported toxicity of heath soil to mycorrhizal fungi found that isolates of *Penicillium genseni* produced the antibiotic gliotoxin, to which *S. bovinus, S. grevillei* (syn. *B. elegans*), and *Cenococcum graniforme* were very sensitive. Levisohn (1957) reported inhibition of *S. granulatus, S. variegatus, S. bovinus, Leccinum scabrum* (syn. *B. scaber*), and *R. luteolus* by *Alternaria tenuis* in laboratory media. Although care must be taken in extrapolating from such antagonisms in laboratory media to behavior in the field, Levisohn found that *L. scabrum* could be introduced successfully as a mycorrhizal fungus into arable soils in which the other mycorrhizal fungi, which were more markedly affected by *A. tenuis*, failed to grow. Handley (1963) suggested that endomycorrhizal fungi of *Calluna* may be antagonistic toward ectomycorrhizal fungi for coniferous trees planted in heathland soils.

There are few published experimental data on microbial effects on colonization of the rhizosphere by ectomycorrhizal fungi. The easier establishment of mycorrhizal fungi in fumigated soils could be due to possible stimulation of mycorrhizal fungi by microorganisms recolonizing such soils or to elimination of antagonists. Ridge and Theodorou (1972) found that the major bacterial recolonizers of fumigated soils are fluorescent pseudomonads, and we have examined the effect of these and other bacteria on colonization of *R. luteolus* on *P. radiata* roots in sterilized forest soil. The results (Table IX) show a highly significant decrease in root colonization in the presence of pseudomonads, whereas a sporing bacterium with populations as high as pseudomonads had no inhibiting effect. Because of the reduction in colonization of glass fibers also in the presence of pseudomonads, production of antagonistic substances by the pseudomonads appears to be a better interpretation of the results than a simple competition for readily available energy sources

TABLE IX

EFFECT OF BACTERIA ON COLONIZATION OF *Pinus radiata*
ROOTS BY *Rhizopogon luteolus*

Treatment	Length colonized (mm)		Root length (mm)
	Seedlings	Fibers	
R. luteolus no bacteria	22.8	9.6	69
R. luteolus + fluorescent *Pseudomonas* R4/F3[a]	10.4	2.4	73
R. luteolus + fluorescent *Pseudomonas* R4/F1[a]	4.0	1.1	87
R. luteolus + *Pseudomonas* R3/AP1[a]	1.2	1.5	80
R. luteolus + sporing bacteria S4/Post 2[a]	19.6	6.0	64
	LSD 3.8 $P = 0.05$		LSD 9 $P = 0.05$
	5.1 $P = 0.01$		11 $P = 0.01$
	6.5 $P = 0.001$		14 $P = 0.001$

[a] Laboratory reference numbers.

in the root exudate. Other studies (Bowen and Theodorou, 1971) have shown a 20–50% reduction in colonization of roots by *R. luteolus* in sterilized soil reinoculated with a general soil microflora. The significant increase in root growth in Table IX with two of the pseudomonads is consistent with findings of both root stimulative and depressive organisms in soil (Bowen and Rovira, 1961).

G. COLONIZATION DYNAMICS

An understanding of the dynamics of the rhizosphere ecosystem in a field soil is necessary to assess the possibility of manipulating the rhizosphere population either for ease of introduction (and for persistence) of mycorrhizal fungi or for reasons such as biological control of plant diseases—of particular value in extensive crops such as forests. The treatment of single factors, as above, especially in a sterile soil is unreal, but it is a first step toward understanding a complex situation and provides quantitative information necessary to building a dynamic picture of the importance of various factors in colonization of roots.

Population biologists in other fields, e.g., zoology, have found the ecosystem or systems analysis approach of considerable use (e.g., Watt, 1968) but it had been largely ignored in the study of microbial colonization of roots. It is not difficult to discern direct analogies between phenomena *thought* to be important in microbial colonization of roots and population phenomena studied in detail in other fields, e.g., competition, dispersal, food supplies, and stability of systems. At this stage of our knowledge, the predictive value of such approaches in any but very broad terms in rhizosphere biology is small, but it forms a useful framework for integrating concepts of colonization, and particularly for focusing attention on large zones of ignorance which need to be studied.

In primary colonization the root grows through the soil at a rate determined by species and environment and comes into contact with potential bacterial and fungal colonizers of the rhizosphere (possibly including pathogens and propagules of mycorrhizal fungi). Under what conditions will these occupy the new root surface and possibly retard colonization by an inoculated fungus, or will existing colonization determine colonization of younger new root tissue? Actively growing seedling roots of *P. radiata* grow at 3–12 mm/day in well-aerated moist soil between 10° and 25°C (Bowen, 1970) depending on soil conditions, but the rate of growth of hyphae of *R. luteolus* (1 strain), *S. granulatus* (2 strains), and *S. luteus* (2 strains) along the root was of the order of 1–1.5 mm/day at 25°C in the absence of other organisms (Theodorou and Bowen, 1971a) and considerably smaller at lower temperatures. At 25% field capacity, growth of *R. luteolus* along roots was reduced to 0.3 mm/day while that of the root was 5 mm/day. Thus under reasonably good conditions for seedling growth, the root will quickly outgrow a mycorrhizal inoculum of growing hyphae. Where basidiospores must first germinate, the fungus is further delayed. A similar situation occurs with regrowth of dormant roots in the spring when new lateral root growth can temporarily break away from a fungal mantle (Robertson, 1954). Mycelial strands however grow along roots at 2–4 mm/day (at 20°C) (Skinner and Bowen, 1972) and once formed, clearly present a method of fairly rapid colonization of roots. This rapid growth is no doubt due to ready translocation of nutrients from an existing food base. Where root growth slows considerably due to age or adverse conditions, fungal growth may be similar to that of the root, e.g., some basidiomycetes and mycorrhizal fungi, such as *Cenococum graniforme*, can grow at very low soil water contents (Bowen, 1964) and can probably colonize new root growth where this is very slow because of low water content of soil. The complete investment of slowly growing short roots by mycorrhizal fungi thus giving the mycorrhizal mantle is, of course, another example of the fungus keeping up with the root.

A vigorously growing seedling root will easily outgrow mycelial inoculum of mycorrhizal fungi and many purely rhizosphere fungi (Taylor and Parkinson, 1961) leaving large amounts of actively exuding young roots available for colonization. Can microorganisms in soil colonize these before spread of organisms from older parts of root? There is very little quantitative information available on growth rates of soil bacteria and fungi on different parts of roots. Experiments by Ridge and Rovira (1970) on bacterial spread along roots indicate that in natural soils fresh colonization from soil is much more frequent than colonization from migration of existing organisms along the root when conditions are conducive to rapid root growth. Taylor and Parkinson (1961) considered lateral colonization of new root surfaces from soil to be more usual than colonization by spread of organisms along the root. Colonization of the new root by organisms in soil will depend on the population of the particular organisms in the soil, their distribution and proximity to the root, sensitivity to root exudates, rate of germination, motility to the root when the root does not actually touch them, and growth rate on the root surface. More information is needed on the rate of germination of fungal spores in the rhizosphere but it is known that spores of many soil fungi, e.g., *Penicillium* and *Helminthosporium*, will germinate in 1–10 hours when fungistasis is relieved, e.g., by root exudates, and therefore they are quite capable of rapidly colonizing new root surfaces approaching or touching them, much more quickly than the spread of mycorrhizal and most other fungi along the root. Thus the fungus with many propagules dispersed through soil and capable of germinating rapidly under the influence of the root has an enormous advantage in primary rhizosphere colonization, and it is almost a matter of chance which species of these the root will touch or approach closely to give them the opportunity of colonizing the root. Almost nothing is known of rates of germination of spores of mycorrhizal fungi in soil but it is difficult to envisage them germinating rapidly enough to be among these first colonizers.

It seems certain that mycorrhizal fungi will not be the first colonizers of recently formed seedling roots where growth of these is at all vigorous. What difficulty then does the mycorrhizal fungus (and late arrivals) have in establishing itself? Harley and Waid (1955) have shown that the numbers of fungi isolated from beech roots increase with age and a poorly defined succession occurs. The root tip was often sterile, fast-growing sporing fungi tended to occur toward the tip, and in older parts slow-growing and sometimes sterile fungi occurred as well. How do these late-developing organisms establish?

With herbaceous plants early microbial colonizers, which are predominantly bacteria, cover most of the root surface (Rovira, 1956b). However with other plants, e.g., *P. radiata*, this is not always the case. We have examined colonization of *P. radiata* roots growing in a sandy forest soil by removing seedling roots of known age, gently agitating them in water to remove sand but as few of the microorganisms as possible, and staining them using the Jones and Mollison method (1948).

Direct measurements of the areas of roots covered by microbial growth were made by weighing cut-outs of tracings of bacterial and fungal growth in representative microscopic fields projected onto a screen. We have no way of telling what percentage of organisms were lost in the washing in this study but even if it was 50% in our very gentle washing procedure, the results of Table X are convincing. Usually the percentage of the root surface covered by microorganisms was less than 10%, even with roots 3 weeks old. This was variable and on one occasion 16% was covered in a portion of the root 4 days old. We have not yet examined whether this generally poor colonization was related to a possibly low microbial population of the soil but it seems unlikely; studies with other soils would be desirable. Another possible reason for the low root surface cover by microorganisms is that exudation is localized and exudates are in lower concentration in the uncolonized parts. It is clear that even with relatively old parts of roots in a field soil there can be considerable areas of *P. radiata* root available for colonization and possible unimpeded paths for fungal growth could be discerned. Old sloughed epidermal cells were usually very heavily colonized.

The spatial considerations above raise the question of the extent to which interaction between microorganisms occurs on roots. The results of Table IX show that antagonism or competition for substrates does occur when organisms are added simultaneously in high concentration, and thus there is clearly scope in rhizosphere microbiology, for consideration of competition between species and stability of microecosystems, possibly along similar lines to those proposed for animals by Watt

TABLE X

MICROBIAL OCCUPATION OF *Pinus radiata* RHIZOPLANE SEGMENTS

Plant age (days)	Distance of segment from apex (cm)	Age of segment (days)	Percent occupation of segment[a]
7	2.5	4	1.2
21	14	21	8.4
28	1	1–2	7.1
28	2.5	4	5.4
28	3	4	15.7
28	3.5	4–5	7.2
90	1	1	0
90	4	4	12.1
90	Base	90	36.6

[a] Percentage of root surface covered by microorganisms with seedlings growing in a forest soil.

(1968) and Garfinkel and Sack (1964). However, it is also likely that at least for *P. radiata* in the soil we used, the main phenomenon was a number of fairly small spatially noninteracting communities. There is great scope for detailed studies of microecology and microbial interactions with general soil microorganisms and mycorrhizal fungi, possibly using techniques such as fluorescent antibody reactions (Beutner, 1961; Schmidt and Bankole, 1965; Trinick, 1970).

It seems in many cases (most cases?) late colonizers will establish not because they displace existing microorganisms but because there is space which they can occupy. In doing this, what do the slow-growing fungi and other late arrivals use as energy sources? There are a number of possibilities: (i) They may use exudates (if any) coming from the uncolonized part of the root. (ii) They may use secondary products of the existing microflora. At this time the first colonizers may be actively growing, resting, or senescing—methods used in microbiological examinations do not elucidate the condition of colonies of different ages on the root. Both this possibility and the previous one may be consistent with late colonizers (including mycorrhizal fungi) having lower metabolic rates and lower demands on energy sources than early colonizing, faster growing organisms. Energy requirements both for growth and maintenance of primary colonizers and slower late colonizers in the rhizosphere are unstudied. (iii) They could achieve energy and growth factor supplies by antagonism toward some early colonizers. A number of mycorrhizal fungi have been shown to produce antibiotics in laboratory culture (Šašek and Musílek, 1967; Marx, 1969). Antibiotic production would also help them maintain their colonization of the root against later entries. (iv) They may alter the permeability or metabolism of the host epidermal cells to increase exudation locally—an unproven hypothesis but one with considerable selective advantage. It is possible that antibiotics are involved in this also. (v) They may have access to energy sources not readily available to other organisms—this is a little explored field and somewhat unusual compounds are likely to be involved.

We have considered only energy sources above but the more general rhizosphere environment of mature parts of roots, i.e., including growth factors, inorganic composition, and physical environment, must also be considered in microbial growth on older parts of roots.

H. Growth around Roots following Infection

Following infection of the root, the fungus can obtain its energy and growth sources in an environment free of competition and then translocate them to the external phase. Except where rapid root growth occurs after a dormant period, growth of mycorrhizal fungi can usually proceed easily in the cortex of long roots of pine (Robertson, 1954; Wilcox, 1968), and therefore growth in the rhizosphere is not so critical for infection of short roots arising from the long root. In other cases, e.g., *Fagus* (Harley, 1969), infection of the long root is less common and external spread will be important. Few quantitative studies have been made on rate of spread of mycorrhizal fungi along the root surface compared with the cortex. Mycelial strands of *R. luteolus* grow along *P. radiata* roots in soil at 15°–20°C at 2–4 mm/day (Skinner and Bowen, 1972).

Growth on the roots following infection is also important when considering pathogen control (Chapter 9), nutrient uptake (Chapter 5), mycorrhizal succession, and spread from tree to tree. Rapid external growth by a mycorrhizal fungus will sometimes enable it to replace an existing mycorrhizal fungus (Marks and Foster, 1967; Wilcox, 1971). Ectendomycorrhizae are usually restricted to young seedlings and are replaced by other types (Mikola, 1965; Wilcox, 1971). These phenomena should be explainable in terms such as growth rate on particular exudates, infection ability of different mycorrhizal fungi, changes in plant susceptibility to infection, competition with other microorganisms, survival from season to season, etc., but there has been little study of such phenomena at this level so far. The spread of fungi from root to root will depend on root contact (especially with fungi where little external mycelium is formed) and the ability of fungi to grow through soils (Robertson, 1954).

V. Conclusions and Future Approaches

Response to mycorrhizal introduction and variation among fungi in their effect on tree growth has been indicated (e.g., Chapter 5) and hence a desirability of selecting fungi for inoculation either to new sites or superimposing them on sites with less beneficial mycorrhizae is apparent. The evidence of this chapter indicates variability among fungi in the ease of introducing them, and selection on this basis and on their ability to persist from season to season as well as on efficiency in tree stimulation is desirable. The rationale for studies of growth of mycorrhizal fungi in the rhizosphere is that an understanding of this will indicate the critical zones for microbial selection and the management practices needed to assist establishment.

Spore inoculation with basidiomycetes has been shown to be an easy practicable way to introduce ectomycorrhizal fungi, and the technology of application has been largely overcome. However this restricts one to the use of fungi with which a ready supply of fruiting bodies is available. Field collection of such sporophores reduces control over inoculum.

More control would be given by the development of methods to induce sporophore formation under laboratory and pilot plant conditions. There is an obvious need also for further study of resistance to adverse conditions and ease of establishment in the field of other types of propagules of mycorrhizal fungi, e.g., oidia, which may be readily produced in laboratory culture. This also imposes some limitation on the fungi which can be used in inoculation and eventually the greater reliability of mycelial forms of inoculum should be aimed at.

Infected roots will generally enable the survival of a fungus from season to season, but studies on survival and spread of different types of propagules under varying soil conditions both in the presence and absence of a host plant will add considerably to our knowledge of persistence of introduced organisms. Development of quick reliable marker techniques for such studies is important. The use of distinctive and reliable morphological markers (see Chapter 2) will be of great use as would be the development of reliable antibody fluorescence techniques for ectomycorrhizal fungi, in a similar way to that used in study of other soil fungi (Schmidt and Bankole, 1965). The development of highly selective media for particular groups of organisms has been of great value in studying microbial ecology (Tsao, 1970; Sands and Rovira, 1970, 1971) and will also be useful in mycorrhizal studies. Similarly, in rhizosphere colonization studies, use could be made of techniques involving labeling of hyphae, e.g., by radioactive isotopes (Robinson and Lucas, 1963).

Growth in the rhizosphere can be considered at two main levels, the descriptive level, where colonization of particular organisms and interactions with other organisms are examined in relation to environmental conditions, and the ecological–physiological level, where the mechanisms of these observations are studied. The study of root exudates and definition of the environment for root growth comes into the second of these categories. Exudates can be viewed from purely plant physiological aspects or attempts can be made to relate them to microbial phenomena on the root surface. Documentation of root exudates in studies performed up to now have provided useful knowledge of some of the main types of substrates in the rhizosphere and the effects of environment on these. They have pointed out, for example, that we should pay more attention to the role of organic acids in nutrition of microorganisms around the root. However, one must ask where do root exudate studies go from here? In the field of pure plant physiology it is obvious that more should be done to relate exudation to the composition of root cells and the mechanisms of loss, and in this way to integrate exudate amount and composition with general plant physiology and known effects of environment on plant behavior. Biophysical and physiological approaches to nutrient uptake and loss mechanisms and compartmentation in cells are fundamentally important in such studies of root exudation. Detailed root exudation analyses on some major groups of compounds would now appear to be sufficient to provide a realistic background for many rhizosphere microbiology studies. Some other groups of compounds need much more study, e.g., volatiles, to lay a similar type of background there. Further detailed analyses of sugars, organic acids, and amino acids should be related to specific realistic questions, e.g., exudation of citric acid for its role in desorption of phosphate from soils or for fluctuation in key compounds known to be important in the ecology of a particular microorganism. Further pure data collections without specific questions on the commonly occurring compounds seem unwarranted. Occurrence of compounds likely to be used by a wide range of microorganisms is not likely to answer questions of microbial specificity—the more unusual compounds would be more likely to do that. An illustration of this point was given by Gunner *et al.* (1966) who found that the dominant bacterium from roots of bean plants given foliar applications of the organo phosphate insecticide "diazinon" used this as a source of sulfur, phosphorus, carbon, and nitrogen. It may be possible, similarly, to stimulate the secretion of specific energy sources (if such exist) for selected mycorrhizal fungi in the rhizosphere by foliar application of the energy source.

In relating root exudates to microbial growth around roots, not only must specificity in microbial response be considered but also the important question is whether certain microorganisms themselves selectively enhance exudation—such a property by some mycorrhizal fungi could enhance their establishment in a highly competitive environment.

There are tremendous gaps to be filled in understanding microbial colonization of roots in such matters as the dynamics of growth on surfaces, the dynamics of competition, energy relations of organisms for growth and maintenance, production of antibiotics under rhizosphere conditions, definition of ecological "niches" on roots, quantitative studies on growth rates at different parts of roots under different environmental conditions, migration rates along roots, and germination of soil propagules and establishment on the root. Growth of mycorrhizal fungi on the young root must be integrated with this knowledge. A really quantitative comprehensive understanding of microbial colonization of roots in natural soils is not attainable at the moment. However, (i) since so many organisms have the ability to colonize roots and whichever do so is very much a chance occurrence depending on placement of propagules in soil, and (ii) since it seems probable that many spatially separated partially independent communities could be involved, it is doubtful if a very detailed predictive statement of microbial composition at all parts of a root under various conditions is ever likely to be possible. It is more realistic to use information gained by such studies as those above to help specify the general characters of microorganisms one should select for ready colonization, persistence, and spread. Certainly it would seem that one has a greater chance of partially controlling the rhizosphere microflora of plants by the use of organisms which also infect the roots, e.g., mycorrhizal fungi, but first the infection must be established.

In this chapter we have not considered the mechanisms of replacement of one mycorrhizal fungus with another, and superimposition of selected mycorrhizal fungi on existing, possibly less desirable ones. This important problem awaits further study.

References

Agnihotri, V. P., and Vaartaja, O. (1967). Root exudates from red pine seedlings and their effects on *Pythium ultimum*. *Can. J. Bot.* **45**, 1031.

Barber, S. A. (1962). A diffusion and mass-flow concept of soil nutrient availability. *Soil Sci.* **93**, 39.

Barber, S. A., and Ozanne, P. G. (1970). Autoradiographic evidence for the differential effect of four plant species in altering the calcium content of the rhizosphere soil. *Soil Sci. Soc. Amer., Proc.* **34**, 635.

Barber, S. A., Walker, J. M., and Vasey, E. H. (1963). Mechanisms for the movement of plant nutrients from the soil and fertilizer to the plant root. *J. Agr. Food Chem.* **11**, 204.

Beutner, H. E. (1961). Immuno-fluorescent staining—the fluorescent antibody method. *Bacteriol. Rev.* **25**, 49.

Björkman, E. (1942). Uber die Bedingungen der Mykorrhiza bildung bei Kiefer und Fichte. *Symb. Bot. Upsal.* **6**, No. 2, 1.

Boulter, D., Jeremy, J. J., and Wilding, M. (1966). Amino acids liberated into the culture medium by pea seedling roots. *Plant. Soil* **24**, 121.

Bowen, G. D. (1961). The toxicity of legume seed diffusates toward rhizobia and other bacteria. *Plant Soil* **15**, 155.

Bowen, G. D. (1963). The natural occurrence of mycorrhizal fungi for *Pinus radiata* in South Australian soils. *CSIRO Div. Soils, Adelaide, Aust., Div. Rep.* **6/63**.

Bowen, G. D. (1968). Chloride efflux along *Pinus radiata* roots. *Nature (London)* **218**, 686.

Bowen, G. D. (1969). Nutrient status effects on loss of amides and amino acids from pine roots. *Plant Soil* **30**, 139.

Bowen, G. D. (1970). Effects of soil temperature on root growth and on phosphate uptake along *Pinus radiata* roots. *Aust. J. Soil. Res.* **8**, 31.

Bowen, G. D. (1964). Unpublished data.

Bowen, G. D., and Kennedy, M. (1959). Effect of high soil temperatures on *Rhizobium* spp. *Queensl. J. Agr. Sci.* **16**, 177.

Bowen, G. D., and Rovira, A. D. (1961). The effects of micro-organisms on plant growth. I. Development of roots and root hairs in sand and agar. *Plant Soil* **15**, 166.

146   *G. D. Bowen and C. Theodorou*

Bowen, G. D., and Rovira, A. D. (1969). The influence of micro-organisms on growth and metabolism of plant roots. *In* "Root Growth" (W. J. Whittington, ed.), pp. 170–201. Butterworth, London.

Bowen, G. D., and Theodorou, C. (1971). Unpublished data.

Brian, P. W., Hemming, H. G., and McGowan, J. C. (1945). Origin of a toxicity to mycorrhiza in Wareham Heath. *Nature* (*London*) **155**, 637.

Briggs, G. E., Hope, A. B., and Robertson, R. N. (1961). "Electrolytes and Plant Cells." Blackwell, Oxford.

Cartwright, J. B. (1967). A study of the interaction between *Cylindrocarpon radicicola* Wr. and the roots of blackbutt (*Eucalyptus pilularis* Sm.). Ph.D. Thesis, University of Sydney.

Chilvers, G. A., and Pryor, L. D. (1965). The structure of Eucalypt mycorrhizas. *Aust. J. Bot.* **13**, 245.

Cirillo, V. P. (1961). Sugar transport in micro-organisms. *Annu. Rev. Microbiol.* **15**, 197.

Dobbs, C. G., and Hinson, W. H. (1953). A widespread fungistasis in soils. *Nature* (*London*) **172**, 197.

Ferenczy, L. (1956). Occurrence of antibacterial compounds in seeds and fruits. *Acta Biol.* (*Budapest*) **6**, 317.

Foster, R. C., and Marks, G. C. (1967). Observations on the mycorrhizas of forest trees. II. The rhizosphere of *Pinus radiata* D. Don. *Aust. J. Biol. Sci.* **20**, 915.

Frenzel, B. (1960). Zur Ätiologie der Anreicherung von Aminosäuren und Amiden im Wurzelraum von *Helianthus annus* L.: Ein Beitrag zur Klärung der Probleme der Rhizosphäre. *Planta* **55**, 169.

Fries, N. (1943). Untersuchungen über Sporenkeimung und Mycelentwicklung bodenbewohnender Hymenomyceten. *Symb. Bot. Upsal.* **6**, No. 4, 1.

Fries, N. (1966). Chemical factors in the germination of spores of Basidiomycetes. *In* "The Fungus Spore" (M. F. Madelin, ed.), pp. 189–99. Butterworth, London.

Fries, N., and Forsman, B. (1951). Quantitative determination of certain nucleic acid derivatives in pea root exudate. *Physiol. Plant.* **4**, 410.

Garfinkel, D., and Sack, R. (1964). Digital computer simulation of an ecological system, based on a modified mass action law. *Ecology* **45**, 502.

Garrett, S. D. (1970). "Pathogenic Root-Infecting Fungi." Cambridge Univ. Press, London and New York.

Greenwood, D. J. (1969). Effect of oxygen distribution in soil on plant growth. *In* "Root Growth" (W. J. Whittington, ed.), pp. 202–223. Butterworth, London.

Gunner, A. B., Zuckerman, B. M., Walker, R. W., Miller, C. W., Deubert, K. H., and Longley, Ruth E. (1966). The distribution and persistence of Diazonin applied to plant and soil and its influence on rhizosphere and soil microflora. *Plant Soil* **25**, 249.

Hacskaylo, E., Palmer, J. G., and Vozzo, J. A. (1965). Effect of temperature on growth and respiration of ectotrophic mycorrhizal fungi. *Mycologia* **57**, 748.

Handley, W. R. C. (1963). Mycorrhizal associations and *Calluna* heathland afforestation. *Forest. Commun. Bull.* **36**.

Harley, J. L. (1969). "The Biology of Mycorrhiza." Leonard Hill, London.

Harley, J. L., and Waid, J. S. (1955). A method of studying active mycelia on living roots and other surfaces in the soil. *Plant Soil* **7**, 96.

Harmsen, G. W., and Jager, G. (1963). Determination of the quantity of carbon and nitrogen in the rhizosphere of young plants. *In* "Soil Organisms" (J.

Doeksen and J. van der Drift, eds.), pp. 245–51. North-Holland Publ., Amsterdam.

Highinbotham, N. (1968). Cell electropotential and ion transport in higher plants. *In* "Transport and Distribution of Matter in Cells of Higher Plants" (K. Mothes *et al.*, eds.), pp. 167–77. Akademie-Verlag, Berlin.

Hiltner, L. (1904). Über neuere Erfahrungen und Probleme auf dem Gebiet der Bodenbakteriologie und unter besonderer Berücksichtigung der Gründüngung und Brache. *Arb. Deut. Landwirtschaftsges.* **98**, 59.

Jenny, H., and Grossenbacher, K. (1963). Root-soil boundary zones as seen in the electron microscope. *Soil Sci. Soc. Amer., Proc.* **27**, 273.

Jones, P. C. T., and Mollison, J. E. (1948). A technique for the quantitative estimation of soil micro-organisms. *J. Gen. Microbiol.* **2**, 54.

Katznelson, H., Rouatt, J. W., and Peterson, E. A. (1962). The rhizosphere effect of mycorrhizal and non-mycorrhizal roots of yellow birch seedlings. *Can. J. Bot.* **40**, 377.

Krupa, S., and Fries, N. (1971). Studies on ectomycorrhizae of pine. I. Production of volatile compounds. *Can. J. Bot.* **49**, 1425.

Laiho, O., and Mikola, P. (1964). Studies on the effect of some eradicants on mycorrhizal development in forest nurseries. *Acta Forest. Fenn.* **77**, 1.

Lamb, R. J., and Richards, B. N. (1971a). Effect of mycorrhizal fungi on the growth and nutrient status of slash and Radiata pine seedlings. *Aust. Forest.* **35**, 1.

Lamb, R. J., and Richards, B. N. (1971b). Unpublished data.

Levisohn, I. (1957). Antagonistic effects of *Alternaria tenuis* on certain root-fungi of forest trees. *Nature (London)* **179**, 1143.

Levisohn, I. (1960). Root forking of pine seedlings growing under non-sterile conditions. *New Phytol.* **59**, 326.

Lewis, D. G., and Quirk, J. P. (1967). Phosphate diffusion in soil and uptake by plants. III. ^{31}P movement and uptake by plants as indicated by ^{32}P autoradiography. *Plant Soil* **26**, 445.

Lewis, D. H., and Harley, J. L. (1965a). Carbohydrate physiology of mycorrhizal roots of beech. I. Identity of endogenous sugars and utilization of exogenous sugars. *New Phytol.* **64**, 224.

Lewis, D. H., and Harley, J. L. (1965b). Carbohydrate physiology of mycorrhizal roots of beech. II. Utilization of exogenous sugars by uninfected and mycorrhizal roots. *New Phytol.* **64**, 238.

Lewis, D. H., and Harley, J. L. (1965c). Carbohydrate physiology of mycorrhizal roots of beech. III. Movement of sugars between host and fungus. *New Phytol.* **64**, 256.

Lockwood, J. L. (1964). Soil fungistasis. *Annu. Rev. Phytopathol.* **2**, 341.

Lundeberg, G. (1970). Utilisation of various nitrogen sources, in particular bound soil nitrogen, by mycorrhizal fungi. *Stud. Forest. Suec.* **79**, 1.

MacDougall, B. M. (1968). The exudation of C^{14}-labelled substances from roots of wheat seedlings. *Trans. Int. Congr. Soil Sci., 9th, 1968,* Vol. 3, p. 647.

MacDougall, B. M. (1970). Movement of ^{14}C-photosynthate into the roots of wheat seedlings and exudation of ^{14}C from intact roots. *New Phytol.* **69**, 37.

MacDougall, B. M., and Rovira, A. D. (1965). Carbon-14 labelled photosynthate in wheat root exudates. *Nature (London)* **207**, 1104.

MacDougall, B. M., and Rovira, A. D. (1970). Sites of exudation of ^{14}C-labelled compounds from wheat roots. *New Phytol.* **69**, 999.

Marks, G. C., and Foster, R. C. (1967). Succession of mycorrhizal associations or individual roots of radiata pine. *Aust. Forest.* **31**, 193.

Martin, J. K. (1971). ^{14}C-labelled material leached from the rhizosphere of plants supplied with ^{14}CO$_2$. *Aust. J. Biol. Sci.* **24**, 1131.

Marx, D. H. (1969). The influence of ectotrophic mycorrhizal fungi on the resistance of pine roots to pathogenic infections. II. Production, identification, and biological activity of antibiotics produced by *Leucopaxillus cerealis* var. *piceina*. *Phytopathology* **59**, 411.

Marx, D. H., and Bryan, W. C. (1969). Studies on ectomycorrhizae of pine in an electronically air filtered, air-conditioned, plant growth room. *Can. J. Bot.* **47**, 1903.

Marx, D. H., and Ross, E. W. (1970). Aseptic synthesis of ectomycorrhizae on *Pinus taeda* by basidiospores of *Thelephora terrestris*. *Can. J. Bot.* **48**, 197.

Marx, D. H., Bryan, W. C., and Davey, C. B. (1970). Influence of temperature on aseptic synthesis of ectomycorrhizae by *Thelephora terrestris* and *Pisolithus tinctorius* on loblolly pine. *Forest Sci.* **16**, 424.

Melin, E. (1925). "Untersuchungen über die Bedeutung der Baummykorrhiza. Eine ökologische physiologische Studie." Fisher, Jena.

Melin, E. (1953). Mycorrhizal relations in plants. *Annu. Rev. Plant Physiol.* **4**, 325.

Melin, E. (1959). Physiological aspects of mycorrhizae of forest trees. *In* "Tree Growth" (T. T. Kozlowski, ed), pp. 247–63. Ronald Press, New York.

Melin, E. (1963). Some effects of forest tree roots on mycorrhizal basidiomycetes. *Symp. Soc. Gen. Microbiol.* **13**, 125–45.

Melin, E., and Krupa, S. (1971). Studies on ectomycorrhizae of pine. II. Growth inhibition of mycorrhizal fungi by volatile organic constituents of *Pinus sylvestris* (Scots pine) roots. *Physiol. Plant.* **25**, 337.

Mikola, P. (1965). Studies on the ectendotrophic mycorrhiza of pine. *Acta Forest. Fenn.* **79**, 1.

Mitchell, P. (1951). Physical factors affecting growth and death. *In* "Bacterial Physiology" (C. H. Werkman and P. W. Wilson, eds.), pp. 126–77. Academic Press, New York.

Moodie, C. D. (1965). Sites of nutrient exchange in soils. *In* "Microbiology and Soil Fertility" (C. M. Gilmour and O. N. Allen, eds.), pp. 1–16. Oregon State Univ. Press, Corvallis.

Moser, M. (1958). Der Einfluss tiefer Temperaturen auf das Wachstum und die Lebenstätigkeit höherer Pitze mit spezieller Berücksichtigung von Mykorrhizapilzen. *Sydowia* **12**, 386.

Neal, J. L., Bollen, W. B., and Zak, B. (1964). Rhizosphere microflora associated with mycorrhizae of Douglas Fir. *Can. J. Microbiol.* **10**, 259.

Neumann, R. (1959). Relationships between *Pisolithus tinctorius* (Mich. ex. Pers) Coker et Couch. and *Eucalyptus camaldulensis* Dehn. *Bull. Res. Counc. Isr.*, *Sect. D* **7**, 116.

Nightingale, G. T. (1934). Ammonium and nitrate nutrition of dormant Delicious apple trees at 48°F. *Bot. Gaz.* (*Chicago*) **95**, 437.

Norman, A. G. (1955). The effect of polymyxin on plant roots. *Arch. Biochem. Biophys.* **58**, 461.

Norman, A. G. (1961). Microbial products affecting root development. *Trans. Int. Congr. Soil Sci., 7th, 1960* Vol. 2, p. 531.

Oswald, E. T., and Ferchau, H. A. (1968). Bacterial associations of coniferous mycorrhizae. *Plant Soil* **28**, 187.

Palmer, J. G., and Hacskaylo, E. (1970). Ectomycorrhizal fungi in pure culture. I. Growth on single carbon sources. *Physiol. Plant.* **23**, 1187.

Pearson, R., and Parkinson, D., (1961). The sites of excretion of ninhydrin-positive substances by broad bean seedlings. *Plant Soil* **13**, 391.

Picci, G. (1959). Dossagio microbiologico di alcune vitamine e di alcuni aminoacidi ceduti dal seme durante la germinazione. *Ann. Fac. Agr. Univ. Pisa* **20**, 51.

Purnell, H. M. (1958). Nutritional studies of *Pinus radiata* D. Don. I. Symptoms due to deficiency of some major elements. *Aust. Forest.* **22**, 82.

Ramdas, L. A., and Dravid, R. K. (1936). Soil temperatures in relation to other factors controlling the dispersal of solar radiation at the earth's surface. *Proc. Nat. Inst. Sci. India* **2**, 131.

Richards, B. N. (1961). Soil pH and mycorrhiza development in *Pinus*. *Nature (London)* **190**, 105.

Richter, M., Wilms, W., and Scheffer, F. (1968). Determination of root exudates in a sterile continuous flow culture. I. The culture method. *Plant Physiol.* **43**, 1741.

Ridge, E. H., and Rovira, A. D. (1970). Unpublished data.

Ridge, E. H., and Theodorou, C. (1972). The effect of soil fumigation on microbial recolonization and mycorrhizal infection. *Soil Biol. Biochem.* **4**, 295.

Riley, D., and Barber, S. A. (1969). Bicarbonate accumulation and pH changes at the soybean root-soil interface. *Soil Sci. Soc. Amer., Proc.* **33**, 905.

Riley, D., and Barber, S. A. (1970). Salt accumulation at the soybean root-soil interface. *Soil Sci. Soc. Amer., Proc.* **34**, 154.

Robertson, N. F. (1954). Studies on the mycorrhiza of *Pinus sylvestris*. I. The pattern of development of mycorrhizal roots and its significance for experimental studies. *New Phytol.* **53**, 253.

Robinson, R. K., and Lucas, R. L. (1963). The use of isotopically labelled mycelium to investigate the host range and rate of spread of *Ophiobolus graminis*. *New Phytol.* **62**, 50.

Rovira, A. D. (1956a). Plant root excretions in relation to the rhizosphere effect. I. Nature of root exudate from oats and peas. *Plant Soil* **7**, 178.

Rovira, A. D. (1956b). Study of the development of the root surface microflora during the initial stages of plant growth. *J. Appl. Bacteriol.* **19**, 72.

Rovira, A. D. (1965). Plant root exudates and their influence upon soil microorganisms. *In* "Ecology of Soil-borne Plant Pathogens" (K. F. Baker and W. C. Snyder, eds.), Vol. I, pp. 170–86. Univ. of California Press, Berkeley.

Rovira, A. D. (1969a). Plant root exudates. *Bot. Rev.* **35**, 35.

Rovira, A. D. (1969b). Diffusion of carbon compounds away from wheat roots. *Aust. J. Biol. Sci.* **22**, 1287.

Rovira, A. D., and Bowen, G. D. (1970). Translocation and loss of phosphate along roots of wheat seedlings. *Planta* **93**, 15.

Rovira, A. D., and Davey, C. B. (1972). Biology of the rhizosphere. *In* "The Plant Root and its Environment." Southern Reg. Educa. Bd. Inst., Virginia (in press).

Rovira, A. D., and MacDougall, B. M. (1967). Microbiological and biochemical aspects of the rhizosphere. *In* "Soil Biochemistry" (A. D. McLaren and G. H. Peterson, eds.), pp. 418–63. Dekker, New York.

Sands, D. C., and Rovira, A. D. (1970). Isolation of fluorescent pseudomonads with a selective medium. *Appl. Microbiol.* **20**, 513.

Sands, D. C., and Rovira, A. D. (1971). *Pseudomonas fluorescens* Biotype G., the dominant fluorescent pseudomonad in South Australian soils and wheat rhizospheres. *J. Appl. Bacteriol.* **34**, 261.

Šašek, V., and Musílek, V. (1967). Cultivation and antibiotic activity of mycorrhizal basidiomycetes. *Folia Microbiol. (Prague)* **12**, 515.

Schmidt, E. L., and Bankole, R. O. (1965). Specificity of immunofluorescent staining for study of *Aspergillus flavus* in soil. *Appl. Microbiol.* 13, 673.

Skinner, M. F., and Bowen, G. D. (1972). Unpublished data.

Slankis, V., Runeckles, V. C., and Krotkov, G. (1964). Metabolites liberated by roots
of white pine (*Pinus strobus L.*) seedlings. *Physiol. Plant.* 17, 301.

Smith, D., Muscatine, L., and Lewis, D. (1969). Carbohydrate movement from
autotrophs to heterotrophs in parasitic and mutualistic symbiosis. *Biol. Rev.*
44, 17.

Smith, W. H. (1969). Release of organic materials from the roots of tree seedlings.
Forest Sci. 15, 138.

Smith, W. H. (1970). Root exudates of seedling and mature sugar maple. *Phytopathology* 60, 701.

Subba-Rao, M. S., Bidwell, R. G. S., and Bailey, D. L. (1962). Studies of rhizosphere
activity by the use of isotopically labelled carbon. *Can. J. Bot.* 40, 203.

Taylor, G. S., and Parkinson, D. (1961). The growth of saprophytic fungi on root
surfaces. *Plant Soil* 15, 261.

Theodorou, C. (1967). Inoculation with pure cultures mycorrhizal fungi of radiata
pine growing in partially sterilized soil. *Aust. Forest.* 31, 303.

Theodorou, C. (1971). Introduction of mycorrhizal fungi into soil by spore inoculation of seed. *Aust. Forest.* 35, 23.

Theodorou, C., and Bowen, G. D. (1968). Unpublished data.

Theodorou, C., and Bowen, G. D. (1969). The influence of pH and nitrate on
mycorrhizal associations of *Pinus radiata* D. Don. *Aust. J. Bot.* 17, 59.

Theodorou, C., and Bowen, G. D. (1970). Mycorrhizal responses of radiata pine in
experiments with different fungi. *Aust. Forest.* 34, 183.

Theodorou, C., and Bowen, G. D. (1971a). Influence of temperature on the mycorrhizal associations of *Pinus radiata* D. Don. *Aust. J. Bot.* 19, 13.

Theodorou, C., and Bowen, G. D. (1971b). Effects of non-host plants on growth
of mycorrhizal fungi of radiata pine. *Aust. Forest.* 35, 17.

Theodorou, C., and Bowen, G. D. (1973). Inoculation of seeds and soil with
basidiospores of mycorrhizal fungi. *Soil Biol. Biochem.* — submitted.

Trinick, M. J. (1970). Rhizobium interactions with soil micro-organisms. Ph.D.
Thesis, University of Western Australia, Perth.

Tsao, P. H. (1970). Selective media for isolation of pathogenic fungi. *Annu. Rev.
Phytopathol.* 8, 157.

Warcup, J. H. (1967). Fungi in soil. *In* "Soil Biology" (N. A. Burgess and F. Raw,
eds.), pp. 51–110. Academic Press, New York.

Watt, K. E. F. (1968). "Ecology and Resource Management." McGraw-Hill, New
York.

Wilcox, H. E. (1968). Morphological studies of the roots of Red pine, *Pinus resinosa*.
II. Fungal colonisation of roots and the development of mycorrhizae. *Amer.
J. Bot.* 55, 686.

Wilcox, H. E. (1971). Morphology of ectendomycorrhizae in *Pinus resinosa*. *In*
"Mycorrhizae" (E. Hacskaylo, ed.), USDA Forest Serv. Misc. Publ. No. 1189,
pp. 54–68. US Govt. Printing Office, Washington, D. C.

Wilde, S. A. (1946). "Forest Soils and Forest Growth." Chronica Botanica, Waltham,
Massachusetts.

Worley, J. F., and Hacskaylo, E. (1959). The effect of available soil moisture on
the mycorrhizal association of Virginia pine. *Forest Sci.* 5, 267.

CHAPTER 5

Mineral Nutrition of Ectomycorrhizae

G. D. BOWEN

I. Introduction

Growth responses to ectomycorrhizal fungi have now been reported on many occasions, and the response has usually been interpreted to be due to increased nutrient uptake from soils low in one or several nutrients.

Although mycorrhizae differ qualitatively or quantitatively from non-mycorrhizal roots in many plant properties which could affect tree growth, historically the nutritional response has been most noticeable, the most frequent, and certainly the aspect receiving most attention from investigators. Observations of increased tree growth occur at one end of the spectrum of the studies made, and at the other, the far-reaching fundamental physiological studies of ion uptake and transfer by mycorrhizae carried out by Harley and his colleagues at Oxford and the studies of the Swedish schools of Melin and of Björkman (see Harley, 1969) are well known. The task of the ecologist–physiologist is to interpret the physiological work in an ecological framework in order to assess the scope for selection of elite strains of mycorrhizal fungi and to specify the fungal and plant characteristics for which we should select (or breed) to maximize production in a given environment.

The earliest studies of the mycorrhizal phenomenon had a distinct ecological–physiological approach, for Frank (1894) suggested that mycorrhizal fungi could obtain nitrogen more easily from forest humus than could higher plants. A further landmark in analyses of the field responses was the physical bases of mycotrophy enunciated by Hatch (1937) in which the mycorrhizal response was referred to a number of morphological factors considered important in uptake of nutrients. Little quantitative work on the physical factors concerned and on variability between different fungus–root associations appears to have been attempted until the 1960's. The Hatch hypotheses were not a complete explanation of increased nutrient uptake by mycorrhizae for he did not consider factors such as selectivity in uptake, an area of study in which Harley has made considerable contribution. In the last few years, soil chemists and physicists have developed mathematical analyses of ion movements to roots. These, integrated into a quantitative framework of root distribution and ion uptake characteristics of the root, now provide a basis for a quantitative study of mycorrhizal function in soil and the relative importance of various mycorrhizal characteristics in uptake of particular ions and water from soil.

The mycorrhizal phenomenon has interfaces with the disciplines of soil science, plant physiology, and soil microbiology. It forms part of the microbial–plant association spectrum ranging from noninfecting rhizosphere microorganisms to ectomycorrhizae, endomycorrhizae, other symbioses, and plant pathogens, all of which have many common effects (developed to different extents) on ion uptake and plant nutrition; these were discussed by Bowen and Rovira (1969) and Rovira and Bowen (1971). Mycorrhizae themselves normally carry a high population of microorganisms in the "mycorrhizosphere" (Katznelson *et al.*,

1962; Neal *et al.*, 1964; Foster and Marks, 1967; Oswald and Ferchau, 1968; Davey, 1971). Although the mycorrhiza and its associated microflora is an ecological unit, care must be taken in interpreting biological activity of mycorrhizae so that properties ascribed to the mycorrhiza are due to it and not to the associated microorganisms, in the same way as plant properties should not be confused with the properties of rhizosphere microorganisms which invariably occur on plants. Where definition of the mycorrhizal activity itself is needed, pure culture syntheses are required, for which a number of basically similar methods have been described (e.g., Hacskaylo, 1953; Marx and Davey, 1969). Even in these cases checks should always be carried out to ensure that the introduced mycorrhizal fungus is the only organism present.

Nutrient relations of any plant in soil consist of at least two distinct phases, viz., (i) uptake from the soil itself and (ii) use of the absorbed nutrient for plant growth. In plant nutrition these two phases have frequently not been separated. Indeed, relatively little detail is known of the efficiency of nutrient use in plants and its integration with the dynamics of plant growth. The emphasis in this chapter, therefore, is mainly on the uptake phases and movement of ions from the fungus to root. Some indications are given that the efficiency of the use of nutrients in mycorrhizal and nonmycorrhizal trees may be worth considering in future studies.

II. Tree Responses

Two types of evidence have been used to deduce a mycorrhizal response in trees: (i) a correlation between vigorous seedlings and mycorrhizal infection, usually in the absence of inoculation. This does not distinguish between cause and effect, i.e., the seedlings may have been mycorrhizal because they were more vigorous or because of some associated genetic characters. Such observations however, have been confirmed by (ii) controlled inoculation studies. Results of such a study are given in Table I from data of Hatch (1937) with plants grown in soil in pots.

Note (i) large increases in dry weight, percentages of nitrogen, phosphorus, and potassium, and the total amount of these per seedling due to inoculation. (ii) For nitrogen, phosphorus and potassium, respectively, the uptake per gm of root in mycorrhizal plants was 1.8, 3.2, and 2.1 times that of nonmycorrhizal plants, strongly suggesting a more efficient nutrient uptake by mycorrhizal plants.

Most analyses of ectomycorrhizal responses deal with the major nu-

TABLE I

RESPONSE OF *Pinus strobus* TO MYCORRHIZAL INOCULATION[a]

	Inoculated	Uninoculated
Dry wt./seedling (mg)	405	303
Root/shoot ratio	0.78	1.04
Nitrogen % dry wt.	1.24	0.85
Nitrogen per seedling (mg)	5.00	2.69
Phosphorus % dry wt.	0.196	0.074
Phosphorus per seedling (mg)	0.789	0.236
Potassium % dry wt.	0.744	0.425
Potassium per seedling (mg)	3.02	1.38
Uptake, mg/mg dry wt. root		
Nitrogen	0.029	0.016
Phosphorus	0.0045	0.0014
Potassium	0.017	0.008

[a] From data of Hatch (1937).

trients, nitrogen, phosphorus, and potassium. From matters discussed in later parts of this chapter it will be seen that increased uptake can be predicted with many ions, especially with ions of low mobility in soil. These increases in uptake of several ions will arise because of greater root growth following relief of a deficiency of another ion by the mycorrhizae and/or from greater uptake rates by the mycorrhizal association itself. Absorption of Ca^{++}, $H_2PO_4^-$, K^+, Rb^+, Cl^-, $SO_4^=$, Na^+, NO_3^-, NH_4^+, Mg^{++}, Fe^{++}, and Zn^{++} has been demonstrated either with mycorrhizae or mycorrhizal fungi.

Table II (Bowen and Theodorou, 1967) shows two more points about mycorrhizal response: (i) Significant differences between different mycorrhizal fungi can be obtained in plant response and phosphate uptake, and (ii) although a deficiency can be partly overcome by mycorrhizal inoculation, the extent of the response will depend also on the nutrient level of the soil; note that growth increased in the following order: uninoculated and unfertilized, *Suillus granulatus* and unfertilized, uninoculated but fertilized, *S. granulatus* and fertilized. In this experiment the inoculation response disappeared as the available phosphate level in soil was raised further. This elimination of an inoculation response by supplying nutrients is usually interpreted to mean that the mycorrhizal response is a nutrient uptake response only. However in no published experiment concerning the mycorrhizal response in a nutrient-deficient soil have mycorrhizal and nonmycorrhizal plants (with fertilizer added) of the

TABLE II

RESPONSE OF INOCULATED AND UNINOCULATED *Pinus radiata*
SEEDLINGS TO ROCK PHOSPHATE[a,b]

Inoculum	Dry weight per seedling[c] (gm)		P content per seedling[c] (mg)	
	No P	Rock phosphate	No P	Rock phosphate
Suillus granulatus	2.15	4.55	0.75	2.61
Rhizopogon luteolus	2.65	4.42	1.10	1.77
S. luteus	1.45	3.60	0.49	1.79
Cenococcum graniforme	1.58	3.70	0.58	1.42
Uninoculated	1.65	3.65	0.58	1.50
	LSD 0.57	$P = 0.05$	LSD 0.69	$P = 0.05$
	0.77	$P = 0.01$	0.92	$P = 0.01$

[a] From Bowen and Theodorou (1967).

[b] Seedlings were grown for 14 months. Rock phosphate was added at the rate of 0.56 gm/3.5 kg air dry soil (approximately equivalent to 376 kg/ha).

[c] Mean of four replicate pots with 5 seedlings each.

same size been analyzed for nutrient content or have the sizes of mycorrhizal and nonmycorrhizal plants of the same nutrient content been compared. Unless this is done, effects of nutrient uptake cannot be separated easily from nutrient use or other physiological effects. An absolute requirement for mycorrhizal fungi for vigorous tree growth has never been convincingly shown; mycorrhizae appear to be more involved with increasing the efficiency of the tree–soil system.

Differences between mycorrhizal fungi observed in pot studies have also been observed in the field (Moser, 1956; Hacskaylo and Vozzo, 1967; Theodorou and Bowen, 1970). The responses in Table III were obtained at a P-deficient site adjacent to an existing *P. radiata* plantation. Finding a response in such a situation was ascribed by Theodorou and Bowen (1970) to an erratic distribution of naturally occurring mycorrhizal fungi and to a poorer efficiency of phosphate uptake by naturally occurring mycorrhizal fungi. The consequences of early mycorrhizal infection (and therefore vigorous tree growth) are seen also in Table III, for at 28 months all trees in uninoculated plots had become mycorrhizal and growth increments subsequently increased. However the margin between the uninoculated and the better mycorrhizal treatments established after the first 28 months has never decreased and still tends to grow larger. Vigorous seedling growth in the better mycorrhizal treatments in the first season led to significantly fewer deaths from summer

TABLE III

FIELD RESPONSE OF *Pinus radiata* TO DIFFERENT MYCORRHIZAL
FUNGI 60 MONTHS AFTER PLANTING[a]

Inoculum	Mean height[b] (cm)
Suillus granulatus No. 5[c]	482
Rhizopogon luteolus No. 17D[c]	454
Rhizopogon luteolus No. 10C[c]	448
Suillus luteus No. 8[c]	414
Uninoculated	334
	LSD 40 $P = 0.05$
	58 $P = 0.01$

[a] Experimental details and results to 36 months were reported by Theodorou and Bowen (1970).

[b] Mean of 27 trees.

[c] Laboratory accession numbers.

drought. Where mycorrhizal fungi are absent from nutrient poor soils one may reasonably expect much larger relative responses to inoculation than those shown in Table III, e.g., the data of Briscoe (1959) in Puerto Rico shows uninoculated seedlings of *P. elliottii* var. *elliottii*, *P. taeda*, and *P. echinata* grew only 12 cm in 4 years after planting out, but inoculated ones grew 149 cm. Bowen (1965) proposed several criteria governing inoculation response in infertile soils; he emphasized consideration of natural occurrence of mycorrhizal fungi, differences in efficiency of fungal species, competition of introduced fungi with established fungal species, and survival of introduced fungi from season to season.

Two particularly important areas of tree nutrition appear to have received only sparse attention, if any, from mycorrhizal investigators, viz., (i) the uptake of trace elements. Increases in uptake of trace elements by mycorrhizal plants would be expected on quite general grounds. Bowen *et al.* (1972) have recorded zinc uptake by mycelial strands of mycorrhizal fungi and Skinner *et al.* (1972) have found increased uptake of ^{65}Zn by *P. radiata* mycorrhizae. Wilde and Iyer (1962) suggested *P. resinosa* could obtain B, Cu, Mn, Mo, and Zn from soil minerals via mycorrhizae but this was not experimentally demonstrated. (ii) Amelioration of toxicities. Zak (1971) has recently demonstrated the destruction of heat-formed phytotoxins in the soil by mycorrhizal fungi, and this raises the question of their amelioration of effects of naturally occurring phytotoxins (Woods, 1960; Bevege, 1968; McCalla, 1971). It is not impossible that some cases of failure to eliminate the mycorrhizal response by adding fertilizers will be due to the mycorrhizal function at that site being involved in ameliorating organic or inorganic toxicities of soil. Of course, control of pathogens or subclinical pathogens (see Chapter 9 by Marx) is also reason for failure to duplicate mycorrhizal responses by adding fertilizer.

III. Nutrient Uptake from Solutions

Here the purely physiological aspects of nutrient uptake by mycorrhizae and uninfected roots are dealt with, and in Section IV these are related to soil situations. Harley (1969) gave a detailed account of some aspects of ion uptake from solutions by mycorrhizae and his findings will be summarized rather than repeated in detail.

A. GENERAL PRINCIPLES

1. Mycorrhizal Structural Characters

a. The Fungal Compartment. The fungus sheath around mycorrhizae (and the Hartig net) forms a compartment external to the root. A well-developed mantle of beech mycorrhizae can be 40 μm thick (Harley, 1969); Wilcox (1971) has noted a mantle thickness of 70–100 μm on *P. resinosa,* and Chilvers (1968) recorded mantles on eucalypt mycorrhizae to be usually between 30 and 40 μm thick. Considerable variation in mantle thickness occurs between fungi, and in ectendomycorrhizae the mantle may be quite inconspicuous—sometimes absent or only one hypha wide. Mantle thickness is also affected by small temperature changes (Redmond, 1955). Depending on the ionic composition of uptake solutions, on the uptake characteristics of the fungus forming the mantle, and on its thickness and compactness, entry of ions to the higher plant may be entirely via the fungus, partly via the fungus and partly directly to the root, or the fungus may be inconsequential. The position of the mantle determines that an ion will encounter the fungus first and, where the ion is in very low concentration and the mantle is compact, most of the ion in solution will be absorbed by the fungus, very little having direct access to the higher plant cells below. That is, in many cases, the uptake characteristics of the system are not those of the root but those of the fungus, and one of the attractions of mycorrhizal associations is the possibility of manipulating uptake by the choice of the fungus. With the ectendomycorrhizal association and some ectomycorrhizae which barely form mantles, the physical effects of the mantle on entry of ions will be small or nonexistent. However, even in these instances some qualitative and quantitative effects on ion uptake and transfer to the plant may occur because ions penetrating between cortical cells will encounter the Hartig net.

The mycorrhizosphere microorganisms may add their own uptake characteristics to those of the mycorrhiza. Bowen and Rovira (1966) showed rhizosphere organisms could increase apparent phosphate uptake by roots by 50–77%, owing to microbial capture of phosphate and effects on host metabolism (Rovira and Bowen, 1966; Barber and Loughman, 1967). Similarly, microbial effects on uptake of rubidium, iron, zinc, potassium, and sulfate have been shown (Epstein, 1968; Barber and Frankenburg, 1969, 1970a,b). There are often larger numbers of bacteria associated with mycorrhizae than with the rhizosphere (Katznelson *et al.*, 1962) and the effects of these will be greatest with uptake studies over short periods from very dilute solutions.

b. Volume. Another general feature of uptake of nutrients from solutions is that uptake is more closely related to volume than to root length or surface area (Russell and Newbould, 1969). As well as the mantle, radial extension of cortical cells which sometimes accompanies mycorrhizal transformation will increase volume of the root over that of uninfected laterals and this alone will increase uptake from solutions over short periods considerably. The radius of the host tissue has been recorded to increase from 120 to 160 μm with beech mycorrhizae (from illustrations of Clowes, in Harley, 1969) and 55 μm radius to 80 μm radius with eucalypt mycorrhizae (Chilvers and Pryor, 1965). This radial extension of the cortex is not a constant feature of mycorrhizae (Wilcox, 1971). The relative volumes of cylinders of radii of 55, 80, and 100 μm (corresponding to eucalypt uninfected lateral roots, root tissue of eucalypt mycorrhizae, and root plus sheath of eucalypt mycorrhizae) are 1, 2.1, and 4, respectively, and similar measurements of 120, 160, and 200 μm for beech give relative volumes of 1, 1.8, and 2.8.

2. Mechanisms of Uptake

A detailed treatment of mechanisms of ion uptake by plants is given by Briggs *et al.* (1961), Jennings (1963), Dainty (1969), and MacRobbie (1971) and for fungi by Jennings (1963) and Burnett (1968). As an ion moves to a plant cell it first encounters the cell wall with a net negative charge, the "Donnan free space," which correlates with the carboxyl groups of polyuronic acids of the cell wall. This can lead to adsorption and/or exchange of cations which leads to some modification of the ionic composition of the solution. The adsorption of ions by cell walls of fungi has been little studied but the existence of polyuronic acids in fungal cell wall materials appears unusual (Aronson, 1965). With some fungi the possibility exists for the rather unusual plant phenomenon of anion adsorption because of the presence of chitosans in the cell walls. Harold (1962) has demonstrated the adsorption of polyphosphate to the cell walls of *Neurospora crassa*. There has been no detailed study on cation or anion adsorption by mycorrhizal fungi.

Before absorption into the cell, ions may enter the "water free space" which can be demonstrated by penetration of radioactive ions into cellular tissue at $0°C$ and their ready removal by washing.

Uptake proper of ions into cytoplasm of the cell, i.e., across the plasmalemma is broadly referred to as "passive" or "active." The net uptake (or *net flux*) is the balance between uptake (or *influx*) and loss (or *efflux*) and mechanisms for these two processes may differ for the one ion. It is at this stage where specificity in uptake occurs most effectively, i.e., uptake of ions at different rates, rejection of some ions, and competition and antagonism between ions either by competing for a common uptake site or by modifying uptake characteristics of the site. The charge properties of the plasmalemma (a net negative charge) will also modify the ionic environment immediately around it. There is almost invariably a difference in electrical potential between the outside of a cell and the cytoplasm and often between the cytoplasm and the vacuole, i.e., across the tonoplast. An ion in the cell has an electrochemical potential from the concentration of the ion and from the electrical potential difference. Ion movement without expenditure of energy can only be downhill from a higher to a lower electrochemical potential and movement in this case is regarded as passive. The rate of movement across a membrane (flux) is determined by the difference in electrochemical potential for that ion (the driving force) and the permeability of the membrane to the ion in question. Work must be expended for an ion to move against an electrochemical potential gradient and the process is referred to as active. In all higher plant cells so far studied (e.g., Dainty, 1969) and fungal cells so far studied (Slayman, 1962; Bowen, 1963) the cytoplasm is electrically negative relative to the outside solution, so where the internal chemical concentration of an anion is higher than the external concentration, influx must have been active. However cations could be in greater chemical concentration in the cytoplasm but be at a lower electrochemical potential than the outside solution; in this case their entry would be passive, i.e., to consider active movement to always occur where the chemical concentration is greater inside the cell can be quite erroneous. Uptake studies in conjunction with inhibiting or reducing metabolism, by lowering the temperature to 1°–

2°C or using metabolic inhibitors, will tell whether uptake of the ion is metabolically mediated but will not distinguish between the ion under study being itself taken up actively or the ion passively following one of the opposite charge which has been actively transported. Where ions are actively transported they may be moved by specific chemical "pumps."

MacRobbie (1971) has proposed entry of some ions into vacuoles in large algae may occur by a micropinocytosis mechanism, i.e., transfer directly from the outside solution to the vacuole by means of microvesicles. Pinocytotic mechanisms have been suggested for uptake of large molecules, e.g., proteins, by higher plants (Jensen and McLaren, 1960) but if it occurs, it appears to be a minor phenomenon. Pinocytosis has not been observed with fungi (Burnett, 1968).

Uptake of many ions from solution has sometimes been described in terms similar to those of the Michaelis–Menten kinetic analyses of enzyme systems, where the substrate is analogous to the ion being absorbed and the "enzyme" is the ion carrier. Such analyses on a number of ions have indicated the possibility of two or more "systems" operating, one for low solute concentrations (e.g., 0.1–1 mM for cations) and the other for much higher concentrations (Epstein, 1966). It has been suggested that these are both at the plasmalemma (Welch and Epstein, 1968) or at the plasmalemma and tonoplast for low and high concentration mechanisms, respectively (Laties, 1969). However, the analogy to enzyme kinetics is by no means the only interpretation possible, and other possible explanations include (i) probable effects on membrane properties as solution concentrations increase and (ii) the progressive exposure of internal root cells to ions as exterior cells become saturated by increasing solution concentration; the interpretation of kinetics of uptake for tissues (as opposed to homogeneous systems) is complicated.

3. Integration with the General Physiology of the Plant

There is a need to integrate ion uptake with the general physiology of the plant. One of the first considerations is the relationship between photosynthesis, top growth, root growth, and ion uptake; even in higher plants uncomplicated by symbioses, there is very little quantitative understanding of these relationships. The intensity of uptake is a function of the previous nutritional history, and a feedback to decrease uptake of many ions occurs as internal concentrations of these ions rise, e.g., phosphate in *P. radiata* (Bowen, 1970a) and potassium in the fungus *Neocosmospora vasinfecta* (Budd, 1969). As uptake of many ions is metabolically mediated, a crude relationship might be expected between sites of accumulation of translocated photosynthate in roots and ion uptake. This has been confirmed with wheat and pine seedlings by Bowen and Rovira (1970). Pitman *et al.* (1971) have pointed out a relation between high salt plants, low salt uptake rates, and low sugar levels in barley roots. Eventually, the broad fields of nutrition and photosynthesis will also have to integrate effects of growth hormones.

B. Phosphate Uptake

1. Orthophosphate

a. Uptake by Infected and Uninfected Roots. Figure 1 (from Bowen, 1968) shows the sites of uptake of phosphate along mycorrhizal and non-mycorrhizal roots of *P. radiata* from a 5×10^{-6} M solution of potassium dihydrogen phosphate in 5×10^{-4} M calcium sulfate over 15 minutes, using an automatic scanning technique (Bowen and Rovira, 1967). The concentration of phosphate was selected to approximate that in solution in a soil moderately well supplied with phosphate. Intact roots were used, thus avoiding the errors which can be associated with excised roots in some plants (Bowen and Rovira, 1967). The short uptake periods ensured that sites of uptake were recorded quite uncomplicated by redistribution of ions following uptake, which could occur with experiments over the longer times. There was a peak in uptake behind the apex of the growing root, i.e., in the area of cell elongation, after which uptake declined along the main axis. Apparent uptake along the older part of the main axis was mainly due to a passive adsorption to the suberized surfaces (Bowen, 1960) although Addom's studies on water (1946) suggest entry to roots could also occur through cracks in the suberized layer. Increases in uptake occurred where uninfected short roots emerged but very large increases in uptake occurred with mycorrhizae. These results are similar to those obtained semiquantitatively by Kramer and Wilbur (1949) on *P. taeda* for uptake periods of 3–4 hours and by Lundeberg (1961) with *P. sylvestris* with a 6 hour uptake.

The experiment providing data for Fig. 1 showed that uptake by the rapidly elongating uninfected part of the main root was approximately the same as that of mycorrhizae on a weight basis. However the absorbing power of the elongating cells of the uninfected roots declines with root age (usually a matter of days) whereas mycorrhizae several months old maintain a high absorbing power. Harley and McCready (1950) showed excised beech mycorrhizae with well-developed sheaths had 2.3–8.9 times the uptake of carrier-free ^{32}P per unit surface area than did uninfected roots. Bowen and Theodorou (1967) found mycorrhizae of pine to have 2–9 times the total absorbing power of uninfected short roots; ectendomycorrhizae with a poorly developed sheath had approxi-

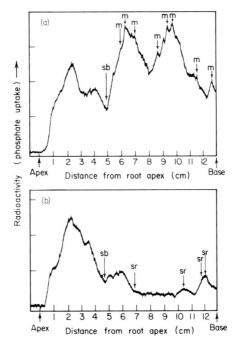

Fɪɢ. 1. Uptake of phosphate along (a) *P. radiata* roots bearing mycorrhizae and (b) uninfected *P. radiata* roots during 15-minutes immersion in $5 \times 10^{-6} M$ phosphate containing ^{32}P; m: positions of mycorrhizae, sr: position of uninfected short roots, and sb: commencement of suberization. From Bowen and Theodorou (1967), by permission of International Union Forest Research Organisations.

mately the same uptake/mg as uninfected short roots while mycorrhizae with well-developed sheaths had considerably greater uptake/mg (see Table IV). Ectendomycorrhizae absorbed more phosphate than uninfected short roots mainly because they were larger. Harley and Mc-Cready (1960) found a much lower uptake of phosphate by mycorrhizae of beech with localized sheaths than by those with well-developed sheaths.

A second approach to uptake by infected and uninfected tissue is to compare performance of fungal and host tissue of mycorrhizae, using the elegant dissection techniques of Harley and McCready (1952). They demonstrated that when the fungus sheath and the underlying host "core" (including the Hartig net) of beech mycorrhizae were exposed to a $1.6 \times 10^{-4} M$ KH$_2$PO$_4$ solution the sheath had 5–8 times more uptake/mg than did the core. Striking though these differences are, their interpretation on a cell physiology basis is difficult because not all cells of the core are active in phosphate uptake, much of the weight of the core probably being made up of vascular tissue. As for all multicellular systems, the microenvironment of inner cells, and therefore their absorbing power, may differ from outer layers (e.g., Laties and Budd, 1964). A comparison of uptake by the core of a mycorrhiza and an uninfected lateral could indicate if the fungus sheath is affecting the absorbing power of the higher plant tissue itself in a similar way to the rhizosphere effect described by Bowen and Rovira (1966) for clover and tomato roots.

Harley and McCready (1952) found that after 7 hours in 4 μM KH_2PO_4, 93% of the absorbed ^{32}P of beech mycorrhizae remained in the fungal sheath; it is not known if the remaining 7% was in the Hartig net or had passed to the higher plant. At 3230 μM, 21% was in the mycorrhizal core after 7 hours. The extent to which the core was supplied by the fungus (because of its external position) was obtained by comparing ^{32}P in the core in intact mycorrhizae with that when the ^{32}P had ready access to the core following slicing of the mycorrhiza into sections 0.5–1.5 mm thick. The effect of slicing diminished with increasing concentration of phosphate from 30 μM, until at 32,300 μM it was negligible, i.e., ecologically, unrealistically high concentrations of phosphate were needed to saturate absorption by the sheath so that it did not significantly affect the phosphorus concentration reaching the core by diffusion through the interhyphal spaces. The situation is analogous to that of transfer of ions to vascular tissue of plant roots reviewed by Russell and Barber (1960), where ionic composition of the transpiration stream approaches that of the surrounding solution only as solution concentrations and transpiration rates increase. As solutes move through the mantle of mycorrhizae, the solution concentration will be depleted depending on the initial concentration of the solution, the avidity of the fungus for the solute (which may change with time as the fungus is charged with the ion), the interhyphal distances, the thickness of the mantle, and the rate of movement of the solution. The experiments by Harley and McCready used excised roots, and with very actively transpiring plants one may expect phosphate to pass between the mantle cells somewhat more readily at lower concentrations than their 30,000 μM KH_2PO_4. At almost all field concentrations of phosphate, however, it can be predicted that the route of phosphate will be via the fungus where the sheath is well developed. It is obvious that these considerations will have to be modified considerably for ectendomycorrhizae. The situation with ectomycorrhizae is an extreme case of the more general one, including noninfecting rhizosphere microorganisms and endomycorrhizae where, however, the microbial compartment is thinner and more discontinuous; concentration effects on phosphate uptake by rhizosphere organisms external to the root have been observed (e.g., Barber and Loughman, 1967).

b. Differences between Mycorrhizae. Bowen and Theodorou (1967) showed that uptake of phosphate from solution by *P. radiata* mycorrhizae differed with the mycorrhizal type (Table IV). Mejstřik (1970) showed the temperature optimum of phosphate uptake by two mycorrhizal types on *P. radiata* and uninfected roots was 25°C, but that above this the mycorrhizal types differed in their response to temperature. In the studies of Table IV (from Bowen and Theodorou, 1967) both types of mycorrhizae in the Kuitpo material were more efficient in phosphate uptake (i.e., uptake/mg) than were uninfected roots. There was a fivefold difference in total uptake but only a twofold difference in efficiency between the mycorrhizal types. The sheaths of these two types were comparable in thickness and structure, but the mean mycorrhizal size of type (ii) was much greater than that of the other. It was concluded that in this instance differences between mycorrhizae in uptake were due both to size and physiological differences. The greater uptake of type (iv) (ectendomycorrhizae) than of uninfected roots in the Belair material was due solely to their larger size.

c. Phosphate Uptake Mechanisms of Mycorrhizae. The inhibition of

TABLE IV

Tнε Uptake of Phosphorus by Excised Mycorrhizae of *P. radiata*[a]

Collection site	Type	Phosphorus uptake[b,c] (10^{-8} mole P)	
		Average per root	Average per mg dry wt.
Kuitpo Forest	(i) Sheath well developed	0.70	7.5
	(ii) Sheath well developed	3.5	15.5
	Uninfected short lateral root	0.4	3.5
Belair Nursery	(iii) Sheath well developed	2.5	15.0
	(iv) Sheath poorly developed	1.5	5.5
	Uninfected short lateral root	0.7	5.5

[a] From Bowen and Theodorou (1967).

[b] Uptake from 1×10^{-3} *M* dihydrogen potassium phosphate for 2 hours at 25°C, pH buffered at 5.5.

[c] Uptake by mycorrhizae expressed as uptake per forked root.

phosphate uptake by low temperatures and by metabolic inhibitors (see Harley, 1969) shows orthophosphate uptake to be metabolically mediated; the process is almost certainly an active one. The fungal hypha can store considerable phosphate (see Section V,B), probably in the vacuole, which moves to the plant in times of phosphate deficiency. The storage of phosphate in the vacuole maintains low inorganic concentrations of phosphate in the cytoplasm and therefore allows sustained high uptake rates by circumventing a phosphate feedback inhibition of uptake (Bowen, 1970a).

The molecular biology of phosphate uptake and identification of pumps in mycorrhizal fungi have not been examined in detail although extensive studies have been made of oxidative and phosphorylative processes in respiration by Harley and his colleagues (see Harley, 1969).

Bowen and Theodorou (1967) examined the uptake kinetics of two types of ectomycorrhizae of *P. radiata* [Kuitpo types (i) and (ii) in Table IV]. Curves obtained with the Hofstee plot (Hofstee, 1952), were similar to those obtained with barley, millet, and lucerne by Noggle and Fried (1960) and could be interpreted to suggest that uptake of phosphate from orthophosphate with both mycorrhizal types was through at least two carrier systems. Hagen and Hopkins (1955) considered these could possibly correspond to uptake of $H_2PO_4^-$ and $H_2PO_4^=$ ions, respectively, by barley roots but Edwards (1968) considered a dual mechanism uptake of $H_2PO_4^-$ a more likely interpretation, as proposed for other ions (Epstein, 1966). A description of differences in phosphate uptake between mycorrhizal types is possible in terms of number of carriers, affinity between carriers and phosphate, and of phosphate release inside the cell. However, is this the correct interpretation of the kinetics (see Section III,A,2, above)?

2. Organic Phosphates

One of the attractive hypotheses on mycorrhizal function is that the fungi may be able to use forms of nutrients not used by the higher plant. Paterson and Bowen (1968) have demonstrated that ectomycorrhizal fungi in culture can use sugar phosphates and nucleotides as sources of energy and phosphate and, using the Gomori reaction, that mycorrhizae of *P. radiata* have surface phosphatases. Harley (1969) reported that Woolhouse had recorded high surface phosphatase activity of beech mycorrhizae. Theodorou (1968) showed *R. luteolus, S. granulatus, S. luteus,* and *Cenococcum graniforme* could obtain phosphate from calcium and sodium phytates by means of phytase. *R. luteolus* possesses two phytase actions (Theodorou, 1971), which give rise to two pentaphosphate isomers from *myo*-inositol hexaphosphate, the major one being similar to that obtained from phytase of *Neurospora crassa* and pseudomonads, and the minor one corresponding to the product of bran phytase.

One of the important questions on mycorrhizal use of organic phosphates is whether this is a property of the fungus and not the host plant. One difficulty in interpreting much plant physiological work is that the plant is rarely grown aseptically and plant properties cannot be distinguished from those of associated surface microflora. However literature cited by Wild and Oke (1966) clearly shows that many sterile plants can use phytate as well as a wide range of other organic phosphates. Using a sensitive bioassay for phosphate deficiency (Bowen, 1970a), Paterson and Bowen (1968) have shown that sterile *P. radiata* seedlings with sugar phosphates, nucleotides, or phytate as the sole source of phosphorus, behaved as phosphorus-sufficient plants and thus were using the phosphorus in these organic sources. A positive Gomori test for surface phosphatases and sugar phosphates was obtained with the apices of long roots, young laterals, and mycorrhizae. The relative phosphatase activity of mycorrhizae and nonmycorrhizal short roots has not been studied quantitatively, but Ridge and Rovira (1971) reported an approximate 25–50% reduction in surface phosphatase activity of wheat roots when microorganisms were present.

It is obvious that both mycorrhizae and uninfected roots can obtain phosphate from a range of organic forms which are probably broken down at the surface and the phosphate released. It is not known if the organic phosphates can also be absorbed intact but this should be amenable to study using substrates radioactively labeled with both ^{14}C and ^{32}P.

C. NITROGEN UPTAKE

1. Nitrate and Ammonium Ions

Carrodus (1966) found that excised beech mycorrhizae could readily absorb ammonium from ammonium chloride but had almost no ability to absorb nitrate. He raised the possibility that uninfected roots of *Fagus sylvatica* might absorb nitrate while the mycorrhizae would be concerned primarily with ammonium-N which is the main source of nitrogen in the litter and humus layers where mor development occurs. There is little doubt however that most higher plants can use both nitrate-N and ammonium-N although there may be preferences for one or the other. Lundeberg (1970) subsequently showed many (but not all) of 27 mycorrhizal fungi in pure culture could use nitrate as the sole source of nitrogen and some made as good growth on this as on ammonium sources. The use of nitrate varied even between isolates within the one species. Trappe (1967) reported that two out of eight mycorrhizal fungi he tested possessed nitrate reductase and thus presumably had the ability to absorb and use nitrate. Although a number of mycorrhizal fungi may not be able to use nitrate because of absence of nitrate reductase, a generalization that no mycorrhizae can absorb nitrate efficiently appears unsound. Furthermore, there do not appear to be any published studies in which nitrate and ammonium uptake of uninfected roots and mycorrhizae have been studied in the one experiment, and there have been no studies on differences between types of mycorrhizae in uptake of ammonium and of nitrate. We therefore have no satisfactory basis for direct comparison of the relative power of uninfected roots and of different mycorrhizae to absorb inorganic nitrogen sources from solution.

In acid forest soils, nitrification is usually very low, but Corke (1958) detected small numbers of nitrifying organisms in highly acid forest soils and suggested that some nitrification occurred in mor litter. A number of forest situations occur where nitrate-N is almost certainly plentiful, e.g., forest nurseries (especially on previously agricultural soil), mull forest soils (Bornebusch, 1930), and situations where nitrate fertilizers are added. In such cases, nitrate assimilation by the mycorrhiza would be advantageous, and selection of mycorrhizal fungi on their ability to use both nitrate and ammonium ions appears worthwhile. In situations where ammonium ions are abundant they may inhibit nitrate reductase.

The mechanisms of ammonium absorption by mycorrhizae have been little studied. Carrodus (1966) reported the expected occurrence of ammonium ions in water free space in the tissue and adsorption of a small amount in the Donnan free space; the extent to which these were associated with the fungal mantle and the host tissue is not known. Sodium azide and dinitrophenol reduced ammonium-N loss from solution (i.e., net uptake) by 70–75% over 4 hours and he concluded ammonium-N uptake was associated with normal respiratory metabolism and oxidative phosphorylation. In such nontracer experiments on net uptake it is often difficult to distinguish between effects of inhibitors on influx and those on efflux (and failure to retain absorbed ions) but the stimulation of ammonium uptake by addition of sugars (Carrodus, 1966) confirms an association with cell metabolism. In the absence of electrical data and concentrations of NH_4^+-N in the cell, is it not possible to determine on present information whether NH_4-N entry into mycorrhizae is active or passive. The incorporation of NH_4^+ into amino acids probably maintains NH_4^+ at a low concentration in the cell, assisting continued diffusion if this is the entry process. It is unknown whether the ammonium ion enters the fungus in the dissociated form as in higher plants (Becking, 1956) or undissociated; Morton and MacMillan (1954) and MacMillan (1956) have advanced evidence for absorption of undissociated ammonium salts by the fungus *Scopulariopsis versicolor,* but existing data are not adequate to resolve this question with mycorrhizae. The removal of keto acids by amination with ammonium ions stimulates dark carbon dioxide fixation by beech mycorrhizae (Harley, 1964).

Where nitrate absorption by mycorrhizae occurs it will probably be an active process, since other fungi studied have been electrically negative with respect to simple external nutrient solutions (Slayman, 1962; Bowen, 1963). Nitrate enters higher plants by active processes (Highinbotham *et al.,* 1967; Highinbotham, 1968).

2. *Organic Nitrogen Compounds*

The known ability of microorganisms to use organic forms of nitrogen has led to the idea that mycorrhizal fungi could "short circuit" the de-composition—mineralization cycle for plant nutrients (e.g., Frank, 1894).

In synthetic laboratory media, ectomycorrhizal fungi can use a wide range of amino acids and more complex organic nitrogen compounds (e.g., Norkrans, 1953; Lundeberg, 1970; Harley, 1969) but differences exist between them in their growth on a range of these substances. Carrodus (1966) observed uptake of glutamic acid, aspartic acid, glutamine, and asparagine by excised beech mycorrhizae and Melin and Nilsson (1953a) showed hyphae of mycorrhizae of *P. sylvestris* could absorb ^{15}N-glutamic acid and transfer it to the mycorrhiza and higher plant. No study has been made into the mechanisms of absorption of organic nitrogenous compounds by mycorrhizal fungi (see Burnett, 1968, for a discussion of amino acid uptake by fungi). However the absorption of organic nitrogenous compounds is by no means restricted to the fungus partner; Hatch (1937) showed that sterile *P. strobus* could use nucleic acids and peptone as the sole source of nitrogen. The absorption of a wide range of amides and amino acids by higher plant roots is well documented (Jennings, 1963).

3. *Fixation of Atmospheric Nitrogen*

By analogy with some other microbial symbioses with plants, e.g., legume, alder, and casuarina symbioses, early mycorrhiza workers proposed atmospheric nitrogen fixation by the mycorrhizal symbiosis. Extensive plant growth studies by Melin failed to show any nitrogen fixation by mycorrhizal seedlings, by mycorrhizal or other fungi (Melin, 1959). The advent of more sensitive assays for nitrogen fixation, such as the use of ^{15}N and the acetylene reduction test, revived interest in the possibility of nitrogen fixation, especially as Kjeldahl determinations of total nitrogen in some forest systems showed an unexplained increase of some 50 kg N/ha annually (Stevenson, 1959; Richards, 1964; Richards and Voigt, 1965; Richards and Bevege, 1967). Bond and Scott (1955) could not detect ^{15}N enrichment with mycorrhizal *P. sylvestris* but Stevenson (1959), Richards and Voigt (1964), and Richards *et al.* 1971) have shown ^{15}N enrichment of mycorrhizal *P. radiata*, *P. elliottii*, and *P. caribaea* var. *hondurensis*. However, in every instance other microorganisms were present. Richards *et al.* also showed that acetylene reduction by mycorrhizal seedlings of *P. elliottii* produced by pure culture synthesis was small and was approximately that of nonmycorrhizal seedlings, even when glucose was added. Ethylene production in the absence of acetylene was not measured. Addition of the nitrogen-fixing bacterium *Beijerinckia indica* and glucose increased acetylene reduction by up to three orders of magnitude with nonmycorrhizal plants. They have concluded that the main agents of nitrogen fixation are microorganisms associated with mycorrhizae. This is consistent with the conclusion that atmospheric nitrogen fixation appears to be restricted to prokaryotic organisms (Nutman, 1971) and has not been conclusively demonstrated in filamentous fungi. Katznelson *et al.* (1962) recorded up to ten times the number of bacteria in the mycorrhizosphere as in the rhizosphere.

Of necessity, incubation times of several days duration under conditions somewhat conducive to anaerobiosis have been used with most ^{15}N studies with mycorrhizal plants, and although nitrogen fixation by the rhizosphere microorganisms and mycorrhizosphere organisms occurs, it is difficult to relate this directly to nitrogen gains in the field. Even though some ^{15}N fixation occurs in the mycorrhizosphere it apparently does not always occur (Bond and Scott, 1955), and significant nitrogen accretion recorded in some field studies has yet to be reconciled with the frequent occurrence of N deficiency in well-established conifer forests (Mustanoja and Leaf, 1965). Stone and Fisher (1969) reported increased availability of soil nitrogen under conifers but this does not explain accretion of N with some conifer systems. More work is needed on microorganisms in the rhizosphere as opposed to those in litter to assess their possible relative importance in the nitrogen fixation reported from field studies. Nonsymbiotic nitrogen fixation in soil and plant systems is discussed by Moore (1966). It is interesting to note that Parker (1957) considered that large numbers of *Clostridium* in soil could fix appreciable amounts of nitrogen and that Corke (1958) has recorded large numbers of these organisms in forest podzols.

D. UPTAKE OF OTHER IONS

Almost the only detailed studies of uptake of potassium, sodium and rubidium by mycorrhizae have been those of Wilson (1957) and of

Harley and Wilson (1959). Potassium and rubidium uptake by beech mycorrhizae were approximately double that of nonmycorrhizal roots. Some 60% of ^{86}Rb fed to beech mycorrhizae in short-term experiments was in the fungal sheath but almost equal distribution of alkali-metals between sheath and core occurred quickly. As for ammonium, adsorption of these cations to the mycorrhizae also occurred. Inhibitor and temperature studies showed the uptake to be metabolically mediated. Although potassium, rubidium, and sodium moved against concentration gradients, a final assessment of active or passive movement cannot be made in the absence of electrical potential data. Harley and Wilson reported the rapid leakage of potassium from beech mycorrhizae at temperatures a little above 20°C.

As with uninfected plant roots, mycorrhizae show considerable selectivity in ion uptake. Wilson (1957) showed sodium to have little effect on rubidium uptake by beech mycorrhizae but potassium depressed uptake of rubidium and sodium. Uptake versus concentration data given by Harley and Wilson (1959) indicate approximately the same relationships described by Epstein (1966) for a number of plants at solution concentrations of cations up to 1–2 mM, and which he designated as high affinity mechanisms showing considerable specificity compared with a mechanism of much lower specificity in the range 10–50 mM, an interpretation subject to the assumptions in kinetic analyses outlined above.

Dijkshoorn (1969) and Lips *et al.* (1971) have proposed a key regulatory role of potassium in nitrate uptake by higher plants via its influence on malate production and exchange of bicarbonate or malate ions for nitrate outside the root. This may be worth examining with mycorrhizal systems.

Zinc uptake by mycorrhizae of *P. radiata* has been studied by Skinner *et al.* (1972) because of the importance of this trace element in forest nutrition in southern Australia. Both actual uptake and zinc adsorption occurred; mycorrhizae absorbed more zinc than uninfected short lateral roots.

Morrison (1962b) demonstrated no sulfate uptake differences between mycorrhizal and nonmycorrhizal *P. radiata* seedlings of high, medium, and low sulfur status. Very rapid movement of sulfate through the mantle of beech mycorrhizae occurred with a sulfate concentration of $3 \times 10^{-5} M$, i.e., lower levels of sulfate than normally found in soil (Fried and Shapiro, 1961). Separation of sheath and core before uptake had little effect on uptake by the core. Except for very short absorption periods, when a large proportion of the ^{35}S was in the sheath, only some 50% was usually retained in the mantle; in contrast to phosphate, the sheath had very little storage capacity for sulfate.

Obviously rapid transfer of sulfate to the core occurs both via the fungus and by diffusion direct to the host tissue.

IV. Uptake from Soils

A. Principles of Solute Transfer and Ion Uptake by Plants in Soil

The preceding section dealing with nutrient uptake from stirred solution basically shows the plant's absorbing power. In soil, however, distances between roots and the movement of ions from soil to roots must also be considered, since the supply of ions to the root may limit uptake rather than the root's absorbing power. A detailed treatment of ion movement in soils is given by Olsen and Kemper (1968), and a general introduction to the concepts involved in ion uptake from soils is given by Nye (1969). Most theoretical analyses so far treat individual roots under ideal conditions, but the problems of roots competing with each other and of irregular distribution of roots in contrast to an idealized uniform distribution have recently been considered by Sanders *et al.* (1971) using an electrical analog of diffusion of ions to roots. A detailed review of the importance of root configuration in ion uptake was given by Barley (1970).

In brief, ions may move to roots by *diffusion* and by *convection* in water ("mass flow"). As an ion is absorbed by the root, its concentration at the root decreases thus leading to a concentration gradient causing diffusion from the bulk of the soil. As the plant transpires water containing solutes moves towards the root surface, i.e., solutes are moved in the water by convection. Soil phenomena affecting rates of diffusion and convection include diffusivity of the ion in soil, dispersion of ions, effects of other ions on diffusion, and soil factors such as water content, tortuosity of diffusion paths, and interaction of ions with soil surfaces. High absorbing power of the root for a particular ion will deplete it at the root surface but low uptake of particular ions by the plant, i.e., selectivity, especially if coupled with large arrival at the root surface with high transpiration, may lead to accumulation of some ions around roots of some species (e.g., Ca^{++}, see Barber and Ozanne, 1970), as can direct pumping of an ion (e.g., Na^+, Highinbotham *et al.*, 1967) out of a root. These ions can then be partly redistributed in the soil by back-diffusion to the soil bulk.

In situations where plants absorb ions from soil solution as rapidly as they arrive at the surface, the transfer of the ion through soil, will control the rate of uptake more than differences in absorbing power between plants and actively absorbing parts of roots, but where supply to the root is usually high, differences in absorbing power between plants and between different organs of the same root will be important in uptake. The uptake of phosphate has been most studied and is a good example of diffusion through soil controlling uptake and of the way in which plant parameters other than absorbing power affect uptake.

B. TREE ROOT CHARACTERISTICS

Morphological features of the mycorrhizal root system are particularly relevant to the physical bases for the mycorrhizal response proposed by Hatch (1937). He proposed the nutritional response was due to increase in absorbing surface of the roots caused by increased diameter and branching of mycorrhizae, to the growth of hyphae into soil, and to the greater longevity of mycorrhizae. These parameters have rarely been measured with different mycorrhizal fungi with a view to assessing their importance in differences between mycorrhizal types and uninfected roots in uptake of nutrients from soil. Such an assessment should aid in selection of mycorrhizal fungi with particular characteristics leading to increased mycorrhizal responses seen in Section II.

The characteristics of the root or mycorrhiza important in uptake are (i) their absorbing power for ions and water, including maintenance of these properties; (ii) abundance and distribution of roots; and (iii) the effective radius of the root or mycorrhiza, i.e., the cylinder of root plus root hairs—or mycorrhiza plus hyphal extensions. Organized aggregates of fungi (as mycelial strands and rhizomorphs) growing from mycorrhizae are treated as analogous to root growth. Individual hyphae projecting from mycelial strands along their length and from mycorrhizae have been treated as analogous to root hairs forming an annulus around uninfected plant roots.

1. Absorbing Power

Absorbing power of mycorrhizae and uninfected roots has been dealt with in Section III. Generally, there is a sparsity of information from soil grown plants, but Bowen (1970b) and Rovira and Bowen (1968) using *P. radiata* seedlings and wheat seedlings have found patterns of uptake along soil and solution grown roots are usually very similar. The distribution of absorbing power along roots is affected markedly by soil conditions, e.g., soil temperature (Bowen, 1970b), but one of the striking features of mycorrhizae is their sustained absorbing power in one place in soil, whereas the high absorbing power of individual parts of uninfected roots is transient. Figure 1 showed mycorrhizae well behind the root apex, i.e., several months old, to be sites of high phosphate absorbing power. Harley (1969) cited the longevity of mycorrhizae in pine trees to vary between several months to a year and in some cases renewed growth occurs each year for three or more years. This longevity is not only a property of the fungal component, since the activity of cortical cells of mycorrhizae is also prolonged (Wilcox, 1971), and one may speculate on the roles of growth factors produced by the mycorrhizal association in this (e.g., cytokinins, Miller, 1971).

The distance over which solutes diffuse to a root in soil is approximately proportional to the square root of the product of the diffusion coefficient of the ion in the soil and the time for which the root is functional in uptake of that ion, although this can be modified by the various soil factors mentioned above and by interaction between diffusion and convection. Profiles of ion removal from soil with time are given in Fig. 2 (from Lewis and Quirk, 1967b). Thus an organ functioning for a long period, such as a mycorrhiza, can feed from a much larger volume of soil than one with a short functional life such as the elongating portion of an uninfected root. Olsen *et al.* (1962) calculated that in a silty clay loam, phosphate in soil 1 mm from a maize root would contribute to its nutrition over 10 days, but that this would be 5 mm had the root been functional in that position for 100 days. These increases in radius of diffusion with time lead to very great increases in the volume of soil which could contribute phosphate to the root system. Note that not all of the increased soil volume will contribute phosphate equally to the

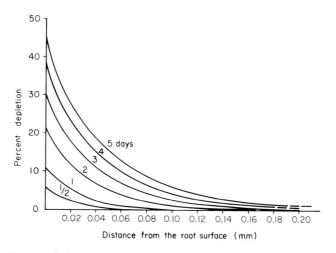

FIG. 2. Computed changes in soil phosphorus concentration with time and distance from the root surface. Root radius 0.2 mm, 150 μg P/gm soil was added to a highly phosphorus fixing soil; soil solution phosphorus was approximately 0.1 ppm. From Lewis and Quirk (1967a), by permission of *Plant and Soil*.

root, since the amount of ion contributed from the soil decreases with distance from the root and with the ion. The effect of diffusivity of different ions on the radius of diffusion is shown in Fig. 3 (from Nye, 1966). Greater depletion of soil away from the root will occur with the rapidly diffusing ion; only narrow depletion zones will occur around roots with slowly diffusing ions.

Where solutes can move significantly by convection, e.g., nitrates and sulfates, the implication of longevity of mycorrhizae is much the same as for diffusion, but there is the added requirement that mycorrhizae are water-uptake organs as well. This is almost certainly the case (see below). The distance solutes move by convection is approximately proportional to the square root of time in unsaturated soils, the usual situation (Todd, 1959), and proportional to time in saturated soils (Gardner, 1965). Some ions can be moved for relatively large distances by convection.

2. Abundance and Distribution of Roots and Mycelial Strands

Barley (1970) referred to abundance as length of root per plant and related this to the problem of the minimum length of root required to meet plant demands for nutrients. However, nutrient status of the soil, and the radius of different members of the root system must be considered as well as length in such deliberations. Because soil offers a resistance to ion transfer, and because roots may compete in a soil, greater

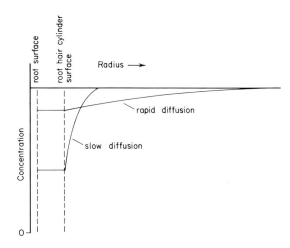

FIG. 3. Nutrient concentration around the root when diffusion is relatively slow (e.g., potassium) or rapid (e.g., nitrate). From Nye (1966), by permission of *Plant and Soil.*

lengths of root are needed to meet the plant's nutrient requirements in soil than in stirred solution.

It is strikingly obvious that root length/volume of soil is very much smaller for tree species than for herbaceous plants. Andrews and Newman (1970) indicated 0.08–3.6 cm/cm^3 for woody plants in the surface horizons of soil and 1.6–340 cm/cm^3 with herbs. Data of Bowen (1964) for *P. radiata* plantations 14–26 years old in a lateritic podzol soil gave values of 2 cm/cm^3 in the surface 8 cm, 0.8 cm/cm^3 at 25–45 cm, 0.4 cm/cm^3 at 91–106 cm, and considerably lower beyond that. It is difficult, however, to compare root lengths required by older stands of woody species with those of herbaceous species as an important source of mineral nutrients for new growth of the tree is redistribution of nutrients from older parts of the plant.

A detailed examination of root density in the surface 10–12 cm of seedlings and plantation trees was reported by Bowen (1968). The data for 20–27-year-old plantations in two soils are given in Table V. With 7-month-old seedlings in a nursery, relatively great distances between long roots and their members were also found; at the tap root the mean distance between long roots was 1.1 cm and this increased greatly with increasing distance from the tap root. The mean distance between mycorrhizae was 2.1 mm, and their average length (and those of uninfected

TABLE V

Root Distribution of *P. radiata* in Plantations[a]

Soil	Long roots			Short lateral roots		
	Mean distance apart[b,c] (mm)	Closest root[d]		Closest short root[e]		
		Range (mm)	Mean (mm)	Range (mm)	Mean (mm)	Length (mm)
Lateritic podzol[f]	12.8	2–25	7.6	1–52	9.0	4.4
Sandy solodized solonetz[g]	14.2	2–31	9.0	1–44	5.9	4.5

[a] From Bowen (1968).
[b] Means of at least 220 roots.
[c] Based on number of long roots in 15 × 15 × 12-cm samples.
[d] Distance to the nearest neighboring long root.
[e] Including uninfected short roots and mycorrhizae, and measured along individual long roots.
[f] 0–12 cm under 27-year-old *P. radiata* stand.
[g] 0–12 cm under 20-year-old *P. radiata* stand.

short roots) was 2.6 mm. Although considerable emphasis is usually placed on the occurrence of mycorrhizae in the litter and surface layers of soil this is partly because the highest concentration of roots is in the surface soil. Lobanow (1960) recorded mycorrhizae of pine at 1–1.5 m depth; Bowen (1964) recorded mycorrhizae from *P. radiata* at 2 m.

The conclusion that very large interroot distances occur with woody species is inescapable. This may make interroot competition less severe in a pure stand and thus represents an aspect of efficiency of the design of tree root systems. Large interroot distances may be inconsequential in a soil well supplied with nutrients, but where the soil is low in nutrients, it imposes long distances for ions to move to roots to satisfy the tree's requirements. In a nutrient deficient soil, this means that much of the nutrient which occupies the large interroot volumes may not normally be available to the plant, especially ions of very low mobility. The nutrition response with mycorrhizae strongly suggests they can use this interroot soil more effectively than uninfected roots. How do they do it?

The longevity of mycorrhizae will effectively increase the radii of diffusion of ions to the root; on a three-dimensional scale even small increases in radii of diffusion greatly increase effective volumes of soil around the root. Fungus growth into soil (Hatch, 1937; Björkman, 1949) may also greatly increase effective volumes of soil. Melin and his co-workers (Melin and Nilsson, 1950, 1952, 1953a,b, 1955; Melin *et al.*, 1958) elegantly showed the ability of mycelia of mycorrhizae to absorb and translocate a wide range of inorganic anions, cations, and glutamic acid to pine seedlings, and Skinner *et al.* (1972) have shown this for the trace element, zinc.

The growth of individual hyphae from mycorrhizae into soil is usually relatively small (Bowen, 1968; see Section IV,B,3, below) but the growth of mycelial aggregations as mycelial strands or as rhizomorphs is usually large. These two terms have frequently been used, incorrectly, as synonyms. Mycelial strands are the type most frequently formed by mycorrhizae and are branching aggregations of hyphae quite distinct from highly organized rhizomorphs (see Garrett, 1970); intermediate structures do sometimes occur however. The variation in detailed structure of mycelial strands of mycorrhizae is seen in illustrations of Schramm (1966).

Few quantitative studies of mycelial strand growth have been made with mycorrhizae. Bowen (1968) examined mycelial strand growth from mycorrhizae into four soils; mycelial strands frequently extended for up to 3.5 cm from the mycorrhiza, branching frequently. Along their lengths individual hyphae grew into soil for up to 350 μm—a feature which is probably more conducive to soil exploitation than the more compact, structured rhizomorphs, of such fungi as *Armillaria mellea*, with smooth surfaces and few individual hyphal outgrowths (Garrett, 1970). There is a tendency for very great development of mycelial strands along surfaces, along cleavage lines in clay and into minute interlamellar cracks of shale (e.g., Schramm, 1966) but their growth can be extensive through most of the soil and excellently suited to permeate efficiently and exploit the large interroot distances occurring with woody plants.

Figure 4 illustrates the removal of poorly diffusible nutrients from large interroot distances by mycorrhizal and nonmycorrhizal roots, by means of longevity of mycorrhizal roots and growth of mycelial strands (see also Figs. 5 and 6).

Extensive soil exploration by mycelial strands of *R. luteolus* mycorrhizae of *P. radiata* are shown in Fig. 6, and in Fig. 5 an illustration of portion of a mycelial strand in soil. The major mycelial strands permeating the soil, varied from 15 to 50 μm radius with hyphae projecting from these for 60–120 μm and approximately 40 μm apart. Usually mycelial strands terminated in wefts of smaller strands grading down to single hyphae. The fresh weight of the length of 22.8 mm of strand illustrated was 0.125 mg. By contrast, the same total length of uninfected short roots of *P. radiata* weighed 1.2 mg.

The growth of mycelial strands of fungi has usually been viewed by plant pathologists as a method of inoculum spread and this function occurs with mycorrhizae also. The nutrient uptake role of mycelial strands has been confirmed by Skinner and Bowen (see Bowen *et al.*, 1972) by

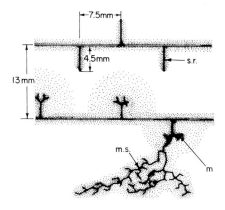

FIG. 4. Representation of removal of a poorly diffusible ion (e.g., phosphate) from interroot soil with mycorrhizal and nonmycorrhizal roots. s.r.: uninfected short root, m.: mycorrhiza, and m.s.: mycelial strand (see Figs. 5 and 6). Intensity of spotting indicates depletion of phosphate due to longevity in uptake and uptake by mycelial strands. From data of Bowen (1968).

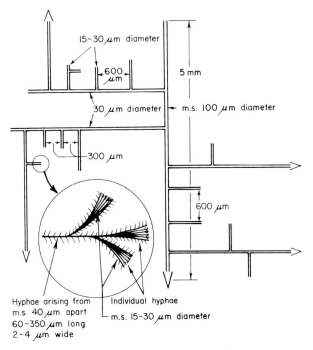

FIG. 5. Representation of part of a mycelial strand growing into soil from a mycorrhiza of *P. radiata* formed by *R. luteolus*. m.s.: mycelial strand.

feeding them up to 1 cm from the tree root with ^{32}P solutions and noting the appearance of ^{32}P in the tree root. Nutrient uptake appears to occur for most of the length of the mycelial strand. By contrast, the highly structured and suberized rhizomorphs of *Armillaria mellea* absorb phosphate and chloride mainly in the unsuberized apical parts (Morrison and Bowen, 1970).

The production of mycelial strands occurs only with some mycorrhizal fungi, and Bowen (1968) has suggested selecting mycorrhizal fungi which produce mycelial strands readily, for obtaining increases in uptake of nutrients from soil. This hypothesis is currently under test. Fungi may differ in the readiness with which they produce mycelial strands, but little is really understood about their formation and growth in soil. Garrett (1970) observed that strand formation occurs typically where a mycelium grows over a surface or through a medium having a negligible content of free nutrients, and he considered that some stimulus from the leading hyphae must occur for following hyphae to grow along the same line. The loss of organic compounds from intact older hyphae of mycelial strands suggests the nutritional value of these exudates to be at least

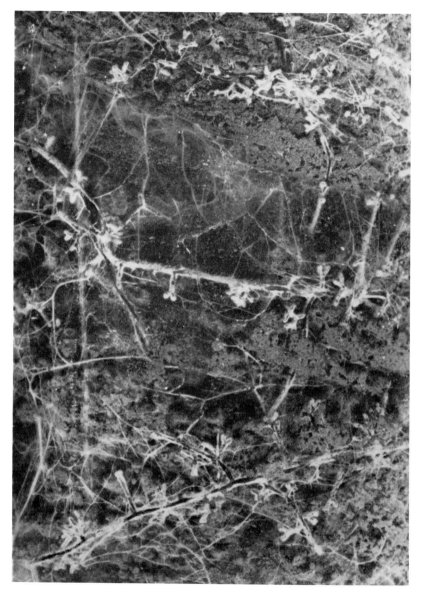

Fig. 6. Mycorrhizae and mycelial strands of *P. radiata* (magnification approximately ×2).

one of these stimuli (Day, 1969). It should be noted in passing that such a hypothesis could also explain sheath formation by mycorrhizal fungi. Day also noted that the form of the nitrogen source was important in mycelial strand formation of the wood rot fungus *Merulius lacrymans.*

Rhizomorph formation can be stimulated in laboratory culture by volatile compounds (Weinhold, 1963) and may also be influenced by soil microorganisms; it is not known whether such factors affect mycelial strand formation in soil. The intense competition from other soil organisms for energy sources in soil and the apparent difficulty many mycorrhizal fungi have in using complex energy sources make it apparent that most of the energy sources for mycelial strand growth must come from the higher plant and that the fungus must be an efficient translocator of carbohydrate. The possibility of the mycorrhiza acting as a sink for photosynthate (Shiroya *et al.*, 1962; Nelson, 1964; Lister *et al.*, 1968; Bevege *et al.*, 1972) obviously has consequences for mycelial strand growth.

Björkman (1949) pointed out that the energy requirement for fungal growth into soil would be considerably less than that required to produce roots to permeate the soil. In the material of Fig. 5 the weight of mycelial strands over 30 μm in diameter was one-tenth that of the same length of short lateral roots. The relationships between root size and nutrient uptake are complex for the rate of uptake/cm² of root increases as the radius increases, but the surface area of a given weight of roots varies inversely with the radius. For ions of low diffusivity in soil, many fine absorbing elements throughout the soil would have considerable advantages. If one considers uninfected short roots of 175 μm radius, root hairs of 150 μm length, and depletion of soil in the root hair "annulus" and for 50 μm around it, uptake/cm of root will be only 2.2 times that of a mycelial strand of 25 μm radius, with a 50 μm depletion zone out from hyphae projecting 150 μm into soil but which weighs only about one-tenth of the root. More extensive permeation of soil near the terminals of finer branches of the mycelial strand will increase the efficiency of uptake from soil still further.

Other functional aspects of growth of fungi into soil include the binding of soil grains together, thus affecting soil structure and associated properties (e.g., *Inocybe* sp., Schramm, 1966; Thornton *et al.*, 1956) and by affecting wettability of soil and litter (Thornton *et al.*, 1956). Went and Stark (1968) speculated that the extensive growth of mycelia of mycorrhizal fungi through soil could serve as an efficient trapping system for nutrients being leached from foliage or soil especially under high rainfall, thus leading to a "closed" nutrient circulation system. This almost certainly occurs to some extent—the efficiency of these trapping systems could be studied experimentally with some ease. No estimates of interception area of mycorrhizal hyphae in litter or soil have been made but it may be rather low and, as for fungal growth through soils, great variation between mycorrhizal fungi is predictable.

3. Root Hair and Hyphal Growth

Mycelial strand growth through soil has been treated above as analogous to root extension through soil. Consideration of uptake by the main axis and by root hairs in uninfected roots is replaced in mycorrhizae by consideration of the mycorrhiza plus fungal mantle and individual hyphal growth radiating from it and from mycelial strands. As mentioned in Section III,A,1, the radius of host tissue can increase from 120 to 160 μm with beech mycorrhizae and 55 μm radius to 80 μm with *Eucalyptus* mycorrhizae but this does not always occur. The fungus mantle may be absent or can be up to 40 μm thick with *Eucalyptus* (Chilvers, 1968) and beech (Harley, 1969) and up to 100 μm with some *Pinus* species (Wilcox, 1971), thus adding considerably to the "root" radius. Mycelial strands may grow into soil from this mantle as do individual hyphae, which will be considered here.

Both theoretical and experimental studies have shown the importance of root hairs in ion uptake from soil (Lewis and Quirk, 1967a,b; Drew and Nye, 1969; Barley and Rovira, 1970). Where root hairs (or hyphae of mycorrhizae) are close enough to deplete soil volumes between them, the effective radius is the root axis plus the root hair (see Fig. 3) or the mycorrhiza (or mycelial strand) plus radiating hyphae. The radius of *Eucalyptus* root plus root hair depicted by Chilvers and Pryor (1965) was approximately 100 μm, i.e., still 20 μm less than the mycorrhiza plus sheath. Bowen (1968) reported very little growth of individual hyphae from ectendomycorrhizae into the soil and large variation between ectomycorrhizal types in growth of their hyphae into soil. In four soils examined, individual hyphae from mycorrhizae of a *Rhizopogon* type penetrated only 120 μm with occasional penetration to 350 μm. Hyphal connection between adjacent branches of mycorrhizae occurred frequently, leading to nodules of soil and mycorrhizae. Photographs by Mikola (1948) show hyphae from *Cenococcum graniforme* penetrating natural soil away from mycorrhizae for more than 250 μm but growth of about 1500 μm with mycorrhizae synthesized in sterile soil; obviously other soil organisms affect hyphal growth. Chilvers noted growth of individual hyphae for 1–2 mm in natural soil. The excellent photographs of Schramm (1966) who studied *P. virginiana* colonization of anthracite wastes show individual hyphae of mycorrhizae formed by *Inocybe* sp. penetrating soil for several millimeters, but in Chilver's and Schramm's studies such growth was not in all directions and should not be regarded as analogous to the root hair cylinder of uninfected roots. Much more information on hyphal growth through soils and factors affecting it is required. It is clear that much variation exists depending on the fungus and the environment. In many, if not most cases, the growth of individual hyphae into soil is relatively small and the most extensive exploitation of soil occurs from mycelial strands and their "cylinder" of projecting individual hyphae.

The diameters of root hairs are commonly in excess of 10 μm (Barley, 1970) and those of *P. radiata* are 15–20 μm diameter compared with 2–4 μm diameter of hyphae of mycorrhizal fungi (Bowen, 1968; see also Chilvers, 1968, with eucalypt mycorrhizae). The narrowness of the hypha will enable entry into pores in soil and organic matter that cannot be entered by root hairs. This leads to greater soil–plant continuity in exploiting soil water and nutrients in pores.

I have emphasized the relatively much greater role of mycelial strands than of individual hyphal growth from mycorrhizae, for soil (and litter) exploration. This is supported by the data of Reid and Woods (1969) who, in considering the role of shared mycorrhizal fungi in transfer of nutrients and metabolites between trees of the same or different species, found movement of ^{14}C-labeled compounds in mycelial strands of *Thelephora terrestris* to be over considerably greater distances than with *Pisolithus tinctorius,* which did not form mycelial strands.

C. WATER UPTAKE

The movement of ions to mycorrhizae by convection in water assumes that mycorrhizae have a sustained water uptake function. The physiology of water absorption by mycorrhizae has been little studied, but it would be most surprising if they did not have this function. Lobanow (1960) observed very low root absorption surface/leaf transpiring surface ratios of some woody plants compared with herbaceous plants and nonmycorrhizal tree species and suggested that the deficit of surfaces required to absorb water were made good by external hyphae of ectomycorrhizae. A parallel situation probably occurs to that with the vesicular arbuscular endomycorrhizal association (with the phycomycete, *Endogone* sp.); Safir *et al.* (1971) have shown approximately 50% less resistance to water movement from soil to plant foliage with mycorrhizal soybeans, probably due to hyphal penetration of soil shortening transfer paths to the root. This increases the supply rate for water and ions and also bypasses any root epidermis resistance to water flow.

Studies by Cromer (1935), Goss (1960), Lobanow (1960), and Theodorou and Bowen (1970) indicate a greater drought resistance of mycorrhizal seedlings. This may be due to greater drought resistance itself (Cromer, 1935) or to generally greater root growth of mycorrhizal seedlings. Pure culture studies by Bowen and Theodorou (1967) have shown large differences between mycorrhizal fungi in their tolerance of low water potential in pure culture. Worley and Hacskaylo (1959) noted that the predominant mycorrhizal fungus on *P. virgiana* differed with moist and dry soils. When roots of *P. taeda* and *P. strobus* are subjected to severe water stress the tips become dormant (Kaufmann, 1968). The mycorrhiza does not seem to react in this way for Cromer (1935) noted mycorrhizae appeared to renew growth within 3 days after relief of water stress, whereas uninfected roots renewed growth only after approximately 14 days.

D. Phosphate Uptake

1. Readily Soluble Phosphates

Soil solution concentrations of phosphate are insufficient to account for convection supplying the phosphate needs of most plants, and diffusion to the root is the main transfer mechanism. A number of mathematical analyses show the limiting factor in phosphate uptake from soil is the slow rate of phosphate diffusion to roots (Nye, 1966; Lewis and Quirk, 1967b). It seems, therefore, that differences in absorbing power of different mycorrhizae will usually play only a minor role in differences between mycorrhizal fungi in stimulation of tree growth in phosphate-deficient soils. However, differences in the absorbing power may well be important in rapid uptake of large amounts of phosphate when flushes of these arise either by flushes of decomposition, e.g., with wetting and drying (Birch and Friend, 1961) or from leaf leachates (Will, 1959). Increased diameter of the mycorrhiza over that of the uninfected root should contribute to increased phosphate uptake/cm of root (see Nye, 1966). Lewis and Quirk (1967a) have demonstrated by autoradiography that the effective radius of roots for phosphate uptake is the root hair cylinder; depending on the particular fungus concerned, the mycorrhiza plus radiating hyphae can provide an effective root radius of approximately the same size or considerably greater than that of the uninfected root plus root hair. Some scope exists for selection of fungi on these bases in addition to selection on the production of mycelial strands.

The importance of longevity in uptake of ions diffusing over short distances has been mentioned above, and this will be of general significance in increased phosphate uptake by ectomycorrhizae but is particularly important where hyphal growth into soil is small or nonexistent (e.g., ectendomycorrhizae). Where extensive hyphal growth from the mycorrhiza occurs this will supplement the effects of longevity for the new mycorrhiza "boundary" will be the hyphal perimeter.

However, Bowen and Theodorou (1967) and Bowen (1968) pointed out that the best way to exploit the large interroot volumes for poorly mobile ions is by mycelial strand penetration of soil. In the field responses to different mycorrhizal fungi reported by Theodorou and Bowen (1970) in a phosphate-deficient soil, the best mycorrhizal strains were ones producing mycelial strands, suggesting the correctness of the above assessment.

2. Poorly Soluble Phosphates

a. Mineral Forms. Although Jahn (1934), Routien and Dawson (1943), and Schatz et al. (1954) proposed that excretion of acid, hydrogen ions, or chelating compounds, respectively, by mycorrhizae could have a solubilizing effect on poorly soluble phosphates, these ideas have not been critically tested experimentally. Solubilization of mineral phosphate by mycorrhizal fungi can readily be shown in laboratory media (e.g., Rosendahl, 1943; Bowen and Theodorou, 1967). In such laboratory studies, however, where carbohydrates are readily available, it is probable that solubilization is due solely to high acid production and it is unwise to extrapolate such laboratory results to a field situation.

Stone (1950) attempted to show solubilization of mineral phosphates by mycorrhizae by growing Sudan grass in association with mycorrhizal P. radiata. The grass did not show an increase in phosphate uptake but this cannot be taken as proof of nonsolubilization, as any solubilized phosphate may have been rapidly absorbed by the mycorrhizae. Bowen and Theodorou (1967) reported a response of mycorrhizal and non-mycorrhizal P. radiata to rock phosphate (see Table II), a finding also made by Stone (1950) with four mineral phosphates. These responses could be due to solubilization by rhizosphere and by mycorrhizosphere microorganisms.

b. Adsorbed Phosphates. In many soils total phosphorus is high but soluble phosphorus is low because of strong specific phosphorus adsorption to clay surfaces (Hingston et al., 1969). These phosphorus sources can be desorbed by certain acids, principally hydroxycarboxylic acid anions, e.g., citrate (Nye, 1968). No examination for exudation of such acids has been made with ectomycorrhizae.

No detailed examination has been reported of phosphorus uptake from phosphorus sorbed strongly onto the soil. An approach to such a study would be to label soil with ^{32}P and to allow it to equilibrate with soil solution. If the specific activity of phosphate absorbed by the plant is the same as that of soil solution, it is likely that phosphate is absorbed from soil solution or from forms in rapid equilibration with soil solution and not from nonlabile specifically sorbed inorganic phosphorus.

Some phosphorus in the solid phases in soil is in equilibrium with a very low concentration of phosphate in soil solution. Longevity of mycorrhizae should assist in some uptake of phosphorus from these solid phases by continued removal of phosphorus from solution over a long period. The studies of Henderson and Stone (1970) strongly suggest that the axenically produced mycorrhizal *P. radiata* can use phosphate fixed on perlite but the mechanism of this is not known.

3. Organic Phosphates

In most mineral soils one-half to two-thirds of the total phosphorus is organic but proportions varying from 4% for a podzol to 90% for an alpine humus have been reported (Cosgrove, 1967). Bowen and Theodorou (1967) reported that 80–93% of phosphorus in decomposing *P. radiata* litter was in various organic forms; much of this may have been bound up in microbial cells and is not available to other microorganisms until death of the cell.

Not only mycorrhizae but also sterile higher plants can use a range of organic phosphates (Section III,B,2). However, it is questionable whether higher plants ever get the chance to express this potential; organic phosphates will be used readily by a wide range of soil microorganisms as carbohydrate and energy sources, and because of their positional advantage in being spread throughout the soil, microorganisms will almost certainly absorb most of the readily available organic phosphates rather than the higher plant. The advantage of the mycorrhizal association, in contrast to uninfected plants, in use of organic phosphate is the ability of mycelia to penetrate soil pores and soil organic matter at distances from the root, thus competing positionally with other soil microorganisms and then translocating absorbed phosphate to the plant. The ability of certain mycorrhizal fungi to produce antibiotics against soil microorganisms (e.g., Šašek and Musílek, 1966; Marx, 1969) and the ability of some to grow at low water potentials, when other organisms are senescing and releasing organic phosphates, are particular cases where some mycorrhizal fungi may have special advantage over many other microorganisms in the competitive soil situation.

E. NITROGEN UPTAKE

1. Nitrate and Ammonium Ions

The processes of nitrate and ammonium uptake from soils are not as well defined as that of phosphate, but have been analyzed to some extent in agricultural soil with herbaceous plants. Where soil nitrate is maintained at a high level and transpiration is rapid, nitrate arriving at the root by convection appears adequate for plant demand (Olsen and Kemper, 1968). In such a case it could be expected that differences between mycorrhizae in rate of nitrate absorption would lead to some differences in nitrate uptake. Even here, however, the growth of hyphae and mycelial strands into soil will shorten movement paths considerably and increase the rate of transfer from soil. However, where nitrate is low in soils (the probable case of many forest soils) it appears that diffusion to roots will be the dominant factor (Olsen and Kemper, 1968; Nye, 1969). Nitrate is not adsorbed by soil and will therefore diffuse more freely than most other major plant nutrients, and over much greater distances than ions such as potassium and phosphate (see Fig. 3). Note also in Fig. 3 the relatively smaller importance of root hairs (or hyphae from mycorrhizae) in the case of a quickly diffusing ion. For nitrate, the need for extensive penetration of interroot volumes by mycelial strands will be useful but is by no means as critical as for phosphate. Andrews and Newman (1970) found when wheat plants were not competing in soil, root pruning did not affect nitrogen uptake, but when roots were competing, pruning substantially reduced uptake by the pruned plant. Extrapolating this to forestry it would seem that in a pure stand, high intensity of soil exploration by mycelial strands will not be as critical for growth as in mixed stands and forest–grassland mixtures, where the vast increase in tree rooting density given by mycelial strands will be significant in competition for nutrients and water with other species probably with high rooting densities. The importance of nitrate in a great many forest situations should not be overlooked despite the tendency for inhibition of nitrification in some forest litters (see Section III,C,1).

Studies by Clarke and Barley (1968) on wheat plants showed that although ammonium uptake was greater than nitrate uptake from solution, nitrate uptake from soil was 2–8 times faster than ammonium uptake because the diffusivity of nitrate through soil is an order of magnitude higher than that of ammonium ions. This clearly shows the limitations of ion uptake studies from solutions in predicting uptake from soil. The uptake of ammonium ions from soil by mycorrhizae will be aided considerably by longevity and fungal growth into soil.

2. Organic Nitrogen Compounds

The ability of both mycorrhizal fungi and higher plants to absorb organic forms of nitrogen was indicated above (Section III,C,2). Miller (1967; in Voigt, 1971) found mycorrhizal root systems of *Pseudotsuga menziesii* and of *P. radiata* had greater ability to absorb organic nitrogen compounds from soil than did nonmycorrhizal roots and that differences occurred between mycorrhizal fungi in use of amino acids. As for organic phosphates, the benefit from mycorrhizal fungi in a soil or litter situation is likely to lie in their location through the soil and in their spatial competition with other soil organisms.

Indeed it seems that the emphasis on mineralization of nitrogenous compounds being important for higher plant nutrition, though real, may have been misplaced. There is no doubt that many higher plants can easily use organic forms of nitrogen but they rarely get the chance, as simple organic forms of nitrogen are in ready demand as sources of energy by soil microorganisms, and it is not until the energy and carbohydrate content of the nitrogen source available to most microorganisms is very low (e.g., NH_4^+ and NO_3^-) that any is available to the plant.

F. UPTAKE OF OTHER IONS

Drew and Nye (1969), working on ryegrass, calculated that when root demand for potassium was high, diffusion of potassium through soil limited uptake to between 71 and 59% of that which would be expected from a stirred solution of the same concentration as soil solution. The presence of root hairs was calculated to have enhanced uptake by up to 77% because they increased the effective root diameter. However when root demand was low, soil supply was not a limiting factor and root hairs added little to the efficiency of the root. This illustrates the dynamic nature of ion uptake from soil depending on ion concentration in the soil, soil transmission properties, and plant demand. In the first case above, mycorrhizae should assist uptake by increase in size of the infected root and fungal growth into soil, but these would be of limited consequence in the second situation. The first situation is the more likely in potassium-deficient forest soils.

Solubilization of potassium minerals by mycorrhizae is erratic but apparently can occur (Voigt, 1971). Rosendahl (1943) found increases of potassium in *P. resinosa* growing in the presence of orthoclase and inoculated with *Tylopilus felleus* (syn *B. felleus*) but not *S. granulatus* or *Amanita muscaria*—no mycorrhizae were formed in the experiment. Boyle (1967; in Voigt, 1971) found mycorrhizal *P. radiata* inoculated with *A. rubescens, R. roseolus,* and *T. felleus* had increased potassium uptake from biotite and muscovite. Routien and Dawson (1943) found calcium, magnesium, iron, and potassium were absorbed in much larger quantities by mycorrhizal plants than by nonmycorrhizal plants from clay with low levels of base saturation. They suggested this was due to greater hydrogen ion production by mycorrhizae and exchange for cations from clay surfaces. Respiration of mycorrhizae was 2–4 times that of nonmycorrhizal short roots. None of the studies above have distinguished between the mycorrhizae and their associated microorganisms.

Sulfate demands by agricultural plants in soils containing gypsum can be met adequately by convection, but in low sulfate soils diffusion of sulfate may be limiting and physical mycorrhizal features would again be important in uptake. Morrison (1962b) found that mycorrhizal and nonmycorrhizal *P. radiata* seedlings growing in perlite, had the same absorption of sulfate. Thus, in soil with ready movement of sulfate to roots, mycorrhizal features are unlikely to increase uptake much beyond that of the uninfected root provided it is capable of absorbing sulfate. However, studies on wheat roots (Rovira and Bowen, 1968) suggest that the sulfate absorbing power of a root declines very rapidly with age and, if this occurs also with uninfected short roots of tree species, the longevity of mycorrhizae in the uptake of sulfate may make a significant difference to total uptake from soils over an extended period.

V. The Use of Absorbed Nutrients

A. Translocation in Hyphae and Mycelial Strands

Translocation in fungi has been discussed by Burnett (1968). Translocation of nutrients and carbohydrates over considerable distances is not a uniform property of fungi for some will move radioactive isotopes for only short distances in the extending hypha tips. Melin and Nilsson (1950, 1952, 1953a,b, 1955, 1957) and Melin *et al.* (1958) have demonstrated the movement of phosphate, ammonium-N, glutamic acid, calcium, and sodium through mycelia to mycorrhizae and the movement of ^{14}C-labeled photosynthate from pine seedlings to mycorrhizal fungi. Reid and Woods (1969) found ^{14}C-labeled compounds from glucose or sucrose applied to foliage of *P. taeda* seedlings in mycelial strands 12 cm from mycorrhizae produced by *Thelephora terrestris* but only slight movement occurred in the hyphae of mycorrhizae formed by *Pisolithus tinctorius*, which did not produce mycelial strands. The greater penetration of soil by mycelial strands was discussed in Section IV,B,2 above. Skinner and Bowen (in Bowen *et al.*, 1972) have shown ^{32}P and [65]Zn translocation by mycelial strands of *R. luteolus* to mycorrhizae of seedlings in soil, for distances in excess of a centimeter. The presumed method of translocation in fungi is by protoplasmic streaming but this mechanism has not been examined in the same biophysical and biochemical detail as for slime molds and coenocytic algae (Kamiya, 1959). The rates and frequency of cytoplasmic streaming in fungi, 2–3 cm/hr in hyphae and 6–15 cm/hr in subterranean mycelia attached to basidiocarps, are quite adequate to explain observed rates of translocation in fungi and bi-directional movement of ions. Note that this rate of movement far exceeds that of diffusion of ions in soil and convective flow through soil (Olsen and Kemprer, 1968; Nye, 1969) and highlights the efficacy of fungal growth from mycorrhizae into soil as a transfer mechanism. Although cytoplasmic streaming and translocation are temperature and oxygen sensitive (thus indicating participation of metabolic energy), they are greatly increased by transpiration when the mycelia are attached to fungus fruiting bodies; Burnett (1968) suggested this may be associated with internal turgor pressure gradients. By analogy we might expect translocation through mycelial strands to the sheath and Hartig net to be stimulated by transpiration. However, this is separate from the effects of transpiration on transfer of nutrients from fungus to host (see Section V,C), although it is likely to have some effect on transfer to the plant. From observation on mycelial strands of *Thelephora terrestris*, Schramm (1966) suggested movement of absorbed nutrients by mycorrhizal fungi could be by retraction of protoplasm from projecting hyphae. Although real, this is likely to be a minor mechanism compared with cytoplasmic streaming.

B. Storage of Nutrients

Most of the existing information on nutrients in the sheath of mycor-rhizae is on phosphate, a detailed study of which was performed by Harley and his associates at Oxford. Their studies with excised roots (see Harley, 1969) showed that some 90% of phosphate remained in the sheath, in a pool, not mixing with incoming phosphate, and that this stored phosphate was released to the plant over a period when external phosphate supplies were removed. Their studies and those of Jennings (1964) showed there are two main phosphorus "pools," a small pool integrated with metabolism of the cell and a larger nonmetabolic storage pool. Storage of phosphate could theoretically be as organic phosphates (e.g., inositol polyphosphates), inorganic polyphosphates (metachromatic granules are commonly found in microorganisms), or as orthophosphate physically separated from the metabolic sites in the cytoplasm (the most probable place being the vacuole). No evidence for high accumulation of inorganic or organic polyphosphates has yet been forthcoming from mycorrhizal fungi despite efforts to detect them (Bowen, 1963). Jennings (1964) suggested storage as orthophosphate. Using the method devised by Loughman (1960) for plant tissues, Bowen (1966) found that some 70–90% of the phosphate of *R. luteolus* and *S. luteus*, grown as pure cultures in a nutrient solution with $3 \times 10^{-5} M$ phosphate for 3 weeks, is in a nonmetabolic orthophosphate pool. As plant cells become rich in phosphate, further phosphate uptake is often depressed (Bowen, 1970a) and the removal of phosphate to a nonmetabolic site would keep cytoplasmic pools low and help maintain high phosphate uptake for long periods, as observed with ectomycorrhizae. Consistent with this is the stimulation of phosphate uptake by ammonium reported by Jennings (1964), since, as he pointed out, ammonium will lead to high rates of synthetic reactions and removal of phosphate from the cytoplasmic pool.

Harley pointed out that maintenance of a high phosphate uptake and storage of phosphate was likely to be ecologically important in situations where large phosphate accessions occurred (e.g., with flushes of decomposition and in leaf leachates). The reality of the storage function of mycorrhizae in soil has been shown by Morrison (1954, 1957, 1962a) and Clode (1956) with *P. radiata* seedlings; with a period of phosphate feeding followed by phosphate stress, nonmycorrhizal plants absorbed less phosphate and translocated it more readily to the tops but only for a limited period. With mycorrhizal plants translocation to the top was initially slower but was sustained and eventually more phosphate moved to the tops than with nonmycorrhizal plants.

It seems however, that the storage of nutrients observed with phosphate may be an exception rather than the rule. Morrison's studies with sulfate (1962b) indicated very little sulfate storage ability of pine and beech mycorrhizae, and an accumulation of sulfate in the cytoplasm leading to depression of further sulfate uptake would explain why he observed sulfate uptake rates to diminish markedly with time when feeding sulfate continuously. The data of Harley and Wilson (1959) suggest little storage capacity for potassium.

The storage of organic acids which are correlated with high cation uptake (Briggs *et al.*, 1961) has not been studied in detail with mycorrhizae. Harley (1964) showed $^{14}CO_2$ fixation by beech mycorrhizae led to 52–75% of the label of organic acids being in malate. It is probable therefore that malate is the main storage form of organic acids and, as with higher plants, it is stored in the vacuole.

C. Movement from Fungus to Host

Details of nutrient movement from fungus to the higher plant have been studied mainly with phosphate. In *P. radiata* where the Hartig net abuts the endodermis, microradiography studies (Bowen, 1966) have shown little movement occurs to adjoining cortical cells, and release of phosphate for the higher plant is at the endodermis. However, the restricted Hartig net in other systems suggests movement first to cortical cells and then to the stele. Metabolic inhibitors prevent the movement of phosphate from fungus to the host plant (Harley and Brierley, 1954, 1955). Phosphate passes to the root tissue as inorganic orthophosphate (Harley and Loughman, 1963). The storage of phosphorus and its slow translocation to the host plant have been discussed above. Melin and Nilsson (1958) suggested transpiration may affect rate of transfer considerably for 28–63% of the absorbed phosphate over 24 hours was retained in the mycorrhizal sheath of intact seedlings of *P. sylvestris* compared with 71–91% when the seedlings were decapitated. This aspect needs further study preferably preventing transpiration by less severe methods than decapitation. Such data tend to be at variance with the phosphate storage by mycorrhizae which has been demonstrated with intact seedlings under probable high transpiration conditions (see Section V,B).

Gunning and Pate (1969) have indicated the frequent occurrence in plant cells of ingrowths of wall material, leading to protoplasts with an unusually high surface-to-volume ratio which would enhance transfer of solutes. A similar situation occurs in the lichen symbiosis between fungi and algae (Brown and Wilson, 1968). Examination of electron micrographs of ectomycorrhizae of Foster and Marks (1966, 1967) and Hofsten (1969) does not indicate their presence in ectomycorrhizae but it must remain a possibility.

D. Efficiency of Nutrient Use

Since applied nutrients can eliminate the mycorrhizal response in deficient soils, it has been concluded that the nutritional mycorrhizal response is simply one of increased nutrient uptake. Earlier claims for growth-promoting substances being involved in the response (e.g., Rayner and Neilson Jones, 1944) have, correctly, been criticized on the basis that manurial treatments used in application of mycorrhizal fungi themselves contained sufficient nutrient to give a plant response. Although increased uptake may be the major part of nutrient response in mycorrhizae it is still not clear that it is the only response, i.e., it is possible that the mycorrhizal plant may use its nutrients more efficiently than the nonmycorrhizal plant. The efficiency of plant use of a limiting nutrient after absorption is a little explored field. The necessary experiment to examine efficiency of nutrient use (or other physiological parameters of mycorrhizal and nonmycorrhizal plants) is not to compare mycorrhizal and nonmycorrhizal plants of different sizes and nutrient content but to examine the growth and physiology of mycorrhizal and nonmycorrhizal plants with either the same total nutrient content or the same growth, achieved by adding the limiting nutrient to uninoculated plants. Circumstantial evidence is accumulating indicating a need for a serious reexamination of this problem: (i) Note in Table II that the phosphorus contents of seedlings inoculated with R. *luteolus* and S. *luteus* in the rock phosphate treatment were identical but highly significant differences occurred in dry weight of the seedlings. (ii) Auxin and cytokinin production by mycorrhizal fungi is known to occur and these may profoundly affect host metabolism. Cytokinin production may well explain prolongation of life of the cortex of mycorrhizae. Hormones may affect hydrolysis of starch in infected cells (Meyer, 1966); evidence is now accumulating that patterns of translocation of assimilates and nutrients in higher plants are affected largely by auxin and cytokinins (Letham, 1967; Davies and Wareing, 1965). Effects on metabolism of plants by noninfecting rhizosphere organisms and by plant pathogens have been recorded (see Bowen and Rovira, 1969). Highinbotham (1968) recorded that auxin lowers cell potential difference from 100 to 70 mV and significantly increases membrane permeability toward potassium and probably affects active transport systems. The possibility of ethylene production by mycorrhizal infection and associated metabolic effects does not appear to have been examined yet.

Björkman (1942) indicated relationships between carbohydrate physiology of tree species, mineral nutrient status, and mycorrhizal development. This and subsequent studies have been discussed by Harley (1969). Growth substances may also be involved, since Moser (1959) found high concentrations of nitrogen compounds reduced auxin production by mycorrhizal fungi in culture. A discussion of physiological interactions in mycorrhizae is given by Harley and Lewis (1969).

Obviously nutrient uptake and use cannot be divorced from other aspects of the physiology of the plant.

VI. Conclusions and Future Questions

Although maximum mycorrhizal nutritional responses are to be expected in relatively low fertility soils devoid of mycorrhizal fungi, it has been shown that responses can be obtained also in the presence of naturally occurring mycorrhizal fungi and that differences exist between mycorrhizal fungi in stimulation of plant growth. Thus there are grounds for inoculation of ectomycorrhizal tree species with specially selected fungi. There is a natural tendency to adopt the terminology of the *Rhizobium*–legume symbiosis and to refer to these as highly effective mycorrhizal fungi or as highly efficient fungi. However, in contrast to legume symbiosis, mycorrhizal fungi have many other roles ranging from nutrient uptake and water uptake, to disease resistance and possible metabolic effects on the plant. Efficiency of a fungus in one respect does not necessarily mean high performance in all respects, and the concept of efficiency or effectiveness in mycorrhizal stimulation of tree growth is rather meaningless unless the property is specified also. Our aim should be to select a fungus with as many high performance characteristics as possible, e.g., nutrient uptake, disease resistance, growth factor production, ease of introduction, and persistence. Introduction is simplest in the absence of other mycorrhizal fungi; superimposition on an existing mycorrhizal flora or supplanting existing mycorrhizae in a forest stand is likely to be more difficult and has been little studied.

Given that differences exist between mycorrhizal fungi in nutrient (and water) relations with tree species, *ad hoc* strain testing for particular situations will undoubtedly be of benefit (and must be done in the first instance) but directed selection based on the mechanisms of their responses should be superior in the long run. For nutrient uptake (and especially phosphate) we are initially selecting on the ability of the fungus to produce mycelial strands under a wide range of conditions; these studies are in the early stages.

For maximum response it is probable that the mycorrhizal fungus will have to be matched to the tree species and possibly even to the provenance for particular soils. Lamb and Richards (1971) found different fungi to give greatest stimulation of *P. elliottii* var. *elliottii* and *P. radiata*, respectively, Lundeberg (1968) found differences between provenances of *P. sylvestris* in mycorrhizal formation, and Marx and Bryan (1971b) recorded genotype of *P. elliottii* var. *elliottii* to influence the amount of mycorrhizal development and the growth response to inoculation with mycorrhizal fungi. An integration of mycorrhizal selection into plant introduction, plant selection, and plant breeding programs must eventually come about, since it is obvious that assessment of the species or provenance suitability for a particular areas could be markedly affected by the associated mycorrhizal fungus. The failure to study mycorrhizal requirements of new introductions and to provide suitable fungi for them may lead to seriously lowered production and discarding of a potentially important species.

The two broad areas of study of mycorrhizal nutrient relations, viz., physiological studies of uptake from solution, and studies on uptake of nutrients from soil are closely related. The physiological studies form a base line from which to investigate the soil situation and a number of aspects of phosphate uptake and storage studied initially from the physiological aspect by Harley and his colleagues have now been shown to have field relevance. A large amount of useful information now exists for phosphate but our information is fragmentary for most other elements. Nitrogen nutrition, in particular, needs a great deal more study, especially the question of ammonium versus nitrate nutrition. Nitrate uptake is certainly relevant in many (if not most) forest situations and selection of mycorrhizal fungi with ability to absorb nitrate as well as ammonium ions appears well worthwhile. The question of nitrogen fixation by mycorrhizae themselves appears to be being answered in the negative, recorded fixation being due not to the fungus but to associated bacteria. However, can fixation by mycorrhizosphere microorganisms really account for the reported accessions of 50 kg N/ha/year under some conifers, or are saprophytic organisms in litter the main agents? Related to the question of nitrogen is the increased nitrification under some conifers; this soil microbiology–plant phenomenon needs further elucidation. Trace element relationships of mycorrhizae have been all but neglected. Another neglected aspect, in view of the selectivity of mycorrhizal fungi for ions (as with higher plants) and the storage capacity of mycorrhizae for some ions, is the possibility that mycorrhizae could ameliorate marginal soil toxicities due to inorganic ions. The detoxication of organic phytotoxins in forest soils also awaits further study.

A more general aspect of the nutrient physiology of mycorrhizae is that linking mycorrhizae to other microorganism–plant associations which form a spectrum from noninfecting rhizosphere associations to ecto- and endomycorrhizae, nitrogen-fixing symbioses, and finally plant pathogens. Rhizosphere microorganisms have been shown to absorb significant amounts of nutrient; to affect uptake by the plant, translocation, and incorporation of phosphate in the plant; and to produce hormonal-type compounds (see Bowen and Rovira, 1969). Stimulation of plant growth by mycorrhizal fungi restricted to the rhizosphere has also been recorded. Many of these phenomena, and effects on membrane properties have also been noted in plant pathogenic associations. Smith *et al.* (1969) have indicated unifying concepts in carbohydrate transfers between associated organisms, and common mechanisms almost certainly occur in nutrient uptake and transfer with different microbe–plant associations (see Bowen and Rovira, 1969; Rovira and Bowen, 1971). While considering other plant–microorganism associations, it is appropriate to point out the enormous advantage accruing to plants in nutritionally poor situations if they are both mycorrhizal and nitrogen-fixing, e.g., *Alnus* and ectomycorrhizae (Mejstřik and Benecke, 1969), legumes and endomycorrhizae.

Despite the value of physiological studies on ion uptake and use from solution, usually over short periods, results cannot be extrapolated directly to a soil situation. They have to be placed in a quantitative framework of root activity, distribution and morphology in soil, transfer processes from soil to root, and plant demand for nutrients at different stages of growth. Considerable reference was made above to analyses of ion uptake from soils performed on agricultural plants under certain specified conditions, e.g., adequate soil moisture. To completely understand uptake by plants in the field it will be necessary to consider more complex situations, e.g., drying soils, and also to carry out the same approaches with forest species and forest soil conditions as has been done with agricultural plants. A basis is being laid by measurement of root and mycorrhizal growth, distribution, and function in soil. The analyses so far suggest that especially where adequate nutrients reach the mycorrhiza or root surface, principally by convection flow in water, differences in absorbing power between different mycorrhizae and uninfected roots will lead to differences in uptake. However, convective transfer seems to be of prime importance where the particular ion is abundant, and in nutrient poor soils, i.e., those giving a mycorrhizal response, it is likely that diffusion of the ion through soil is limiting uptake. Under these conditions the physical attributes of mycorrhizae assume particular importance.

One of the striking features about roots of woody species is their sparsity relative to herbaceous roots. It may be that tree species do not normally require a large root volume to supply their nutritional needs because much of the mineral requirement for young growing parts comes from retranslocation within the plant. In a closed system it may be necessary for the plant to take up little more than is lost by leaching and litter fall, and trapping these nutrients in competition with other microorganisms, suggested by Harley (1969) and by Went and Stark (1968), may be an important ecological function. Most efficient trapping will of course arise with mycorrhizal fungi with extensive mycelial growth from the mycorrhiza. Interception of leached nutrients in the soil and litter by hyphae has yet to be measured but it is unlikely to be by any means complete. Therefore both before significant litter fall commences, i.e., in the critical seedling stage, and first few years of a forest, and even in a closed stand, uptake of nutrients from soil and decomposing litter will be necessary.

The uptake of nutrients from soil (and litter) from the relatively large interroot distances of trees is achieved by longevity of mycorrhizal roots and particularly by growth of mycelial strands into soil. Both properties circumvent the extensive growth needed by nonmycorrhizal roots for full use of soil which occurs in many agronomic plants. The implications of longevity for transfer in the soil are the same for both diffusive and convective flow. Mycelial strand production (a greatly variable property between mycorrhizal fungi) (i) confers high penetration of large inter-root distances, (ii) confers positional advantage for competition with other microorganisms for both inorganic and organic nutrients, and (iii) in mixed stands or open forest and grassland is a factor giving large advantage in competition with other species for water and nutrients. Their role in uptake of nutrients by extensive soil colonization is particularly obvious for ions of low mobility, e.g., phosphorus, and probably mineral and sorbed forms of ions. Even when ions are transferred mainly by convection in water, rates of translocation in hyphae and mycelial strands far exceed water movement in soil, and thus rate of entry of ions to the higher plant–mycorrhiza system will probably be enhanced. Rate of transfer of water from soil to plant is likely to be increased considerably by fungal growth from mycorrhizae into soils (Safir *et al.*, 1971); water relations of ectomycorrhizae and mycorrhizal fungi are a field awaiting far more study.

Relatively little is known about possible metabolic effects on the plant by mycorrhizal fungi, e.g., effects on the distribution and use of nutrients in the higher plant and effects on uptake and transfer parameters of plant membranes, and this will no doubt be a field for increasing study. Marx and Bryan (1971a) found 45% survival of nonmycorrhizal *P. taeda* seedlings at 40°C root temperatures but 70 and 95% survival of plants inoculated with *T. terrestris* and *P. tinctorius*, respectively, unrelated to growth responses. Other circumstantial evidence for metabolic effects due to the fungus were discussed in Section V,D. Perhaps selection of fungi on metabolic properties will be possible in the future.

Finally, it should be pointed out that increasing nutrient uptake and efficiency of nutrient use is only one aspect of tree growth. It has to be integrated into such other aspects as tree demand for nutrients under a particular situation. Increasing nutrient uptake is not likely to be of much relevance in a situation where some other factor is limiting growth! By focusing on soil processes and roots one is sometimes in danger of forgetting that the prime product is the above the ground part of the plant, and therefore, this should be a prime point of focus. One of the more recent moves in nutritional physiology of plants is to attempt to construct physiological models integrating distribution of nutrients to various parts of the plant, effects on net assimilation, and redistribution of assimilates for growth of above the ground parts and of roots and other absorbing organs. Eventually effects of growth substances will have to be integrated into these models (e.g., Sweet and Wareing, 1966). Mycorrhizae influence nutrient uptake, carbohydrate distribution, and growth substance production, and our eventual aim should be to understand how these factors are integrated in plant growth in soil.

References

Addoms, R. M. (1946). Entrance of water into suberized roots of trees. *Plant Physiol.* **21**, 109.

Andrews, R. E., and Newman, E. I. (1970). Root density and competition for nutrients. *Oecol. Plant.* **5**, 319.

Aronson, J. M. (1965). The cell wall. *In* "The Fungi" (G. C. Ainsworth and A. S. Sussman, eds.), Vol. 1, pp. 49–76. Academic Press, New York.

Barber, D. A., and Frankenburg, U. C. (1969). The absorption of ions by excised barley roots grown under sterile and non-sterile conditions. *Annu. Rep. Letcombe Lab. A.R.C. (U.K.)* **19**, 40.

Barber, D. A., and Frankenburg, U. C. (1970a). The contribution of microorganisms to the apparent absorption of ions by excised roots grown under non-sterile conditions. *Annu. Rep. Letcombe Lab. A.R.C. (U.K.)* **20**, 32.

Barber, D. A., and Frankenburg, U. C. (1970b). The effect of microorganisms on the absorption of metal chelates. *Annu. Rep. Letcombe Lab. A.R.C. (U.K.)* **20**, 35.

Barber, D. A., and Loughman, B. C. (1967). Effect of microorganisms on absorption of inorganic nutrients by intact plants. II. Uptake and utilization of phosphate by barley plants grown under sterile and non-sterile conditions. *J. Exp. Bot.* **18**, 170.

Barber, S. A., and Ozanne, P. G. (1970). Autoradiographic evidence for the differential effect of four plant species in altering the calcium content of the rhizosphere soil. *Soil Sci. Soc. Amer., Proc.* **34**, 635.

Barley, K. P. (1970). The configuration of the root system in relation to nutrient uptake. *Advan. Agron.* **22**, 159.

Barley, K. P., and Rovira, A. D. (1970). The influence of root hairs on the uptake of phosphate. *Commun. Soil Sci. Plant Anal.* **1**, 287.

Becking, J. H. (1956). The mechanism of ammonium ion uptake by maize roots. *Acta Bot. Neer.* **5**, 1.

Bevege, D. I. (1968). Inhibition of seedling hoop pine (*Araucaria cunninghamii*) on forest soils by phytotoxic substances from the root zones of *Pinus, Araucaria* and *Flindersia. Plant Soil* **29**, 263.

Bevege, D. I., Bowen, G. D., and Skinner, M. F. (1972). Unpublished data.

Birch, H. F., and Friend, M. T. (1961). Resistance of humus to decomposition. *Nature (London)* **191**, 731.

Björkman, E. (1942). Uber die Bedingungen der Mykorrhizabildung bei Kiefer und Fichte. *Symb. Bot. Upsal.* **6**, 1–191.

Björkman, E. (1949). The ecological significance of the ectotrophic mycorrhizal association in forest trees. *Sv. Bot. Tidskr.* **43**, 223.

Bond, G., and Scott, G. D. (1955). An examination of some symbiotic systems for fixation of nitrogen. *Ann. Bot. (London)* [N.S.] **19**, 67.

Bornebusch, C. H. (1930). The fauna of forest soil (Skovbundens dyreverden). *Forstl. Forsogsv. Danm.* **11**, 1.

Bowen, G. D. (1960). Unpublished studies.

Bowen, G. D. (1963). Unpublished studies.

Bowen, G. D. (1964). Root distribution of *Pinus radiata. Div. Rep. 1/64, CSIRO Div. Soils, Adelaide* pp. 1–26.

Bowen, G. D. (1965). Mycorrhiza inoculation in forestry practice. *Aust. For.* **29**, 231.

Bowen, G. D. (1966). Unpublished studies.

Bowen, G. D. (1968). Phosphate uptake by mycorrhizas and uninfected roots of *Pinus radiata* in relation to root distribution. *Trans. Int. Congr. Soil Sci., 9th, 1968* Vol. 2, p. 219.

Bowen, G. D. (1970a). Early detection of phosphate deficiency in plants. *Commun. Soil Sci. Plant Anal.* **1**, 293.

Bowen, G. D. (1970b). Effects of soil temperature on root growth and on phosphate uptake along *Pinus radiata* roots. *Aust. J. Soil Res.* **8**, 31.

Bowen, G. D., and Rovira, A. D. (1966). Microbial factor in short-term phosphate uptake studies with plant roots. *Nature (London)* **211**, 665.

Bowen, G. D., and Rovira, A. D. (1967). Phosphate uptake along attached and excised wheat roots measured by an automatic scanning method. *Aust. J. Biol. Sci.* **20**, 369.

Bowen, G. D., and Rovira, A. D. (1969). The influence of micro-organisms on growth and metabolism of plant roots. *In* "Root Growth" (W. J. Whittington, ed.), pp. 170–201. Butterworth, London.

Bowen, G. D., and Rovira, A. D. (1970). Unpublished studies.

Bowen, G. D., and Theodorou, C. (1967). Studies on phosphate uptake by mycorrhizas. *Proc. Int. Union Forest Res. Organ., 14th, 1967* Vol. 5, p. 116.

Bowen, G. D., and Theodorou, C. (1967). Unpublished studies.

Bowen, G. D., Theodorou, C., and Skinner, M. F. (1973). Towards a mycorrhizal inoculation programme. *Proc. Amer.-Aust. Forest Nutr. Conf., Canberra, Aust. 1971* (in press).

Boyle, J. R. (1967). Biological weathering of micas in rhizospheres of forest trees. Ph.D. Dissertation, Yale University, New Haven, Connecticut.

Briggs, G. E., Hope, A. B., and Robertson, R. N. (1961). "Electrolytes and Plant Cells." Blackwell, Oxford.

Briscoe, C. B. (1959). Early results of mycorrhizal inoculation of pine in Puerto Rico. *Carib. Forest.* **20**, 73.

Brown, R. M., and Wilson, R. (1968). Electron microscopy of the lichen *Physcia aipolia* (Ehrh.) Ngl. *J. Phycol.* **4**, 230.

Budd, K. (1969). Net transport of potassium by non-growing fungal mycelium. *Abstr. Int. Bot. Congr., 11th, 1969* p. 24.

Burnett, J. H. (1968). "Fundamentals of Mycology." Arnold, London.

Carrodus, B. B. (1966). Absorption of nitrogen by mycorrhizal roots of beech. I. Factors affecting the assimilation of nitrogen. *New Phytol.* **65**, 358.

Chilvers, G. A. (1968). Some distinctive types of eucalypt mycorrhiza. *Aust. J. Bot.* **16**, 49.

Chilvers, G. A., and Pryor, L. D. (1965). The structure of eucalypt mycorrhizas. *Aust. J. Bot.* **13**, 245.

Clarke, A. L., and Barley, K. P. (1968). The uptake of nitrogen from soils in relation to solute diffusion. *Aust. J. Soil Res.* **6**, 75.

Clode, J. J. E. (1956). As micorrizas na nigraçao do fosforo, estudo com O ^{32}P. *Publ. Serv. Flor. Aquic. Portugal* **23**, 167.

Corke, C. T. (1958). Nitrogen transformations in Ontario forest podzols. *In* "First North American Forest Soils Conference" (T. D. Stevens and R. L. Cook, eds.), pp. 116–21. Michigan State Univ. Press, East Lansing.

Cosgrove, D. J. (1967). Metabolism of organic phosphates in soil. *In* "Soil Biochemistry" (A. D. McLaren and G. H. Peterson, eds.), pp. 216–28. Dekker, New York.

Cromer, D. A. N. (1935). The significance of the mycorrhiza of *Pinus radiata*. *Bull. Forest. Bur. Aust.* **16**, 1–19.

Dainty, J. (1969). The ionic relations of plants. *In* "Physiology of Plant Growth and Development" (M. B. Wilkins, ed.), pp. 455–85. McGraw-Hill, New York.

Davey, C. B. (1971). Non-pathogenic organisms associated with mycorrhizae. *In* "Mycorrhizae" (E. Hacskaylo, ed.), USDA Forest Serv. Misc. Publ. No. 1189, pp. 114–21. US Govt. Printing Office, Washington, D. C.

Davies, C. R., and Wareing, P. F. (1965). Auxin-directed transport of radiophosphorus in stems. *Planta* **65**, 139.

Day, S. C. (1969). The morphogenesis of mycelial strands in the timber dry rot fungus, *Merulius lacrymans* (Wulf.) Fr. Ph.D. Thesis, University of Cambridge.

Dijkshoorn, W. (1969). The relation of growth to the chief ionic constituents of the plant. *In* "Ecological Aspects of the Mineral Nutrition of Plants" (I. H. Rorison, ed.), pp. 201–213. Blackwell, Oxford.

Drew, M. C., and Nye, P. H. (1969). The supply of nutrient ions by diffusion to plant roots in soil. II. The effects of root hairs on the uptake of potassium by roots of rye grass (*Lolium multiflorum*). *Plant Soil* **31**, 407.

Edwards, D. G. (1968). The mechanism of phosphate absorption by plant roots. *Trans. Int. Congr. Soil Sci., 9th, 1968* Vol. 2, p. 183.

Epstein, E. (1966). Dual pattern of ion absorption by plant cells and by plants. *Nature (London)* **212**, 1324.

Epstein, E. (1968). Microorganisms and ion absorption by roots. *Experientia* **24**, 616.

Foster, R. C., and Marks, G. C. (1966). The fine structure of the mycorrhizas of *Pinus radiata* D. Don. *Aust. J. Biol. Sci.* **19**, 1027.

Foster, R. C., and Marks, G. C. (1967). Observations on the mycorrhizas of forest trees. II. The rhizosphere of *Pinus radiata* D. Don. *Aust. J. Biol. Sci.* **20**, 915.

Frank, A. B. (1894). Die Bedeutung der Mykorrihzapilze für die gemeine Kiefer. *Forstwiss. Zentralbl.* **16**, 185.

Fried, M., and Shapiro, R. E. (1961). Soil-plant relationships in ion uptake. *Annu. Rev. Plant Physiol.* **12**, 91.

Gardner, W. R. (1965). Movement of nitrogen in soil. *Agron. Monogr.* **10**, 550.

Garrett, S. D. (1970). "Pathogenic Root-Infecting Fungi." Cambridge Univ. Press, London and New York.

Goss, R. W. (1960). Mycorrhizae of ponderosa pine in Nebraska grassland soils. *Nebr. Agr. Exp. Sta., Res. Bull.* **192**, 1–47.

Gunning, B. E. S., and Pate, J. S. (1969). "Transfer cells"—plant cells with wall ingrowths, specialized in relation to short distance transfer of solutes—their occurrence, structure and development. *Protoplasma* **68**, 107.

Hacskaylo, E. (1953). Pure culture syntheses of pine mycorrhizae in terralite. *Mycologia* **45**, 971.

Hacskaylo, E., and Vozzo, J. A. (1967). Inoculation of *Pinus caribaea* with pure cultures of mycorrhizal fungi in Puerto Rico. *Proc. Int. Union Forest Res. Organ., 14th, 1967* Vol. 5, p. 139.

Hagen, C. E., and Hopkins, H. T. (1955). Ionic species in orthophosphate absorption by barley roots. *Plant Physiol.* **30**, 193.

Harley, J. L. (1964). Incorporation of carbon dioxide into excised beech mycorrhizas in the presence and absence of ammonia. *New Phytol.* **63**, 203.

Harley, J. L. (1969). "The Biology of Mycorrhiza." Leonard Hill, London.

Harley, J. L., and Brierley, J. K. (1954). Uptake of phosphate by excised mycorrhizal roots of the beech. VI. Active transport of phosphorus from the fungal sheath into the host tissue. *New Phytol.* **53**, 240.

Harley, J. L., and Brierley, J. K. (1955). Uptake of phosphate by excised mycorrhizal roots of the beech. VII. Active transport of ^{32}P from fungus to host during uptake of phosphate from solution. *New Phytol.* **54**, 296.

Harley, J. L., and Lewis, D. H. (1969). The physiology of ectotrophic mycorrhizas. *Advan. Microbial Physiol.* **3**, 53.

Harley, J. L., and Loughman, B. C. (1963). The uptake of phosphate by excised mycorrhizal roots of the beech. IX. The nature of the phosphate compounds passing into the host. *New Phytol.* **62**, 350.

Harley, J. L., and McCready, C. C. (1950). Uptake of phosphate by excised mycorrhizal roots of the beech. *New Phytol.* **49**, 388.

Harley, J. L., and McCready, C. C. (1952). Uptake of phosphate by excised mycorrhizal roots of the beech. III. The effect of the fungal sheath on the availability of phosphate to the core. *New Phytol.* **51**, 342.

Harley, J. L., and Wilson, J. M. (1959). The absorption of potassium by beech mycorrhizas. *New Phytol.* **58**, 281.

Harold, F. M. (1962). Binding of inorganic polyphosphate to the cell wall of *Neurospora crassa. Biochim. Biophys. Acta* **57**, 59.

Hatch, A. B. (1937). The physical basis of mycotrophy in the genus *Pinus. Black Rock Forest Bull.* **6**, 1–168.

Henderson, G. S., and Stone, E. L. (1970). Growth of mycorrhizal monterey pine supplied with phosphorus fixed on perlite. *In* "Tree Growth and Forest Soils" (C. T. Youngberg and C. B. Davey, eds.), pp. 171–80. Oregon State Univ. Press, Corvallis.

Highingbotham, N. (1968). Cell electropotential and ion transport in higher plants. *In* "Transport and Distribution of Matter in Cells of Higher Plants" (K. Mothes *et al.*, eds.), pp. 167–77. Akademie-Verlag, Berlin.

Highingbotham, N., Etherton, B., and Foster, R. J. (1967). Mineral ion contents and cell transmembrane electropotentials of pea and oat seedling tissue. *Plant Physiol.* **42**, 37.

Hingston, F. J., Atkinson, R. J., Posner, A. M., and Quirk, J. P. (1969). Specific adsorption of ions. *Nature* (*London*) **215**, 1459.

Hofstee, B. H. J. (1952). On the evaluation of the constants V_m and K_m in enzyme reactions. *Science* **116**, 329.

Hofsten, A. (1969). The ultrastructure of mycorrhiza I. *Sv. Bot. Tidskr.* **63**, 455.

Jahn, E. (1934). Die peritrophe Mykorrhiza. *Ber. Deut. Bot. Ges.* **52**, 463.

Jennings, D. H. (1963). "The Absorption of Solutes by Plant Cells." Oliver & Boyd, Edinburgh.

Jennings, D. H. (1964). Changes in the size of orthophosphate pools in mycorrhizal roots of beech with reference to absorption of the ion from the external medium. *New Phytol.* **63**, 181.

Jensen, W. A., and McLaren, A. D. (1960). Uptake of protein by plant cells. The possible occurrence of pinocytosis in plants. *Exp. Cell Res.* **19**, 414.

Kamiya, N. (1959). Protoplasmic streaming. *In* "Handbuch der Protoplasmaforschung" (L. V. Heilbrunn and F. Weber, eds.), Vol. 8, Part 3a. Springer-Verlag, Berlin and New York.

Katznelson, H., Rouatt, J. W., and Peterson, E. A. (1962). The rhizosphere effect of mycorrhizal and non-mycorrhizal roots of yellow birch seedlings. *Can. J. Bot.* **40**, 377.

Kaufmann, M. R. (1968). Water relations of pine seedlings in relation to root and shoot growth. *Plant Physiol.* **43**, 281.

Kramer, P. J., and Wilbur, K. M. (1949). Absorption of radioactive phosphorus by mycorrhizal roots of pine. *Science* **110**, 8.

Lamb, R. J., and Richards, B. N. (1971). Effect of mycorrhizal fungi on the growth and nutrient status of slash and radiata pine seedlings. *Aust. Forest.* **35**, 1.

Laties, G. C. (1969). Dual mechanisms of salt uptake in relation to compartmentation and long-distance transport. *Annu. Rev. Plant Physiol.* **20**, 89.

Laties, G. C., and Budd, K. (1964). The development of differential permeability in isolated steles of corn roots. *Proc. Nat. Acad. Sci. U. S.* **52**, 462.

Letham, D. S. (1967). Chemistry and physiology of kinetin-like compounds. *Annu. Rev. Plant Physiol.* **18**, 349.

Lewis, D. G., and Quirk, J. P. (1967a). Phosphate diffusion in soil and uptake by plants III. ^{31}P movement and uptake by plants as indicated by ^{32}P autoradiography. *Plant Soil* **26**, 445.

Lewis, D. G., and Quirk, J. P. (1967b). Phosphate diffusion in soil and uptake by plants IV. Computed uptake by model roots as a result of diffusive flow. *Plant Soil* **26**, 454.

Lips, S. H., Ben-Zioni, A., and Vaadia, Y. (1971). Potassium recirculation in plants and its importance for adequate nitrate nutrition. *In* "Recent Advances in Plant Nutritions" (R. M. Samish, ed.), Vol. 1, pp. 207–215. Gordon and Breach, New York.

Lister, G. R., Slankis, V., Krotkov, G., and Nelson, C. D. (1968). The growth and physiology of *Pinus strobus* L. seedlings as affected by various nutritional levels of nitrogen and phosphorus. *Ann. Bot. (London)* [N. S.] **32**, 33.

Lobanow, N. W. (1960). "Mykotrophie der Holzpflanzen." V.E.B. Deut. Verlag Wiss., Berlin.

Loughman, B. C. (1960). Uptake and utilization of phosphate associated with respiratory changes in potato tuber slices. *Plant Physiol.* **35**, 418.

Lundeberg, G. (1961). Accumulation and distribution of phosphorus in pine seedlings (*Pinus sylvestris* L.). *Skr. K. Skogshogsk., Stockholm* **38**, 1–26.

Lundeberg, G. (1968). The formation of mycorrhizae in different provenances of pine (*Pinus sylvestris* L.). *Sv. Bot. Tidskr.* **62**, 249.

Lundeberg, G. (1970). Utilization of various nitrogen sources, in particular bound soil nitrogen, by mycorrhizal fungi. *Stud. Forest. Suec.* **79**, 1–95.

McCalla, T. M. (1971). Studies on phytotoxic substances from soil microorganisms and crop residues at Lincoln, Nebraska. *In* "Biochemical Interactions among Plants," pp. 39–43. *National Academy of Sciences*, Washington, D. C.

MacMillan, A. (1956). The entry of ammonia into fungal cells. *J. Exp. Bot.* **7**, 113.

MacRobbie, E. A. C. (1971). Fluxes and compartmentation in plant cells. *Annu. Rev. Plant Physiol.* **22**, 75.

Marx, D. H. (1969). The influence of ectotrophic mycorrhizal fungi on the resistance of pine roots to pathogenic infection. II. Production, identification and biological activity of antibiotics produced by *Leucopaxillus cerealis* var. *piceina*. *Phytopathology* **59**, 411.

Marx, D. H., and Bryan, W. C. (1971a). Influence of ectomycorrhizae on survival and growth of aseptic seedlings of loblolly pine at high temperatures. *Forest Sci.* **17**, 37.

Marx, D. H., and Bryan, W. C. (1971b). Formation of ectomycorrhizae on half-sib progenies of slash pine in aseptic culture. *Forest Sci.* **17**, 488.

Marx, D. H., and Davey, C. B., (1969). The influence of ectotrophic mycorrhizal fungi on the resistance of pine roots to pathogenic infections. III. Resistance of formed mycorrhizae to infection by *Phytophthora cinnamomi. Phytopathology* **59**, 549.

Mejstřik, V. (1970). The uptake of ^{32}P by different kinds of ectotrophic mycorrhiza of *Pinus. New Phytol.* **69**, 295.

Mejstřik, V., and Benecke, U. (1969). The ectotrophic mycorrhizas of *Alnus viridis* (Chaix) D.C. and their significance in respect to phosphorus uptake. *New Phytol.* **68**, 141.

Melin, E. (1959). Mykorrhiza. *In* "Handbuch der Pflanzenphysiologie" (W. Ruhland, ed.), Vol. II, pp. 605–38. Springer-Verlag, Berlin and New York.

Melin, E., and Nilsson, H. (1950). Transfer of radioactive phosphorus to pine seedlings by means of mycorrhizal hyphae. *Physiol. Plant.* **3**, 88.

Melin, E., and Nilsson, H. (1952). Transfer of labelled nitrogen from an ammonium source to pine seedlings through mycorrhizal mycelium. *Sv. Bot. Tidskr.* **46**, 281.

Melin, E., and Nilsson, H. (1953a). Transfer of labelled nitrogen from glutamic acid to pine seedlings through the mycelium of *Boletus variegatus* (Sw). Fr. *Nature* (*London*) **171**, 134.

Melin, E., and Nilsson, H. (1953b). Transport of labelled phosphorus to pine seedlings through the mycelium of *Cortinarius glaucopus* (Schaeff. ex Fr.) Fr. *Sv. Bot. Tidskr.* **48**, 555.

Melin, E., and Nilsson, H. (1955). Ca^{45} used as an indicator of transport of cations to pine seedlings by means of mycorrhizal mycelia. *Sv. Bot. Tidskr.* **49**, 119.

Melin, E., and Nilsson, H. (1957). Transport of C^{14} labelled photosynthate to the fungal associate of pine mycorrhiza. *Sv. Bot. Tidskr.* **51**, 166.

Melin, E., and Nilsson, H. (1958). Translocation of nutritive elements through mycorrhizal mycelia to pine seedlings. *Bot. Notis.* **111**, 251.

Melin, E., Nilsson, H., and Hacskaylo, E. (1958). Translocation of cations to seedlings of *Pinus virginiana* through mycorrhizal mycelia. *Bot. Gaz.* **119**, 241.

Meyer, F. H. (1966). Mycorrhiza and other plant symbioses. *In* "Symbiosis" (S. M. Henry, ed.), Vol. 1, pp. 171–255. Academic Press, New York.

Mikola, P. (1948). On the physiology and ecology of *Cenococcum graniforme. Commun. Inst. Forest. Fenn.* **36**, Part I, 1–104.

Miller, C. O. (1971). Cytokinin production by mycorrhizal fungi. *In* "Mycorrhizae" (E. Hacskaylo, ed.), USDA Forest Serv. Misc. Publ. No. 1189, pp. 168–74. US Govt. Printing Office, Washington, D. C.

Miller, R. J. (1967). Assimilation of nitrogen compounds by tree seedlings. Ph.D. Dissertation, Yale University, New Haven, Connecticut.

Moore, A E. (1966). Non-symbiotic nitrogen fixation in soil and soil-plant systems. *Soils Fert.* **29**, 113.

Morrison, D., and Bowen, G. D. (1970). Unpublished studies.

Morrison, T. M. (1954). Uptake of phosphorus-32 by mycorrhizal plants. *Nature* (*London*) **174**, 606.

Morrison, T. M. (1957). Mycorrhiza and phosphorus uptake. *Nature* (*London*) **179**, 907.

Morrison, T. M. (1962a). Absorption of phosphorus from soils by mycorrhizal plants. *New Phytol.* **61**, 10.

Morrison, T. M. (1962b). Uptake of sulphur by mycorrhizal plants. *New Phytol.* **61**, 21.

Morton, A. G., and MacMillan, A. (1954). The assimilation of nitrogen from ammonium salts and nitrate from fungi. *J. Exp. Bot.* **5**, 232.

Moser, M. (1956). Die Bedeutung der Mykorrhiza für Aufforstungen in Hochlagen. *Forstwisse Centralbl.* **75**, 8.

Moser, M. (1959). Beitrage zur Kenntnis der Wuchsstott-beziehungen im Bereich ectotrophen Mykorrhizen. *Arch. Mikrobiol.* **34**, 251.

Mustanoja, K. A., and Leaf, A. L. (1965). Forest fertilization research, 1957–64. *Bot. Rev.* **31**, 151.

Neal, J. L., Bollen, W. B., and Zak, B. (1964). Rhizosphere microflora associated with mycorrhizae of Douglas Fir. *Can. J. Microbiol.* **10**, 259.

Nelson, C. D. (1964). The production and translocation of photosynthate C^{14} in conifers. *In* "The Formation of Wood in Forest Trees" (M. H. Zimmermann, ed.), pp. 243–257. Academic Press, New York.

Noggle, J. C., and Fried, M. (1960). A kinetic analysis of phosphate absorption by excised roots of millet, barley and alfalfa. *Soil Sci. Soc. Amer., Proc.* **24**, 33.

Norkrans, B. (1953). The effect of glutamic acid, aspartic acid and related compounds on the growth of certain *Tricholoma* species. *Physiol. Plant.* **6**, 584.

Nutman, P. S. (1971). Perspectives in biological nitrogen fixation. *Sci. Progr. (London)* **59**, 55.

Nye, P. H. (1966). The effect of the nutrient intensity and buffering power of a soil, and the absorbing power, size and root hairs of a root, on nutrient absorption by diffusion. *Plant Soil* **25**, 81.

Nye, P. H. (1968). Processes in the root environment. *J. Soil Sci.* **19**, 205.

Nye, P. H. (1969). The soil model and its application to plant nutrition. *In* "Ecological Aspects of the Mineral Nutrition of Plants (I. H. Rorison, ed.), pp. 105–114. Blackwell, Oxford.

Olsen, S. R., and Kemper, W. D. (1968). Movement of nutrients to plant roots. *Advan. Agron.* **20**, 91.

Olsen, S. R., Kemper, W. D., and Jackson, R. D. (1962). Phosphate diffusion to plant roots. *Soil Sci. Soc. Amer., Proc.* **26**, 222.

Oswald, E. T., and Ferchau, H. A. (1968). Bacterial associations of coniferous mycorrhizae. *Plant Soil* **28**, 187.

Parker, C. A. (1957). Non-symbiotic nitrogen fixing bacteria in soil. III. Total nitrogen changes in a field soil. *J. Soil Sci.* **8**, 48.

Paterson, J., and Bowen, G. D. (1968). Unpublished studies.

Pitman, M. G., Mowat, J., and Nair, H. (1971). Interactions of processes for accumulation of salt and sugar in barley plants. *Aust. J. Biol. Sci.* **24**, 619.

Rayner, N. C., and Neilson Jones, W. (1944). "Problems in Tree Nutrition." Faber & Faber, London.

Redmond, D. R. (1955). Studies in forest pathology. XV. Rootlets, mycorrhiza, and soil temperatures in relation to birch dieback. *Can. J. Bot.* **33**, 595.

Reid, C. P. P., and Woods, F. W. (1969). Translocation of C^{14}-labelled compounds in mycorrhizae and its implication in interplant nutrient cycling. *Ecology* **50**, 179.

Richards, B. N. (1964). Fixation of atmospheric nitrogen in coniferous forests. *Aust. Forest.* **28**, 68.

Richards, B. N., and Bevege, D. I. (1967). The productivity and nitrogen economy of artificial ecosystems comprising various combinations of perennial legumes and coniferous tree species. *Aust. J. Bot.* **15**, 467.

Richards, B. N., and Voigt, G. K. (1964). Role of mycorrhiza in nitrogen fixation. *Nature (London)* **201**, 310.

Richards, B. N., and Voigt, G. K. (1965). Nitrogen accretion in coniferous forest ecosystems. *In* "Second North American Forest Soils Conference" (C. T. Youngberg, ed.), pp. 105–116. Oregon State Univ. Press, Corvallis.

Richards, B. N., Bevege, D. I., and Lamb, R. J. (1971). Personal communication.

Ridge, E. H., and Rovira, A. D. (1971). Phosphatase activity of intact young wheat roots under sterile and non-sterile conditions. *New Phytol.* **70,** 1017.

Rosendahl, R. O. (1943). The effect of mycorrhizal and non-mycorrhizal fungi on the availability of difficultly soluble potassium and phosphorus. *Soil Sci. Soc. Amer., Proc.* **7,** 477.

Routien, J. B., and Dawson, R. F. (1943). Some inter-relations of growth, salt absorption, respiration and mycorrhizal development in *Pinus echinata. Amer. J. Bot.* **30,** 440.

Rovira, A. D., and Bowen, G. D. (1966). Phosphate incorporation by sterile and non-sterile plant roots. *Aust. J. Biol. Sci.* **19,** 1167.

Rovira, A. D., and Bowen, G. D. (1968). Anion uptake by plant roots. Distribution of anions and effects of microorganisms. *Trans. Int. Congr. Soil Sci., 9th, 1968* Vol. 2, p. 209.

Rovira, A. D., and Bowen, G. D. (1971). Microbial effects on nutrient uptake by plants. *In* "Recent Advances in Plant Nutrition" (R. M. Samish, ed.), Vol. 1, pp. 307–320. Gordon and Breach, New York.

Russell, R. S., and Barber, D. A. (1960). The relationship between salt uptake and the absorption of water by intact plants. *Annu. Rev. Plant Physiol.* **11,** 127.

Russell, R. S., and Newbould, P. (1969). The pattern of nutrient uptake in root systems. *In* "Root Growth" (W. J. Whittington, ed.), pp. 148–69. Butterworth, London.

Safir, G., Boyer, J. S., and Gerdemann, J. W. (1971). Mycorrhizal enhancement of water transport in soybean. *Science* **172,** 581.

Sanders, F. E., Tinker, P. B., and Nye, P. H. (1971). Uptake of solutes by multiple root systems from soil. I. An electrical analog of diffusion to root systems. *Plant Soil.* **34,** 453.

Šašek, V., and Musílek, V. (1966). Cultivation and antibiotic activity of mycorrhizal basidiomycetes. *Folia Microbiol. (Prague)* **12,** 515.

Schatz, A., Cheronis, N. D., Schatz, V., and Trelawny, G. (1954). Chelation (sequestration) as a biological weathering factor in pedogenesis. *Proc. Pa. Acad. Sci.* **28,** 44.

Schramm, J. R. (1966). Plant colonization studies on black wastes from anthracite mining in Pennsylvania. *Trans. Amer. Phil. Soc.* [N. S.] **56,** 194.

Shiroya, T., Lister, G. R., Slankis, V., Krotkov, G., and Nelson, C. D. (1962). Translocation of the products of photosynthesis to roots of pine seedlings. *Can. J. Bot.* **40,** 1125.

Skinner, M. F., Bevege, D. I., and Bowen, G. D. (1972). Unpublished data.

Slayman, C. L. (1962). Measurement of membrane potentials in *Neurospora. Science* **136,** 876.

Smith, D., Muscatine, L., and Lewis, D. (1969). Carbohydrate movement from autotrophs to heterotrophs in parasitic and mutualistic symbiosis. *Biol. Rev.* **44,** 17.

Stevenson, G. (1959). Fixation of nitrogen by non-nodulated seed plants. *Ann. Bot. (London)* [N. S.] **23,** 622.

Stone, E. L. (1950). Some effects of mycorrhizae on the phosphorus nutrition of Monterey pine seedlings. *Soil Sci. Soc. Amer., Proc.* **14,** 340.

Stone, E. L., and Fisher, R. F. (1969). An effect of conifers on available soil nitrogen. *Plant Soil* **30**, 134.

Sweet, G. B., and Wareing, P. F. (1966). Role of plant growth in regulating photosynthesis. *Nature (London)* **210**, 77.

Theodorou, C. (1968). Inositol phosphates in needles of *Pinus radiata* D. Don and the phytase activity of mycorrhizal fungi. *Trans. Int. Congr. Soil Sci., 9th, 1968* Vol. 3, p. 483.

Theodorou, C. (1971). The phytase activity of the mycorrhizal fungus, *Rhizopogon luteolus*. *Soil Biol. Biochem.* **3**, 89.

Theodorou, C., and Bowen, G. D. (1970). Mycorrhizal responses of radiata pine in experiments with different fungi. *Aust. Forest.* **34**, 183.

Thornton, R. H., Cowie, J. D., and McDonald, D. C. (1956). Mycelial aggregates of sand soil under *Pinus radiata*. *Nature (London)* **177**, 231.

Todd, D. K. (1959). "Ground Water Hydrology." Wiley, New York.

Trappe, J. M. (1967). Principles of classifying ectotrophic mycorrhizae for identification of fungal symbionts. *Proc. Int. Union Forest. Res. Organ., 14th, 1967* Vol. 5, p. 46.

Voigt, G. K. (1971). Mycorrhizae and nutrient mobilization. *In* "Mycorrhizae" (E. Hacskaylo, ed.), USDA Forest Serv. Misc. Publ. No. 1189, pp. 122–131. US Govt. Printing Office, Washington, D. C.

Weinhold, A. R. (1963). Rhizomorph production by *Armillaria mellea* induced by ethanol and related compounds. *Science* **142**, 1065.

Welch, R. M., and Epstein, E. (1968). The dual mechanisms of alkali cation absorption by plant cells: Their parallel operation across the plasmalemma. *Proc. Nat. Acad. Sci. U. S.* **61**, 447.

Went, F. W., and Stark, N. (1968). The biological and mechanical role of soil fungi. *Proc. Nat. Acad. Sci. U. S.* **60**, 497.

Wilcox, H. E. (1971). Morphology of ectendomycorrhizae in *Pinus resinosa*. *In* "Mycorrhizae" (E. Hacskaylo, ed.), USDA Forest Serv. Misc. Publ. No. 1189, pp. 54–68. US Govt. Printing Office, Washington, D. C.

Wild, A., and Oke, O. L. (1966). Organic phosphate compounds in calcium chloride extracts of soils: identification and availability to plants. *J. Soil Sci.* **17**, 356.

Wilde, S. A., and Iyer, J. G. (1962). Growth of red pine (*Pinus resinosa* Ait.) on scalped soils. *Ecology* **43**, 771.

Will, G. M. (1959). Nutrient return in litter and rainfall under some exotic conifer stands in New Zealand. *N. Z. J. Agr. Res.* **2**, 719.

Wilson, J. M. (1957). A study of the factors affecting the uptake of potassium by the mycorrhiza of beech. Ph.D. Thesis, Oxford University.

Woods, F. W. (1960). Biological antagonisms due to phytotoxic root exudates. *Bot. Rev.* **26**, 546.

Worley, J. F., and Hacskaylo, E. (1959). The effect of available soil moisture on the mycorrhizal association of Virginia pine. *Forest Sci.* **5**, 267.

Zak, B. (1971). Detoxication of autoclaved soils by a mycorrhizal fungus. *U. S. Forest Serv., Paci. Northwest Forest Range Exp. Sta., Res. Notes.*

CHAPTER 6

Carbohydrate Physiology of Ectomycorrhizae

EDWARD HACSKAYLO

I. Introduction

Fungi, as do all living plants, require carbohydrates as metabolites in the synthesis of cellular components and as energy sources for metabolic processes; but unlike the autotrophic chlorophyll-bearing plants, they must obtain their carbohydrates from exogenous sources. For mycorrhizal fungi, the carbon source, as well as its role in the symbiotic system, has been the subject of many studies. Because of the complexity of the system, it has been necessary to study fungi and roots separately as well as together as a physiological unit. This has resulted in an assemblage of information which must be supplemented and correlated by future investigators.

207

II. Host Dependence

Frank (1885) first suggested that ectomycorrhizal fungi depend upon trees for carbohydrates which are transferred to fungi through networks of hyphae that encircle and penetrate receptive roots. Others later suggested that under natural conditions this attachment of hyphae to particular roots was requisite to fruiting of ectomycorrhizal fungi.

In 1938, Rommell isolated plots beneath spruce trees in Sweden with sheet iron barriers which severed the roots from the body of the trees. When no sporophores of mycorrhizal fungi appeared outside trenched plots, he assumed that the flow of carbohydrates from host to fungus had been interrupted. This, he concluded, proved that mycorrhizal fungi were dependent upon the autotroph for completion of the life cycle and that they were unable to decompose litter as did saprophytic fungi. Although Rommel's experiments appeared to be conclusive, reservations were recently expressed regarding dependence of the fungi upon hosts for fruiting (Harley, 1969).

Hacskaylo (1965) demonstrated a direct dependence on photosynthesis of pine seedlings for sporophore formation in the mycorrhizal fungus *Thelephora terrestris*. In these greenhouse experiments, seedlings with attached young sporophores were either decapitated or their needles were covered with a black bag. In both instances sporophore development was immediately arrested. However, if pine seedlings that had been covered with a black bag were again exposed to natural daylight, the sporophores continued to develop. In this instance, completion of the life cycle was very sensitive to the supply of food available through the roots of its host plant.

Apparently certain species of *Hebeloma* may be facultative symbionts and fruit in the absence of mycorrhizal hosts (Hacskaylo and Bruchet, 1972). The frequency of such relationships is probably small among ectomycorrhizal fungi.

Actual data are very limited on identification of organic materials released from roots of trees that have ectomycorrhizal associations. Very few give details on secretion of sugars. In one study, Lister *et al.* (1968) chromatographically identified glucose and arabinose in exudates of *Pinus strobus* roots. In another study, Smith (1969) found glucose, rhamnose, sucrose, and in one instance, fructose in root exudates of five tree species, four of which were pines; glucose was the most prevalent sugar detected. Identities of the carbohydrates and quantities available for utilization by mycorrhizal fungi at root surfaces are virtually unknown. It can be assumed, however, that carbon compounds available to the fungus at the surface of the root are primarily those supplied through exudates from the host plant. The compounds must be available in concentrations of sufficient magnitude to support the fungus during colonization of the receptive roots.

III. Translocation of Carbon Compounds

Movement of carbohydrates from the host to ectomycorrhizal fungi was conclusively demonstrated by Melin and Nilsson (1957). In this study, pine seedlings growing axenically in Erlenmeyer flasks were inoculated with the mycorrhizal fungi, *Rhizopogon roseolus* and *Boletus variegatus*. After mycorrhizae had formed, the seedlings were exposed to an atmosphere containing [^{14}C]-labeled carbon dioxide. Labeled carbon compounds produced in photosynthesis were rapidly transported into the associated mycorrhizal mycelia and accumulated in high concentrations in the hyphal mantle. Thus the radioactivity of the mycorrhizae was much higher than that of roots of uninoculated seedlings. It was concluded that carbon compounds which have their origin as photosynthates are translocated to the roots to serve as major energy and food sources for mycorrhizal fungi.

Björkman (1960) found that carbohydrates were transferred, not only from hosts into mycorrhizal hyphae, but from one host plant to another through mycorrhizal hyphae common to both. He traced injected [^{14}C]-labeled glucose and [^{32}P]-labeled orthophosphate from trunks of mature spruce and pine trees to adjacent *Monotropa* plants. Björkman concluded that *Monotropa* was an epiparasite on tree species.

Woods and Brock (1964) suggested that mycorrhizal fungi may perform an important function in cycling of materials in an ecosystem. Although their work did not include carbohydrates, they were able to trace ^{32}P and ^{45}Ca from *Acer* trees which have endomycorrhizae, to a total of 19 different species of plants. They believed that these plants mutually shared mycorrhizal fungi.

The capability of mycorrhizae and associated hyphal systems to transport material between individual pine seedlings was studied by Reid (1971). Reid emphasized that movement of materials from carbon compounds occurred in three ways: mycorrhizal root systems to external hyphae, external hyphae to mycorrhizal root systems, and one root system to another via a mutually shared mycorrhizal fungus.

The results of Björkman (1960) and Reid (1971) strongly suggest that carbohydrates can move in considerable quantities from fungus to host tissue, though the results presented by Lewis and Harley (1965,a,b,c) do not favor this interpretation. In their experiments on carbohydrate movement in excised mycorrhizae, Lewis and Harley concluded that sucrose was absorbed primarily by host tissue and that large quantities of sugar were then translocated laterally into the fungal tissue and accumulated as mannitol, trehalose, and glycogen. There was no reciprocal flow of carbohydrate into the root. The information in the literature thus suggests that movement of carbohydrates may occur either from host to fungus or fungus to host; however, conditions governing the direction of movement are not clearly defined.

IV. Carbohydrates and Mycorrhiza Formation

If sugars exuded from the host plant support metabolic activities of mycorrhizal fungi that are growing on the surface of young root tips, then colonization of the root by the mycorrhizal fungus will be directly related to the internal carbohydrate status of the short roots. This means that light, temperature, photoperiod, translocation rates, availability of nutrients, or other factors that affect the internal concentrations and composition of carbon compound within the root have a direct influence upon establishment of mycorrhizae and on their maintenance.

A. Björkman's Hypothesis

Among many efforts to define the mechanism controlling ectomycorrhiza formations, perhaps the most prominent were those of Björkman (1942). Björkman observed that when he applied ammonium nitrate to forest soil deficient in nitrogen, but not in phosphorus, plant growth increased and the frequency of mycorrhiza formation diminished. However, when additional nitrogen was applied to soils already rich in nitrogen but deficient in phosphorus, neither plant growth nor frequency of mycorrhiza formation changed. Applications of phosphoric acid only to nitrogen-rich soil produced an initial increase in the formation of mycorrhizae. The addition of both nitrogen and phosphorus in various proportions reduced the frequency of mycorrhizae in soils containing either high or low nitrogen levels.

Using several levels of fertility, Björkman then grew seedlings in forest soils at 6, 12, 17, 23, and 49% of full daylight, from May to October. Mycorrhiza development was good in all plants grown at 49 and 23% but was erratic at the 17% level (Fig. 1). At 12%, mycorrhizae formed only in soils deficient in available nitrogen, but not in soils that contained a readily available source of the element. At 6%, mycorrhizae never formed. Analyses of carbohydrates in the roots, expressed in terms of reducing

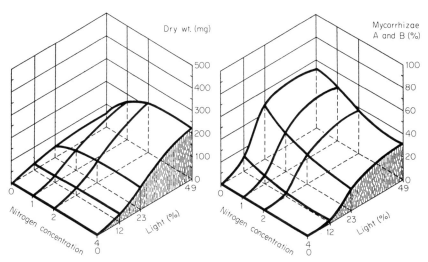

FIG. 1. Growth (left) and mycorrhiza development (right) of 1-year-old pine seedlings in raw humus at various light intensities and NH_4NO_3 additions. After Björkman (1942) by permission of Symbolae Botanicae Upsaliensis.

substances, showed that the sugar contents were very low in the heavily shaded plants, and in the illuminated plants grown at high nitrogen and phosphorus levels. Björkman concluded that when there is an extreme deficiency or a high sufficiency of available nutrients, particularly nitrogen, the sugar concentration in the plant is reduced. The extreme deficiency reduces the plant metabolism, including carbohydrate formation, and high sufficiency favors the rapid conversion of carbohydrates into amino acids. Generally, a deficiency of phosphorus reduces the concentration of sugar by inhibiting photosynthesis. However, if the deficiency is only moderate, the assimilation of carbohydrate is normal with reduced formation of protein, and at times there will be a surplus of sugar.

Björkman suggested that, since mycorrhizal fungi can generally assimilate only soluble carbohydrates, their absence or presence only in very small quantities in roots could be directly related to the complexity of factors that prevent the formation of mycorrhizae. To explain the variations in ectomycorrhiza formation, he hypothesized that mycorrhizal fungi parasitize roots for soluble sugars. They will not enter the roots unless these contain a certain quantity of surplus soluble carbohydrates. Therefore, the rate of photosynthesis should be relatively high, and of nutrient uptake relatively low. These conditions would ensure that assimilation would not tie up most of the available carbohydrates and sufficient quantities would be translocated to the roots for utilization by the mycorrhizal fungi.

In another experiment, Björkman (1944) bound stems of 3-year-old *Pinus sylvestris* plants with a thin wire so that translocation from the needles to the roots was interrupted. The effect was very rapid cessation of mycorrhiza formation. However, those plants that were not strangulated continued to develop mycorrhizae. Long roots grew in unstrangulated plants, but not as much as in strangulated plants. When the wires were removed from the stems, fresh conducting tissue developed, and the ability to form mycorrhizae was restored. The quantities of soluble carbohydrates in the roots were determined and correlated with the frequency of mycorrhizae. Roots of permanently and temporarily strangulated plants contained on the average 0.7 and 1.85% soluble carbohydrate of the dry weight, respectively. The nonstrangulated roots contained 2.2%. Björkman again stated that it appeared that a definite connection existed between the soluble carbohydrate content of the root and the formation of mycorrhizae (Table I).

B. Further Considerations

Hacskaylo and Snow (1959) also showed that high levels of available nitrogen and phosphorus and low light intensity suppressed the formation of mycorrhizae on pine. They favored the hypothesis of Björkman, but stated that conditions that permit and promote formation of

TABLE I

Growth and Development of Mycorrhizae in 4-Year-Old Strangulated and Nonstrangulated Pine Plants[a]

	Strangulated permanently	Strangulated temporarily	Nonstrangulated
Av. length of roots (mm)	3090	5300	5640
Av. % newly formed mycorrhizae	2.3	43	65.2
Av. dry wt. shoots (mg)	7252	11332	8905
Av. dry wt. roots (mg)	1601	3595	4290
Av. content soluble carbohydrates in needles[b]	7.3	6.6	6.2
Av. content soluble carbohydrates in roots[b]	0.7	1.85	2.2

[a] After Björkman (1944).
[b] Expressed as percentage glucose of dry weight.

mycorrhizae are controlled by a complex of both internal and environmental factors, which are now only partially known.

In 1962, Handley and Sanders attempted to duplicate the nutrient status in the substrate and the light intensities that Björkman had used for determining the effects of these factors on the formation of mycorrhizae. However, they used sterile quartz sand substrate rather than soil, and grew the plants during a later part of the year. According to their analysis of root carbohydrates, the concentration of soluble reducing substances increased as a result of mycorrhizal associations. They suggested that the increased amount of reducing substances was the result of accumulations of carbohydrates in the fungal mycelia.

Meyer (1962) also attempted to test Björkman's hypothesis by growing European beech (*Fagus sylvatica*) seedlings in soils of varying fertility. When he added nitrogen and phosphorus to soils of highest fertility, mycorrhiza formation was not reduced, but sometimes it increased. Also, the greatest amounts of reducing substances were found in these roots. Meyer concluded that the increased sugar content of mycorrhizae was a result of the activity of the mycorrhizal fungi and of other microorganisms in the soil. He later suggested that auxins produced by the fungi promote hydrolysis of carbohydrates in the tissues and prevent the accumulation of starch grains in the stele (Meyer, 1968). Supposedly, the fungi absorb carbohydrates from the tissues, thereby increasing the flow of soluble carbohydrates from stems to roots.

To further study the apparent sugar increase in roots after infection, Schweers and Meyer (1970) exposed mycorrhizal and nonmycorrhizal *Pinus sylvestris* seedlings to labeled carbon dioxide. The amount of radioactive carbon dioxide liberated by respiration in the region of the root system was measured after terminating the assimilation period. Forty percent of the assimilated $^{14}CO_2$ was detected in the mycorrhizal roots, and less than 10%, in nonmycorrhizal ones. When other parts of the plants were measured, they found that the assimilative activity increased with an increase in the frequency of mycorrhizae, but the shoot-to-root ratio of activity decreased. They interpreted this as support for the concept that rising levels of sugars in the root are the result of root infection by mycorrhizal fungi and not the cause.

Lister *et al.* (1968) also grew seedlings in the nitrogen and phosphorus regimes that were used by Björkman in 1942. Seedlings were exposed to $^{14}CO_2$ in a closed chamber. The root systems were later removed and analyzed. All the root systems contained sucrose, glucose, fructose, arabinose, and small amounts of unknown soluble carbohydrates; the major carbohydrate was sucrose. The highest amounts of radioactive glucose and fructose were detected in plants grown on moderate to very high phosphorus and nitrogen. In these nutrient groups where Björkman had found few or no mycorrhizae, mycorrhiza formation was completely inhibited.

Lister *et al.* then cultivated white pine (*Pinus strobus*) seedlings on Björkman's medium containing various nitrogen concentrations. They harvested the plants in the fall and collected mycorrhizae from seedlings grown on the two lowest nitrogen levels and nonmycorrhizal short roots from seedlings on highest nitrogen level. Chromatographic analyses showed that sucrose, glucose, and fructose were present in both types of root structures.

In reviewing the influence of auxins and other metabolites on mycorrhiza formation, Slankis (1971) concluded that soluble carbohydrates are not the sole factor in formation of ectomycorrhizae. Although mycorrhizal roots return to a nonmycorrhizal state when nitrogen and phosphorus concentrations in the nutrients are increased, the reverse process does not occur when there is a decrease or deficiency in soluble sugars in the roots. Slankis, however, conceded that there seemed to be a correlation between light intensity and mycorrhiza frequency, but the action of light was still not clear.

Harley (1969) also has not fully accepted Björkman's hypothesis. Citing the work of his student, D. H. Lewis, he doubted the validity of determining carbohydrate content by analyzing for reducing substances: first, because easily soluble reducing substances normally contain substances other than sugars, and second, because preparatory treatments could have resulted in formation of new and different reducing substances. Björkman's failure to estimate the presence of nonreducing disaccharides in mycorrhizae was also questioned. Harley also believed that the ferricyanide reduction and copper reduction techniques used by Björkman in 1942 were not specific to carbohydrates. Harley acknowledged the validity of the findings of many (Hatch, 1937; Björkman, 1942; Hacskaylo and Snow, 1959) that moderate deficiencies of essential major nutrients, especially N and P, promote mycorrhiza formation. However, he also stressed that the whole area of study—correlation of intensity of infection with carbohydrate level—needed reexamination.

As though in rebuttal to his critics, Björkman (1970) again presented data to support his carbohydrate hypothesis. He inoculated *Pinus sylvestris* seedlings with *Boletus bovinus* and *B. subtomentosus* and grew them from May until October in soil provided with different levels of nitrogen, phosphorus, and potassium, or distilled water. Light conditions were varied. One group received full daylight; two others received one-half or one-quarter daylight. Roots were examined for mycorrhizae, and samples of needles and roots were prepared for carbohydrate analyses at the same time of day. Björkman emphasized that samples were taken simultaneously and tested immediately, and stated that his precaution was necessary because the samples of carbohydrates varied during different times of the day. This time he used the reducing substances method described by Meyer (1962).

The level of reducing substances in needles was approximately the same regardless of nutrient conditions; however, it was uniformly lower in most shaded plants. Lower levels of reducing substances were also found in roots of shaded plants and, to some extent, in roots of plants fertilized with nitrogen and grown in full daylight or one-half daylight. Higher values were obtained in roots of plants grown on phosphorus fertilized soils.

Björkman stated that the addition of *B. subtomentosus* had little if any effect on the sugar content of mycorrhizal roots, thus refuting Meyer's (1962) theory. When experiments were repeated the following year, levels of reducing substances in the needles and roots were the same. Björkman stressed that misleading conclusions were probably drawn by comparing his experiments which were started in early spring, on natural forest soil, with those of Handley and Sanders (1962), which were on artificial substrate.

In the same paper, Björkman reported on the effect of varying periods of illumination on mycorrhiza-inoculated pine seedlings growing in fertilized or unfertilized raw humus. Some of the humus had been previously sterilized. The experiments were performed in growth chambers under artificial light. Three different photoperiods were used: 24-, 16-, and 8-hour illuminations. In plants on a 24-hour photoperiod in unfertilized, unsterilized soil, 36% of the short roots were transformed into mycorrhizae. On the 16-hour photoperiod, there was a transformation to 21% mycorrhizae, but on the 8-hour photoperiod, no mycorrhizae formed. Mycorrhizae also did not form on any seedling grown in sterilized humus. Reducing substances were clearly lower in roots grown in the 8-hour series in both sterilized and unsterilized humus.

According to Boullard (1961), increases in photoperiod from 6 to 16 hours or more increased the development of root systems of *Cedrus atlantica, Pinus pinaster*, and *P. sylvestris*. The numbers of short roots and their conversion to mycorrhizae also increased. When Hacskaylo (1969) grew *P. virginiana* seedlings inoculated with *Thelephora terrestris* on a 16-hour photoperiod and switched them to an 8-hour photoperiod, plants became dormant within 4 weeks. Long root growth and short root initiation, however, continued at a rate comparable to those which remained on a 16-hour regime. Short roots were unbranched and uninfected. In this experiment, formation of sporophores of the mycorrhizal fungus, *T. terrestris*, occurred on the long photoperiod, but either did not initiate or, if initiated, ceased to develop further when shifted to the short photoperiod. Although no analyses of carbohydrate content of the roots were made, the suppression of mycorrhizal formation on the short photoperiod may be indicative of low sugar content, as suggested by Björkman (1970).

V. Utilization of Carbon Compounds by Ectomycorrhizal Fungi

Melin (1925) first performed investigations to determine which carbon compounds were actually utilized by root-inhabiting fungi. In controlled laboratory experiments, the mycorrhizal fungi, *Boletus variegatus* and *B. elegans*, and nonmycorrhizal fungi, *Mycelium radicis sylvestris α*, *M.r. sylvestris β*, *M.r. sylvestris*, and *M.r. abietis*, were grown in pure culture on synthetic media containing various carbon compounds (Table II).

All grew well on glucose and to some extent, on maltose. *B. variegatus* was the only species that grew well on dextrin and the only one that did not on starch. *B. variegatus*, *B. elegans*, and *M.r. sylvestris α* utilized mannitol; *B. variegatus* and *M.r. sylvestris α*, lactose; and *B. elegans*, xylose. None of the fungi used inulin or cellulose. This was the only

TABLE II

UTILIZATION OF CARBON COMPOUNDS BY ECTOMYCORRHIZAL
AND OTHER ROOT-INHABITING FUNGI[a]

	Boletus variegatus	*Boletus elegans*	*M. radicis sylvestris α*	*M. radicis sylvestris β*	*M. radicis sylvestris*	*M. radicis abietis*
Xylose	−	+	−	−	−	−
Arabinose	−	−	−	−	−	−
Glucose	+	+	+	+	+	+
Mannitol	+	+	+	−	−	−
Sorbitol	−	−	−	−	−	−
Dulcitol	−	−	−	−	−	−
Glycerol	−	−	−	−	−	−
Lactose	+	−	+	−	−	−
Maltose	+	+	+	−	+	+
Starch	−	+	+	+	+	+
Inulin	−	−	−	−	−	−
Dextrin	+	−	−	−	−	−
Cellulose	−	−	−	−	−	−

[a] Data of Melin, 1925.

TABLE III

RELATIVE GROWTH OF *Tricholoma* SPECIES ON VARIOUS CARBOHYDRATES[a]

Carbohydrate	Saprophytic fungi[b]			Mycorrhizal fungi[b]			
	T. brevipes	T. fumosum	T. nudum	T. flavobr.	T. imbric.	T. pessund.	T. vaccinum
Xylose	4	7	71	57	7	22	1
Glucose	100	100	100	100	100	100	100
Fructose	79	86	104	71	55	120	22
Galactose	25	28	30	13	7	8	6
Mannose	85	103	85	73	121	78	74
Lactose	17	15	104	3	2	2	1
Maltose	36	67	104	21	12	23	11
Saccharose	18	103	112	47	17	6	20
Raffinose	7	97	94	2	2	2	1
"Start-glucose"	49	28	47	64	22	12	12
Starch	18	93	136	11	9	72	13
Starch + "Start-glucose"	54	83	133	93	45	117	50
Inulin	11	105	92	3	2	1	2
Inulin + "Start-glucose"	44	107	101	49	25	13	17
Cellulose	+	+	+	–	–	–	–
Cellulose + "Start-glucose"	+	+	+	–	–	–	+

[a] Data of Norkrans, 1950.
[b] Figures represent mycelial dry weights. Symbols (+ or –) represent visual estimates of mycelial growth.

information available on utilization of carbon compounds by mycorrhizal fungi until nearly 20 years later, when How (1940) confirmed the responses of *B. elegans* to some of the same carbon sources studied by Melin.

As part of an extensive study on *Cenococcum graniforme*, Mikola (1948) found that this distinctively black mycorrhizal fungus utilized sucrose and maltose equally as well as glucose. It grew on lactose to a limited degree. Dextrin, starch, and mannitol promoted mycelial growth, but cellulose did not. Mikola concluded that *C. graniforme* did not seem to be as exacting as many other mycorrhizal fungi.

Also using *C. graniforme*, Keller (1952) found that glucose, mannose, trehalose, cellobiose, and α-dextrin were equally effective as carbon sources, while sorbitol, galactose, β-dextrin, starch, inulin, and cellulose were not effective at all. It was interesting that growth of the fungus increased when maltose was added to β-dextrin, but not when it was added to α-dextrin.

The capacity of four mycorrhizal and three saprophytic species of *Tricholoma* to utilize carbohydrates was examined by Norkrans (1950). The mycorrhiza formers, *Tricholoma flavobrunneum, T. imbricatum, T. pessundatum*, and *T. vaccinum* grew only on simple sugars, primarily glucose, but the saprophytic species used cellulose as well (Table III). With the addition of a start sugar, in this instance glucose, *T. vaccinum* was able to decompose cellulose. Norkrans concluded that species of *Tricholoma* comprise two different ecological types—litter decomposers, and mycorrhiza formers. The difference between the types was considered to be quantitative rather than qualitative.

When critically examined, the data of several of these investigators may be misleading and incomplete. For example, very small quantities of impurities in test solutions could have acted as starters and affected the utilization of starch, cellulose, or other compounds. Sucrose, especially in media with low pH, could have been hydrolyzed, and its moieties, glucose and fructose, utilized instead. Also, the limited number of compounds in experiments and data which were not always directly comparable could have altered results. In view of these uncertainties, Palmer and Hacskaylo (1970) carefully studied the response of various fungi to a large number of carbon sources. Six species of ectomycorrhizal fungi, representing a wide spectrum of genera, were surface-grown on a liquid medium to which 39 soluble and 13 insoluble carbon sources, mainly carbohydrates at equivalents of 2 gm/liter, were added (Tables IV and V). The soluble carbon compounds were sterilized by filtration in aqueous solution to avoid hydrolysis of complex carbohydrates, and then added to an autoclaved basal medium; insoluble or heavily colloidal compounds were autoclaved.

TABLE IV

GROWTH OF ECTOMYCORRHIZAL FUNGI FOR 21 DAYS
AT 21°C ON SOLUBLE CARBON COMPOUNDS[a]

CARBON SOURCE		Amanita rubescens GROWTH (DRY WT. mg)	Cenococcum graniforme GROWTH (DRY WT. mg)	Rhizopogon roseolus GROWTH (DRY WT. mg)	Russula emetica GROWTH (DRY WT. mg)	Suillus cothurnatus GROWTH (DRY WT. mg)	Suillus punctipes GROWTH (DRY WT. mg)
		TRIOSE					
A	DL–Glyceraldehyde	4	5	2	4	3	1
		TETROSE and 4 – CARBON DERIVATIVE					
A	D–Erythrose	1	4	2	3	3	1
O	*i*–Erythritol	1	4	2	2	2	2
		PENTOSES and 5 – CARBON DERIVATIVES					
A	D(−)Arabinose	1	3	2	1	3	1
A	L(+)Arabinose	1	4	2	2	3	2
A	D–Lyxose	2	4	2	3	2	2
A	D(−)Ribose	1	5	2	4	4	3
A	D(+)Xylose	2	5	2	2	2	3
M	L(−)Rhamnose	1	4	2	4	3	2
M	L(+)Rhamnose	1	5	1	2	2	2
D	2–Deoxy–D–Ribose	1	4	2	2	3	1
		HEXOSES and 6 – CARBON DERIVATIVES					
A	D(+)Galactose	6	14	8	3	6	2
A	D–Glucose	40	44	35	33	57	30
A	D(+)Mannose	32	37	43	34	46	18
K	Fructose	42	23	16	29	6	18
K	L(−)Sorbose	1	5	2	2	2	3
D	D(+)Fucose	1	3	2	3	3	2
D	L –Fucose	1	4	2	3	2	2
D	2 Deoxy–D–Glucose	1	3	2	2	3	1
O	Inositol	1	4	2	3	3	2
O	Mannitol	13	4	21	22	5	7
O	Sorbitol	9	4	11	6	4	2
		HOMOGENOUS DISACCHARIDES					
G	Cellobiose	55	33	47	26	47	10
G	D(+)Maltose	5	11	24	5	7	5
G	D(+)Trehalose	34	16	42	7	31	17
		HETEROGENOUS DISACCHARIDES					
BG	α–Lactose	1	4	2	3	3	1
BG	β–Lactose	1	5	2	3	3	1
BG	Melibiose	2	4	2	3	3	3
FG	Sucrose	1	13	2	2	3	3
		HETEROGENOUS TRISACCHARIDES					
GFG	D(+)Melezitose	1	6	2	3	2	2
BFG	Raffinose	2	6	2	2	3	2
		HOMOGLYCANS					
G	Dextrin	4	9	24	3	7	3
G	Glycogen	4	11	10	3	5	2
G	Starch	3	4	15	4	6	3
		HETEROGLYCAN					
BPL	Pectin	43	18	37	31	20	32
		NON–CARBOHYDRATES					
S	Acetate	1	3	2	3	2	2
S	Citrate	1	4	4	2	3	4
S	Tartrate	1	4	2	3	3	2
E	Triacetin	1	4	2	3	3	3
		STANDARDS					
	D–Glucose	40	44	35	33	57	30
	Agar(0.5mg)+Mycelium	1	4	2	2	3	2

[a] Liquid media contain 2 gm carbon/liter. Dry weights for glucose rated 100%. (Symbols in Column 1: A = aldose, D = deoxy sugar, K = ketose, M = methyl sugar, D = alcohol.) (Data of Palmer and Hacskaylo, 1970.)

TABLE V

Growth of Ectomycorrhizal Fungi for 21 Days at 21°C on 13 Insoluble and 5 Soluble Carbon Sources[a]

CARBON SOURCE	Amanita rubescens GROWTH DRY WT. mg	Cenococcum graniforme GROWTH DRY WT. mg	Rhizopogon roseolus GROWTH DRY WT. mg	Russula emetica GROWTH DRY WT. mg	Suillus cothurnatus GROWTH DRY WT. mg	Suillus punctipes GROWTH DRY WT. mg
MONOSACCHARIDE DERIVATIVE						
D-Glucose pentaacetate	1	4	6	2	2	1
OLIGOSACCHARIDE DERIVATIVES						
Cellobiose octaacetate	2	5	4	2	4	3
Sucrose octaacetate	2	8	4	2	2	4
HOMOGLYCANS						
Cellulose (cotton)	1	3	3	2	2	2
Cellulose (ECTEOLA)	3	4	5	2	3	1
Cellulose (paper)	1	3	2	2	2	1
Glycogen	4	7	9	4	4	1
Inulin	6	12	15	10	5	1
Starch	4	7	9	2	3	1
Xylan	2	4	2	2	3	2
HETEROGLYCAN						
Pectin	18	14	20	9	14	7
POLYSACCHARIDE DERIVATIVES						
Cellulose acetate	2	4	4	2	2	1
Chitin	2	6	5	2	3	2
Hesperidin	2	3	6	3	3	1
Polygalacturonic acid	2	4	5	2	2	2
Quercetin	2	4	5	2	2	1
STANDARDS						
D-Glucose	18	54	42	13	23	39
Agar (0.5mg)+Mycelium	2	3	3	2	2	1

[a] Contain 4 gm carbon/liter. Dry weights for glucose rated 100%. (Data of Palmer and Hacskaylo, 1970.)

Individual isolates of the fungi used as many as 22, or as few as 11 of the carbon sources. Among the soluble monosaccharides, only two, D-glucose and D-mannose, supported growth that would substantially increase the dry weight of all isolates above that of the controls. All 6 isolates used the ketohexose fructose, though *Suillus cothurnatus* grew very poorly. Three isolates grew on the hexose alcohol, sorbitol; and four, on mannitol. Only on the latter did dry weight of any isolate exceed 50% of the standard. The three homogeneous disaccharides supported mycelial growth in all isolates, and cellobiose and trehalose were better than maltose (Table IV). *Cenococcum graniforme* was the only isolate to grow on heterogeneous di- or trisaccharides, sucrose, melezitose, and raffinose, and none utilized either form of lactose. Except for *S. punctipes*, each fungus grew on one or more of the soluble homoglycans; *Rhizopogon roseolus* grew well on all three. Of the soluble and insoluble polysaccharides in carbohydrate derivatives, pectin alone increased growth of all fungi (Table V). The only other readily metabolized compounds were the homoglycans, glycogen and starch, and the heteroglycan, inulin. No other compound was used even slightly by more than two fungi.

Utilization of sucrose was poor throughout the experiment. In 21 days, only the dry weight of *C. graniforme* increased; in 28 days, dry weights of *Amanita rubescens, R. roseolus,* and *Russula emetica* also increased, but neither isolate of *Suillus* grew at all.

Some isolates of mycorrhizal fungi grew better on carbon compounds other than glucose than on glucose. For example, compared to growth on glucose, dry weights of *Rhizopogon roseolus* increased 122% on mannose, 134% on cellobiose, 120% on trehalose, and 105% on pectin. However, since the availability of these compounds in natural systems is questionable, certain of these data may have more value in the laboratory than in the forest.

These data seem to indicate that broad generalizations should be avoided concerning utilization of sucrose, fructose, glycogen, and mannitol by all ectomycorrhizal fungi *in vitro.* Although these compounds may have significance in mycorrhizal systems (Lewis and Harley, 1965a,b,c), it is apparent that certain fungi, at least in the absence of their host, do not readily utilize all of them.

It is suspected that as hyphae of ectomycorrhizal fungi grow between the cortical cells, pectinase is secreted, hydrolyzing the pectin. Pectin has been found to be a good carbon source for fungi, at least when no other source is provided (Palmer and Hacskaylo, 1970). How (1940) recorded marked increases in growth of *Boletus elegans* on alcohol-washed pectin. However, availability of pectin in forest litter is probably very small, since it is utilized promptly by soil saprophytes, particularly bacteria. That pectinase has never been detected in assays with mycorrhizal fungi as it has been with lignicolous hymenomycetes might result from suppression of this enzyme in the presence of soluble carbohydrates (Lyr, 1963).

Production of amylase, xylanase, and mannase (Ritter, 1964; Lyr, 1963) and utilization of one or more carbon sources have been recorded for hymenomycetous species that are most probably mycorrhizal (Ferry and Das, 1968; Iwamoto, 1962; Jayko *et al.,* 1962; Meloche, 1961, 1962; Rawald, 1962).

Detection of cellulolytic enzymes in mycorrhizal fungi may depend upon sensitivity of the assays, since repeated results vary considerably. Norkrans (1950) and Rawald (1962) generally found that several saprophytic species of *Tricholoma* produce cellulase and that the mycorrhizal species do not. Ritter (1964), however, reported production of cellulase by all species of *Tricholoma* examined. Lyr (1963), while studying other genera, generally found smaller amounts of cellulolytic enzymes in mycorrhizal fungi than in wood decay and litter-decomposing fungi. Among the mycorrhizal fungi, *Boletus subtomentosus, B. luteus, B. variegatus,* and *Amanita citrina* exhibited some cellulolytic activity, but *B. badius* and *A. muscaria* did not. All were able to attack hemicelluloses.

The ability of *T. fumosum* to produce cellulase and utilize cellulose was demonstrated in Norkrans' (1950) experiments. This fungus was also able to form ecto- and ectendomycorrhizae. Thus, Norkrans proposed that as long as the mycorrhizal mycelium obtains glucose from the host plant, cellulase production is suppressed. However, when the host no longer produces a surplus of simple carbohydrates, the cell wall—a cellulose substrate—induces an increase in cellulase production. This, she hypothesized, may furnish a possible explanation of the formation of ectendomycorrhizae at least in this particular instance. According to Mikola (1965) and Wilcox (1971) most ectendomycorrhizae are the result of infections by fungi that apparently belong to an entirely different group of fungi than those forming ectomycorrhizae. *Tricholoma* may be one example of an exception (Norkrans, 1950). As a group, however, ectomycorrhizal fungi are generally unable to decompose litter and other naturally occurring complex carbohydrates and obtain their required carbon compounds from the break-down products.

A few attempts have been made, with limited success, to show that ectomycorrhizal fungi do not always depend upon carbon from their hosts. Young (1947) found that a species of *Boletus,* which formed mycorrhizae with *Pinus taeda,* digested cellulose. He concluded that soluble reducing substances were released by the fungus which benefited the host, but provided little evidence that an autotrophic host would benefit substantially from it.

Lindeberg (1948) discovered that an isolate of *B. subtomentosus,* which is ectomycorrhizal on pine, could decompose and grow readily on sterilized litter. This isolate, as well as the mycorrhiza-forming fungus *Lactarius deliciosus,* was capable of forming polyphenol oxidase just as litter-decomposing hymenomycetes do. This characteristic, however, does not appear to be general among ectomycorrhizal fungi.

MacDougall and Dufrenoy (1944) claimed that pine roots in soil continued to grow and develop new mycorrhizae following detachment from shoots. They thought that mycorrhizal fungi were absorbing carbohydrates from organic matter in the soil and transferring them to the roots, thereby supplying necessary carbon compounds that otherwise are supplied through photosynthesis. It is more probable, however, that carbohydrates had accumulated in the root before the seedlings were decapitated.

VI. Carbohydrate Transformation

A. Absorption and Metabolic Pathways

A very sophisticated approach to carbohydrate metabolism of intact mycorrhizae was developed at Oxford University in England. Harley and his students and co-workers employed radioactive techniques and biochemical analyses to study the transfer of carbon compounds to mycorrhizal fungi and the nature of the compounds that are metabolized or stored in the host and in the fungus. Primarily using excised beech (*Fagus*) mycorrhizae, they studied the metabolism of the mycorrhiza as a physiological unit.

In an early study, Harley and Jennings (1958) collected beech mycorrhizae at various times of the year and measured their rates of respiration in several sugar solutions. The mycorrhizae preferentially absorbed sugars from solutions containing glucose, fructose, and sucrose. Glucose was most readily absorbed. The sucrose molecule was not absorbed, but was rapidly hydrolyzed. After hydrolysis, the glucose moiety was preferentially absorbed and the fructose tended to remain in the ambient solution (Fig. 2). As the hexoses were absorbed, sucrose and polysaccharides inside the tissues increased, indicating that the monosaccharides were being assimilated into disaccharides and insoluble carbohydrates.

In analyses of excised beech mycorrhizae and nonmycorrhizal roots, trehalose and mannitol were detected in mycorrhizae, but not in uninfected roots (Lewis and Harley, 1965a). When uninfected roots were immersed in solutions of these compounds, very small amounts were utilized. This led to the conclusion that ectomycorrhizal fungi obtain sugars from the host through a concentration gradient which is established and maintained by the conversion of sugars in hyphae into forms that *Fagus* uses poorly (i.e., trehalose) or not at all (i.e., mannitol). The principal carbon sources involved in metabolism of the fungus within the *Fagus* root apparently were sucrose, glucose, and fructose.

In another study, Lewis and Harley (1965b) stripped the mycelial sheath from the core of *Fagus* mycorrhizae and studied the absorption of carbohydrates. The core responded in the same manner as uninfected roots. Analyses showed it contained sucrose synthesized from glucose and fructose, but little in the way of insoluble carbohydrates. The sheath, however, contained only trehalose, mannitol, and insoluble glycogen.

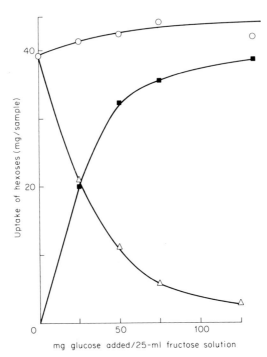

Fɪɢ. 2. Uptake of hexoses by beech mycorrhizae from solutions containing a mixture of fructose and glucose. Total hexose uptake (○), fructose uptake (△), and glucose uptake (■). (Data of Harley and Jennings, 1958; by permission of The Royal Society.)

Lewis and Harley (1965c) studied the rate of absorption of sucrose in mycorrhizal roots. It was approximately two and a half times faster than that of uninfected roots. The presence of active surface invertase on the uninfected root indicated that uptake of sucrose by those roots involved prior hydrolysis. Using [^{14}C]glucose and [^{14}C]fructose, they traced the destination of the glucose to trehalose and glycogen, and of fructose to mannitol. Their suggested diagram of the metabolic pathway for incorporation of glucose and fructose in mycorrhizal roots of beech is illustrated in Fig. 3.

Lewis and Harley's discovery that the fungi in ectomycorrhizae hydrolyze sucrose contrasts to some extent with findings of Palmer and Hacskaylo (1970), who worked only with pure carbon sources *in vitro*. In that study, it was not until the fungi were subjected to sucrose at the concentration of 6 gm of carbon/liter for 21 days that all of these fungi showed indication of growth on this sugar. This suggests possible increased activation of an adaptive enzyme for sucrose utilization. These

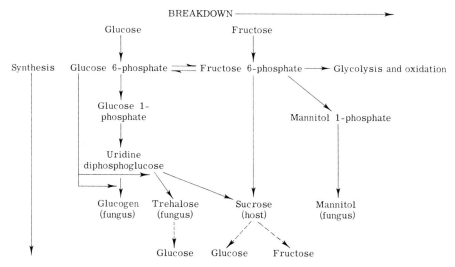

Fig. 3. Possible pathways of hexose utilization by mycorrhizal roots of beech.
(Data of Lewis and Harley, 1965b, by permission of *The New Phytologist.*)

authors have expressed reservations about broad generalizations on utilization of sucrose, fructose, glycogen, and mannitol by all ectomycorrhizal fungi *in vitro*. Whether the metabolic activities within all ectomycorrhizae require stimulation by the host or enzymes from both associates is still unclear.

B. RESPIRATION

The absorption of ions by mycorrhizae is accomplished through the expenditure of energy derived from metabolism of carbohydrates (Harley, 1969). Routien and Dawson (1943) attempted to demonstrate that in pines the increase in the uptake of cations was related to increased respiration of the mycorrhizae.

They postulated that an enhanced capacity of plants for absorption of cations from soil colloids must involve an enhanced capacity for excretion of hydrogen ions. Supposedly these hydrogen ions result from dissociation of carbonic acid which is formed from carbon dioxide in mycelial respiration. Although they were unable to obtain direct evidence to show that there was a relationship between mycorrhizal formation and increased salt absorption, they believed it was theoretically a distinct possibility. McComb (1943) also suggested that increased nutrient absorption by mycorrhizae might be explained by a higher respiration rate.

Hacskaylo *et al.* (1965) studied respiration rates of six ectomycorrhizal fungi at various temperatures in pure culture (Fig. 4). Each species

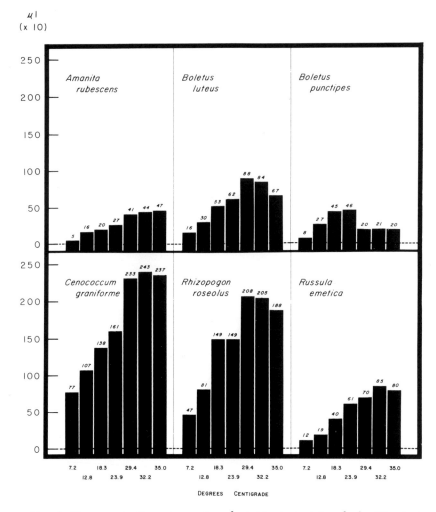

FIG. 4. Mean respiration rates measured at 10-minute intervals for 90 minutes, expressed in μliters of oxygen uptake per gram of fresh weight per hour. (Data of Hacskaylo *et al.*, 1965.)

displayed individual characteristics in respiration patterns over a range of seven temperatures. These authors found that the maximum rate of respiration did not always coincide with the temperature at which the greatest amount of growth occurred. For example, during a 24-day period, the optimum temperature for mycelial development of *Rhizopogon roseolus* was 18°C; however, maximum respiration occurred at 29°C.

In respiration studies on *Pinus taeda* roots, oxygen absorption of

mycorrhizal roots was about 1.4 times greater than that of nonmycorrhizal roots (Kramer and Hodgson, 1954). Harley *et al.* (1956) demonstrated that respiration of mycorrhizal roots is sensitive to temperature
and that oxygen consumption depends upon oxygen supply. Sodium
azide stimulated carbon dioxide production in the presence of oxygen
and inhibited it under anaerobic conditions (Harley and McCready,
1953).

In further studies Harley and ap Rees (1959) found that azide, cya-
nide, carbon monoxide, and dinitrophenol stimulated respiration in beech
mycorrhizae and that these same chemicals, excluding dinitrophenol,
inhibited respiration in isolated host tissue. The authors concluded that
respiratory and electron transport systems in mycorrhizae are mediated
by cytochrome oxidase. Harley and Jennings (1958) reported that as
hexoses are absorbed there is an increase of respiration similar to that
during absorption of nutrient salts. They suggested that respiratory
effects of sugars and salts may be due to increased availability of ADP
from which ATP can be reformed. They concluded that carbohydrate
utilization by mycorrhizae is largely confined to the fungal sheath.

VII. Conclusion

In nature, ectomycorrhizal fungi depend primarily upon the roots of
their hosts for carbohydrates, usually sucrose, glucose, and fructose.
Certain species of fungi may, however, possess enzymes to hydrolyze
cellulose and other complex carbohydrates, but this characteristic does
not appear to be widespread.

Soluble carbohydrates of roots must be considered a major factor in
the formation of ectomycorrhizae, the growth of vegetative hyphae, and
sporophore development. Thus, factors that change the availability of
root sugars, photosynthetic activity, photoperiod of the host, availability
of soil nutrients, particularly nitrogen and phosphorus, auxin production
by associated fungi, and probably other factors, are of major importance
to carbohydrate metabolism of associated fungi.

Translocation of sugars from root tissue to the fungus sheath and
subsequent accumulation there, as mannitol, trehalose, and glycogen,
tend to conserve a supply of carbohydrates for the metabolic processes
in the fungi. Also, movement back to the host has been indicated in some
intact systems.

Respiration studies indicate a cytochrome oxidase system exists in the
fungus tissue as well as the host. In ectomycorrhizae, utilization of carbohydrates apparently is greater in the fungal mantle than in the root tissue.

Variations in experimental procedures used by investigators have made duplication of data and adequate interpretations difficult. There is little doubt that more research is needed to clarify several controversial points of view.

References

Björkman, E. (1942). Über die Bedingungen der Mykorrhizabildung bei Kiefer und Fichte. *Symb. Bot. Upsal.* **6**, 1–191.

Björkman, E. (1944). The effect of strangulation on the formation of mycorrhiza in pine. *Sv. Bot. Tidskr.* **38**, 1–14.

Björkman, E. (1960). *Monotropa hypopitys* L. an epiparasite on tree roots. *Physiol. Plant.* **13**, 308–329.

Björkman, E. (1970). Mycorrhiza and tree nutrition in poor forest soils. *Stud. Forest. Suec.* **83**, 1–24.

Boullard, B. (1961). Influence du Photopériodisme sur le Mycorrhization de Jeunes Conifères. *Bull. Soc. Linn. Normandie, Sér.* **10**, **2**, 30–46.

Ferry, B. W., and Das, N. (1968). Carbon nutrition of some mycorrhizal *Boletus* species. *Trans. Brit. Mycol. Soc.* **51**, 795–798.

Frank, A. B. (1885). Ueber die auf Wurzelsymbiose beruhende Ernährung gewisser Bäume durch Unterirdische Pilze. *Ber. Deut. Bot. Ges.* **3**, 128–145.

Hacskaylo, E. (1965). *Thelephora terrestris* and mycorrhizae of Virginia pine. *Forest Sci.* **11**, 401–404.

Hacskaylo, E. (1969). Unpublished experiments.

Hacskaylo, E., and Bruchet, G. (1972). Hebelomas as mycorrhizal fungi. *Bull. Torrey Bot. Club* **99**, 17–20.

Hacskaylo, E., and Snow, A. G. (1959). Relation of soil nutrients and light to prevalence of mycorrhizae. *Northeast. Forest Serv. Sta., Pap.* **125**, 1–13.

Hackaylo, E., Palmer, J. G., and Vozzo, J. A. (1965). Effects of temperature on growth and respiration of ectotrophic mycorrhizal fungi. *Mycologia* **57**, 748–756.

Handley, W. R. C., and Sanders, C. J. (1962). The concentration of easily soluble reducing substances in roots and the formation of ectotrophic mycorrhizal associations. A re-examination of Björkman's hypothesis. *Plant Soil* **16**, 42–61.

Harley, J. L. (1969). "The Biology of Mycorrhiza," 2nd ed. Leonard Hill, London.

Harley, J. L., and ap Rees, T. (1959). Cytochrome oxidase in mycorrhizal and uninfected roots of *Fagus sylvatica*. *New Phytol.* **58**, 364–386.

Harley, J. L., and Jennings, D. H. (1958). The effect of sugars on the respiratory responses of beech mycorrhiza to salts. *Proc. Roy. Soc., Ser. B* **148**, 403–418.

Harley, J. L., and McCready, C. C. (1953). A note on the effect of sodium azide on the respiration of beech mycorrhiza. *New Phytol.* **51**, 342–344.

Harley, J. L., McCready, C. C., Brierley, J. K., and Jennings, D. H. (1956). The salt respiration of excised beech mycorrhizas. II. *New Phytol.* **55**, 1–28.

Hatch, A. B. (1937). The physical basis of mycotrophy in the genus *Pinus*. *Black Rock Forests Bull.* **6**, 1–168.

How, J. E. (1940). The mycorrhizal relations of larch. I. A study of *Boletus elegans* (Schum.) in pure culture. *Ann. Bot. (London)* [N. S.] **4**, 135–150.

Iwamoto, R. (1962). Pure culture of the mycelia of *Russula pseudodelica* and *Lactarius chrysorrheus*. *Trans. Mycol. Soc. Jap.* **3**, 133–136.

Jayko, L. G., Baker, T. I., Stubblefield, R. D., and Anderson, R. F. (1962). Nutrition and metabolic products of *Lactarius* species. *Can. J. Microbiol.* **8**, 361–371.

Keller, H. G. (1952). Untersuchungen über das Wachstum von *Cenococcum graniforme* (Sow.) Ferd. et. Winge auf verschiedenen Kohlenstoffquellen. Thesis, Juris-Verlag, Zurich.

Kramer, P. J., and Hodgson, R. H. (1954). Differences between mycorrhizal and non-mycorrhizal roots of loblolly pine. *Proc. Int. Bot. Congr., 8th, 1954* Vol. 13, pp. 133–134.

Lewis, D. H., and Harley, J. L. (1965a). Carbohydrate physiology of mycorrhizal roots of beech. I. Identity of endogenous sugars and utilization of exogenous sugars. *New Phytol.* **64**, 224–237.

Lewis, D. H., and Harley, J. L. (1965b). Carbohydrate physiology of mycorrhizal roots of beech. II. Utilization of exogenous sugars by uninfected and mycorrhizal roots. *New Phytol.* **64**, 238–255.

Lewis, D. H., and Harley, J. L. (1965c). Carbohydrate physiology of mycorrhizal roots of beech. III. Movement of sugars between host and fungus. *New Phytol.* **64**, 256–269.

Lindeberg, G. (1948). On the occurrence of polyphenol oxidases in soil-inhabiting Basidiomycetes. *Physiol. Plant.* **1**, 196–205.

Lister, G., Slankis, V., Krotkov, G., and Nelson, C. D. (1968). The growth and physiology of *Pinus strobus* L. seedlings as affected by various nutritional levels of nitrogen and phosphorus. *Ann. Bot. (London)* [N. S.] **32**, 33–43.

Lyr, H. (1963). Zur Frage des Streuabbaus durch ectotrophe Mykorrhizapilze. *Int. Mykorrhiza Symp., 1960* pp. 123–142.

McComb, A. L. (1943). Mycorrhizae and phosphorus nutrition of pine seedlings. *Bull. Iowa Agr. Exp. Sta.* **314**, 582–612.

MacDougal, D. T., and Dufrenoy, J. (1944). Study of symbiosis of Monterey pine with fungi. *Yearb. Amer. Phil. Soc.* pp. 170–174.

Melin, E. (1925). "Untersuchungen über die Bedeutung der Baummykorrhiza. Eine ökologische physiologische Studie." Fischer, Jena.

Melin, E., and Nilsson, H. (1957). Transport of C^{14}-labelled photosynthate to the fungal associate of pine mycorrhiza. *Sv. Bot. Tidskr.* **51**, 166–186.

Meloche, H. P., Jr. (1961). The metabolism of ribose-5-phosphate by cell-free extracts of *Lactarius torminosus*. *Biochim. Biophys. Acta* **51**, 586–588.

Meloche, H. P., Jr. (1962). Enzymatic utilization of glucose by a Basidiomycete. *J. Bacteriol.* **83**, 766–774.

Meyer, F. H. (1962). Die Buchen und Fichtenmykorrhiza in verschiedenen Bodentypen, ihre Beeinflussung durch Mineraldünger sowie für die Mykorrhizabildung wichtige Faktoren. *Mitt. Bundesforschungsant. Forst-Holzwirt.* **54**, 1–73.

Meyer, F. H. (1968). Auxin relationships in symbiosis. "Transport of Plant Hormones," pp. 320–330. North-Holland Publ., Amsterdam.

Mikola, P. (1948). On the physiology and ecology of *Cenococcum graniforme*. *Commun. Inst. Forest. Fenn.* **36**, 1–104.

Mikola, P. (1965). Studies on the ectendotrophic mycorrhiza of pine. *Acta Forest. Fenn.* **79**, 1–56.

Norkrans, B. (1950). Studies in growth and cellulolytic enzymes of *Tricholoma*. *Symb. Bot. Upsal.* **11**, 1–126.

Palmer, J. G., and Hacskaylo, E. (1970). Ectomycorrhizal fungi in pure culture. I. Growth on single carbon sources. *Physiol. Plant.* **23**, 1187–1197.

Rawald, W. (1962). Zur Abhängigkeit des Mycelwachstums höheren Pilze von der Versongung mit Kohlenhydraten. Z. *Allg. Mikrobiol.* **2**, 303–313.

Reid, C. P. P. (1971). Transport of C^{14}-labelled substances in mycelial strands of *Thelephora terrestris. In* "Mycorrhizae" (E. Hacskaylo, ed.), USDA Forest Serv. Misc. Publ. No. 1189, pp. 222–227. US Govt. Printing Office, Washington, D. C.

Ritter, G. (1964). Vergleichende Untersuchungen über die Bildung von Ektoenzymen durch Mykorrhizapilze. Z. *Allg. Mikrobiol.* **4**, 295–312.

Rommell, L. G. (1938). A trenching experiment in spruce forest and its bearing on problems of mycotrophy. *Sv. Bot. Tidskr.* **32**, 87–99.

Routien, J. B., and Dawson, R. F. (1943). Some interrelationships of growth, salt absorption, respiration and mycorrhizal development in *Pinus echinata. Amer. J. Bot.* **30**, 440–451.

Schweers, W., and Meyer, F. H. (1970). Einfluss der Mykorrhiza auf der Transport von Assimilation in die Wurzel. *Ber. Deut. Ges.* **83**, 109–119.

Slankis, V. (1971). Formation of ectomycorrhizae of forest trees in relation to light, carbohydrates, and auxins. *In* "Mycorrhizae" (E. Hacskaylo, ed.), USDA Forest Serv. Misc. Publ. No. 1189, pp. 151–167. US Govt. Printing Office, Washington, D. C.

Smith, W. H. (1969). Release of organic materials from the roots of tree seedlings. *Forest Sci.* **15**, 138–143.

Wilcox, H. E. (1971). Morphology of ectendomycorrhizae in *Pinus resinosa. In* "Mycorrhizae" (E. Hacskaylo, ed.), USDA Forest Serv. Misc. Publ. No. 1189, pp. 54–68. US Govt. Printing Office, Washington, D. C.

Woods, F. W., and Brock, K. (1964). Interspecific transfer of Ca^{45} and P^{32} by root systems. *Ecology* **45**, 886–889.

Young, A. E. (1947). Carbohydrate absorption by the roots of *Pinus taeda. Queensl. J. Agr. Sci.* **4**, 1–6.

CHAPTER 7

Hormonal Relationships in Mycorrhizal Development[1]

V. SLANKIS

[1] The author's unpublished data discussed in this chapter comprise results obtained in Sweden at the Institute of Physiological Botany, University of Uppsala and in Canada at the Forest Research Laboratory, Canada Department of Forestry, Maple, Ontario.

I. Introduction

Nielsen (1930) was the first to show that an auxin which he named "rhizopin," was produced by fungi in pure culture. This metabolite was later identified by Thimann (1935) as indoleacetic acid (IAA). Since then, it has become apparent that numerous fungi and bacteria liberate IAA extracellularly. A comprehensive review on the fungi which produce auxin has been published by Gruen (1959). Subsequently there have been many reports about mycorrhizae-forming fungi and other fungi which produce auxin.

Already in the early 1930's it was known that invasion of higher plant tissues by certain bacteria and fungi creates hyperauxiny and induces cytogenesis in the invaded tissues. Thimann (1936) showed that legume nodules are active auxin-forming centers which produce considerable amounts of IAA. Hyperauxiny in crown galls was observed by Link and Eggers (1941). In ectomycorrhizae of *Pinus radiata*, MacDougal and Dufrenoy (1944) reported an abundance of an indole compound, which they presumed was IAA. The presence of several indole compounds in mycorrhizae of *P. sylvestris* and *P. strobus* was subsequently reported by Subba-Rao and Slankis (1959, partly published data).

From the data accumulated on growth hormones produced by ectomycorrhizae-forming fungi, it is evident that benefits to the higher plant provided by the symbiotic fungus are not limited to supplying inorganic and organic nutrients from soil. The fungal symbiont also provides the host plant with growth hormones, including auxins, cytokinins, gibberellins, and growth-regulating B vitamins. Since these hormones are homologous to those formed endogenously by the host plant, the latter may have excessive amounts of hormonal regulators. The above-normal levels of these potent substances must greatly influence growth and development of the host plant.

Auxins and cytokinins have been shown to mobilize nutrients and control their translocation in higher plants (cf. Thimann, 1969, 1972; Fox, 1969; Skoog and Schmitz, 1972). Cytokinins have also been reported to enhance chloroplast development (Skoog and Schmitz, 1972). In connection with studies on ectomycorrhizae, the influence of superoptimal concentrations of auxin on host plant root development and morphogenesis have been investigated more extensively than have the effects of other growth hormones at increased concentrations.

This chapter will deal mainly with hormone production by symbiotic fungi in pure culture, the influence of hormones on development of roots of the host plant, and the role of hormones in formation of mycorrhizal relationships.

II. Growth Hormones and Growth Regulators Produced by Symbiotic Fungi

The effect of extracellular growth-promoting metabolites liberated by mycorrhizal fungi on host plants has been observed in field experiments, pot cultures, and *in vitro*. Formation of these metabolites by symbiotic fungi has been demonstrated in axenic cultures under strictly controlled laboratory conditions.

A. Field Experiments and Pot Cultures

Rayner (1934, 1939; Rayner and Neilson Jones, 1944; see also Rayner's unpublished data reported by Levisohn, 1956) studied the cause of toxicity on forest tree seedlings in the nutritionally poor Wareham soil at Dorset, England. In these experiments it was established that before the actual mycorrhizal infection occurred, free-living *Rhizopogon luteolus* in the rhizosphere stimulated growth and lateral root initiation of Sitka spruce (*Picea sitchensis*). Levisohn (1953, 1956) observed a similar phenomenon under greenhouse conditions with pot cultures using Wareham soil. In the control pots, 50% of the ash (*Fraxinus excelsior*) seedlings died 4 months after sowing and the survivors were of extremely poor quality. Conversely, no losses occurred in seedlings inoculated with an endotrophic endophyte introduced by means of surface-sterilized mycorrhizal ash roots. Growth and color of the inoculated seedlings improved steadily, despite the fact that only at the end of the growing season were few mycorrhizal infections observed. Similar results were obtained with *Betula verrucosa* seedlings inoculated with a pure culture of *Boletus scaber*. Again, before actual infection, shoot growth of seedlings was stimulated, their leaf areas were enlarged, and color of foliage was improved. Furthermore, Levisohn (1956) convincingly demonstrated that a free-living mycelium of ectomycorrhizal fungi may appreciably stimulate the growth of woody plants known to symbiose with typical endomycorrhizae-forming fungi. Thus, ectomycorrhizae-forming *Rhizopogon luteolus*, without establishing a symbiotic relationship, stimulated the growth of *Chamaecyparis Lawsoniana* and *Thuja plicata*. Similarly, inoculation with *Boletus scaber* changed the pale-green color of *Robinia Pseudoacacia* to dark green within 6 weeks after sowing, and, after an additional 9 weeks, growth stimulation was also observed.

B. AXENIC CULTURE STUDIES

Lindquist (1939) obtained considerable growth promotion in aseptically grown spruce seedlings by supplementing their nutritional medium
with culture filtrates of symbiotic fungi. Compared with controls, seedlings which received 1 ml per flask of the culture filtrate from ectomycor-
rhiza-forming *Mycelium radicis nigrostrigosum* or the filtrate from an
unidentified ectomycorrhizae-forming fungus, produced more and longer
needles and had higher dry weight of shoots and roots. However, the
stimulatory substance(s) in the culture solutions of the fungi was not
identified.

Experimenting with excised root cultures of pine, Slankis (1948)
demonstrated that the addition of 2–8 ml of cell-free culture solution of
Suillus (*Boletus*) *luteus* and *S.* (*B.*) *variegatus* to 20 ml of nutrient solution used to raise excised *Pinus sylvestris* roots, stimulated lateral initiation and induced root morphogenesis, similar to that which nonmycorrhizal pine roots undergo during conversion into mycorrhizae (Fig.
1). Shemakhanova (1962) also reported that culture filtrates of *Suillus*
(*Boletus*) *luteus* and *S.* (*B.*) *bovinus* induce dichotomy resembling
ectomycorrhizae both in excised pine roots and in roots of 10-week-old
pine seedlings. In the older excised root cultures, having established
laterals and sublaterals, the dichotomous branching was more intensive.
A 0.5–1.0 ml supplement of the fungus culture filtrate proved to be most
effective. New lateral initiation was particularly stimulated by *S.* (*B.*)
bovinus filtrate; addition of 1 ml induced earlier root differentiation and
lateral formation in greater numbers than a supplement of 2 ml. Culture
filtrates of fungi isolated from mycorrhizal root surfaces were also rather
stimulatory to lateral initiation but failed to induce dichotomy. By contrast, a filtrate from *Phallus impudicus* was totally ineffective.

Using aseptically grown, excised pine roots at the radicle stage of development and more advanced stages of growth, Turner (1962) tested
the effect of culture filtrates from 7 Phycomycetes, 24 Fungi Imperfecti
and 22 Basidiomycetes of which 7 were mycorrhizal, 13 nonmycorrhizal
and two of unknown status. The excised radicles received 5 ml of the
fungus culture filtrate; the older root cultures received 10 ml. Some 92%
of the fungi tested produced exudates that influenced the roots but great
diversity was observed in the effect on initial elongation of radicles, initiation of laterals, and dichotomous branching. The seven mycorrhizal fungi
tested were no exception in this respect. In experiments with excised
roots of *Quercus alba,* Wargo (1966) inhibited root elongation by adding
exudates of mycorrhiza-forming *Russula amygdaloides* and *Tricholoma
personatum,* but not with *Armillaria mellea* (a suspected symbiotic

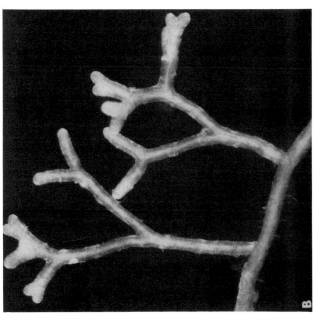

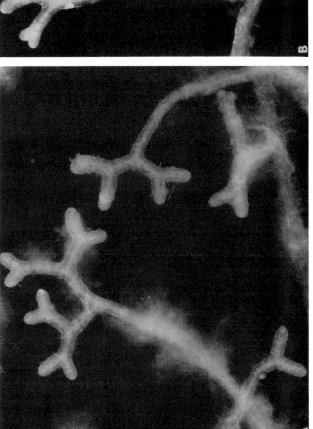

FIG. 1. Mycorrhizae-like morphology induced in excised roots of *Pinus sylvestris* L. by inoculation with mycorrhizal fungi or by addition of their exudates. The roots were cultivated in nutrient solution under aseptic conditions. (A) Dichotomy, caused by inoculation with *Suillus* (*B.*) *luteus*. ×16. (B) Dichotomy, caused by the exudates of *Suillus* (*B.*) *variegatus* added to the nutrient solution. ×16. [From V. Slankis (1948) by permission of *Physiol. Plant.*]

associate with *Q. alba*) or with the ectomycorrhizal *Cenococcum graniforme.*

C. The Nature of Extracellular Growth Hormones and Growth Regulators of Mycorrhizal Fungi

1. Auxins

Nielsen (1932) first reported that extracts from dried fruiting bodies of the mycorrhizal fungus *Boletus edulis* induced positive curvature of *Avena* coleoptiles. MacDougal and Dufrenoy (1944) by cytochemical means demonstrated that excess auxin was present in the stem mycorrhiza of *Corallorhiza maculatum* orchids and in mycorrhizae of *Pinus radiata*. In pine mycorrhizae, they discovered abundant auxin in the fungal hyphae and, particularly, in the fungal mantle. Auxin was also present in the vacuoles of hyphae intermeshed with cortical cells and in the endodermis and pericycle of the roots. On the basis of the color reaction induced by hydrochloric acid and iron perchloride reagent, they assumed that the indole compound present was IAA. The fact that crushed root tissues from fungus-free roots did not produce a color reaction was taken as evidence that this auxin was produced by the fungus and had been translocated to the roots. Subba-Rao and Slankis (1959, partly published data) also observed intensive coloration with Ehrlich's reagent in mycorrhizae and in the swollen terminal regions of mycorrhizae-bearing mother roots of *Pinus resinosa* and *P. strobus* nursery seedlings. However, no IAA was detected in the extracts by paper chromatography. Instead, relatively large quantities of a number of unidentified indole compounds were present. Conversely, short roots of aseptically grown pine seedlings did not produce any color reaction. These findings suggest the possibility that the color reaction observed by MacDougal and Dufrenoy (1944) might have been produced by indole compounds other than IAA. However, it is also possible that, in the root samples tested by Subba-Rao and Slankis, the accumulated auxin was in a bound state as the roots were collected from the nursery rather early in the spring. Eliasson (1971) reported that in root samples of *Populus tremula* auxin was present in samples collected from May until October, whereas in November, roots were auxin-deficient. The highest auxin levels in the samples were found in June and July. Slankis (1949, 1950, 1951, 1958) induced in excised and attached root systems of *Pinus sylvestris* ectomycorrhizae-like root structures (Fig. 2) by adding to the nutrient solution synthetic such auxins as indoleacetic acid (IAA), indolepropionic acid (IPA), indolebutyric acid (IBA), and naphthaleneacetic acid (NAA). Since similar morphological structures in excised pine roots had been previously induced by culture filtrates of mycorrhizae-forming fungi (Slankis, 1948), he postulated that the active factor liberated by the symbiotic fungus was one or more auxins.

Recently, paper chromatography and the *Avena* coleoptile elongation test have been used to demonstrate liberation of auxins in culture solution at a large number of mycorrhizae-forming fungi. Moser (1959) tested 23 species of ectomycorrhizae-forming basidiomycetes for auxin production. They were isolated from mycorrhizae of *Pinus, Larix, Picea, Betula,* and *Fagus* spp. As a nitrogen source and precursor for auxin production Moser added a relatively large amount (2.041 gm/liter) of tryptophan to the nutrient solution. The results were as tabulated below:

Type of auxin produced:	IAA	IPA	IBA
Suillus placidus	+	−	−
S. plorans (7 strains)	+	−	+
S. tridentinus	+	−	−
S. grevillei (3 strains)	+	−	+
Boletinus cavipes	+	−	−
Xerocomus subtomentosus	+	−	−
Lacterius porninsis	+	−	−
Phlegmatium elegantior	+	+	+
Phl. calochroum	+	−	−
Phl. aureopuverulentum	+	+	−
Phl. sulphureum	+	−	−
Phl. caesiocanescens	−	+	−
Phl. varium	−	+	−
Phl. orichalceum	−	+	−

It is apparent that the majority of the fungi tested produced IAA alone or together with IPA and/or IBA. However, three of the *Phlegmatium* spp. produced only IPA. Since the auxin-synthesizing enzymes were present also in mycelia grown in tryptophan-deficient media, it appeared that these enzymes were not adaptive.

Ulrich (1960a) has also shown that mycorrhizae-forming fungi of *Boletus* and *Amanita* spp. produce extracellular auxins in pure culture. The amounts of tryptophan added to the medium as auxin precursor varied from 1.0 to 30 mg/liter. For *Suillus (Boletus) variegatus, S. (B.) granulatus, S. (B.) luteus, S. (B.) bovinus, B. felleus,* and *B. badius* the stable end product of the metabolized tryptophan was IAA. *Boletus badius* was an exception. It produced IAA only during the first week and then the tryptophan was metabolized into some end products other than IAA. In a 200-ml medium containing tryptophan at a concentration of 30 mg/liter during a period of 1 month, *S. (B.) bovinus* had liberated 50 μg of IAA (corresponding to 1.4×10^{-6} M IAA). The

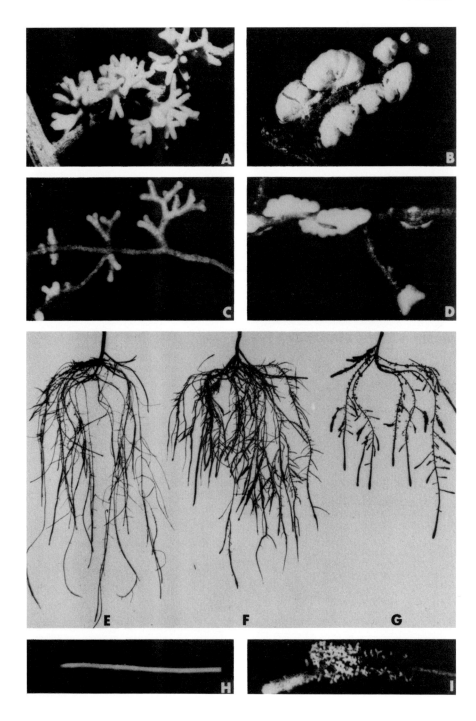

four *Amanita* species—*A. muscaria, A. rubescens, A. frostiana,* and *A. caesaria*—differed from the *Boletus* spp. in that many other compounds were formed simultaneously with IAA. Since Salkovski and Ehrlich's reagent sprays caused color reactions on the chromatograms, these compounds probably were indole derivatives, and, as Ulrich indicated, their presence may suggest a complex pathway of tryptophan metabolism (cf. Horak, 1963, 1964).

More recently, Gogala (1967, 1971) demonstrated that the ecto-mycorrhiza-forming fungus *Boletus edulis* var. *pinicolus* synthesized and liberated several growth hormones and growth regulators. By means of paper chromatography, physicochemical reactions, and classical biological tests, it was found that the fruiting bodies, the mycelium of this fungus in pure culture, as well as the culture medium in which the mycelium had grown for 1 month, contained three indole derivatives, corresponding to IAA, Harada's (1962, cited by Gogala, 1971) compound E, and a compound which seemed to be tryptophan and also growth hormones related to gibberellins and cytokinin (see below).

It is apparent that the amounts of auxins produced, their composition, and time necessary for their production in detectable amounts vary between different species (Moser, 1959; Ulrich, 1960a) and even between strains (Moser, 1959). Thus, *Suillus* (*Boletus*) *variegatus* cultures produced detectable amounts of IAA after 3 days whereas cultures of *Amanita frostiana* required about 7 months. In culture solutions containing 1 gm/liter of tryptophan, auxin production by *S.* (*B.*) *granulatus* was 15 times higher than by *S.* (*B.*) *variegatus* (Ulrich, 1960a). Experimenting with three *Suillus grevillei* strains and twelve strains of *S. plorans,* Moser (1959) found that strains of the same species, grown in identical media produced indole compounds of different quality and quantity.

Furthermore, it appears that some mycorrhizal fungi are able to synthesize IAA in tryptophan-deficient media. A preliminary experiment by Moser (1959) showed that in a nutrient solution without tryptophan,

---

FIG. 2. Mycorrhizae and auxin-induced mycorrhizae-like structures in pine roots. First row: coralloid (A) and tubercle (B) mycorrhizae of *Pinus strobus* L. Second row: structures resembling coralloid mycorrhizae (C) and tubercle mycorrhizae (D) induced in excised roots of *Pinus sylvestris* L. by NAA 2.5 mg/liter and 10 mg/liter. (E)–(G) Root systems of 8-month-old seedlings of *Pinus strobus* L. cultivated in nutrient solutions under aseptic conditions. (E) Control. (F) Indoleacetic acid added, the amount of which is gradually increased from 0.0005 to 0.045 mg/liter. (G) Indoleacetic acid added, the amount of which is gradually increased from 0.0005 to 9 mg/liter (¼ natural size). Inserts: morphology of magnified apical regions of long roots—(H) control; (I) NAA 9 mg/liter. (×2). [From V. Slankis (1958). "Physiology of Forest Trees" (1st ed.) by permission of the Ronald Press Co., New York.]

but with a supplement composed of *dl*-alanine and *l*-asparagine (in an amount corresponding to 280 mg N/liter) and indole, *Suillus plorans,* *S. placidus,* and *S. tridentinus* within 3 weeks liberated several indole compounds, one of which presumably was anthranilic acid (*o*-amino-benzoic acid). After 6 weeks, *S. placidus* and *S. tridentinus* had pro-duced an additional compound, probably IAA. Horak (1963, 1964) demonstrated the ability of *Boletus edulis* to metabolize tryptophan from anthranilic acid and also from a combination of indole and alanine. Of the compounds isolated, the closest to IAA was indolepropionic acid. As shown by Slankis (1951), this indole compound induces mycorrhizae-like morphogenesis in pine roots similar to that induced by IAA and IBA. According to Ulrich (1960), *S. (B.) variegatus,* and particularly *S. (B.) granulatus,* efficiently synthesized IAA in a nutrient solution containing only malt extract, glucose, and mineral salts. However, for other symbiotic fungi tested in the same medium, a supplement of tryp-tophan was necessary.

Of great interest are Moser's (1959) findings that auxin production by ectomycorrhizal fungi is greatly influenced by the source of nitrogen and its concentration. Thus, nitrogenous compounds which can be more readily assimilated than tryptophan and are stimulatory to mycelial growth, hampered tryptophan assimilation and, consequently, auxin synthesis. Thus, alanine, asparagine, aspartic acids, glycine and glu-tamic acid were found to be inhibitory. The inhibition of auxin produc-tion increased with increased dosages. Furthermore, according to Moser (1959), a preliminary experiment indicated that inorganic ammonium salts are also inhibitory.

Reports by some investigators do not confirm auxin production by mycorrhizal fungi. Employing the Salkovski colorimetric method and *Avena* coleoptile tests modified by Boyarskin (for reference, see Shemak-hanova, 1962), Shemakhanova (1962) could not detect any auxin-like substance in culture filtrates of *Suillus (Boletus) granulatus,* *S. (B.) bovinus,* *S. (B.) luteus, Xerocomus (B.) subtomentosus, Amanita rubes-cens,* and *Cenococcum graniforme.* Yet, these filtrates induced root for-mation in bean hypocotyl cuttings and enhanced germination of *Pinus* seeds. Similar negative results were reported by Wargo (1966). In his experiments, chromatographically analyzed culture solutions of *Trichoderma personatum, Russula amigdaloides, Cenococcum grani-forme,* and *Armillaria mellea* (a suspected symbiotic associate with *Quercus alba*) did not contain IAA in detectable amounts. As possible reasons, Wargo mentions too low concentrations of IAA present in the medium, insufficient length of time for culturing the mycelia, and the possibility that these fungi could not produce IAA *in vitro.*

2. Cytokinins and Gibberellins

Since the report of Brushnell and Allen (1962), that extracts of *Puccinia graminis tritici* uredospores and *Erysiphe graminis* spores induce a cytokinin-like effect, it has become increasingly apparent that production and release of cytokinins is rather common by fungi and bacteria which associate with higher plants. Recently, Miller (1967a, 1971) obtained evidence that mycorrhizal fungi also produced and liberated extracellular cytokinins. Using aseptic soybean tissue cultures, known to grow only in the presence of cytokinin, he obtained growth stimulation when pieces of these tissues were placed on agar remote from the mycelial inoculum of *Rhizopogon roseolus* (Fig. 3). The degree of stimulation decreased as distance from the mycelium increased. From 200 liters of the fungus culture solution, Miller isolated in crystalline form about 1 mg each of the most abundant cytokinins, zeatin, and a ribonucleotide of zeatin. Also a third cytokinin, probably zeatin ribonucleotide, was present. All these constituents had been isolated earlier from maize kernels (Letham *et al.*, 1967; Miller, 1967b). In addition to *Rhizopogon roseolus*, the release of cytokinin in culture solution was

Fig. 3. The apparent production of cytokinin by the fungus *Rhizopogon roseolus*. Soybean tissue pieces were planted on the agar surface in all three flasks, and all flasks contained the basal medium. The flask on the left, however, also contained 1 mg/liter kinetin. In the center flask, an inoculum of the fungus was placed in the center of the four soy bean pieces. [From C. O. Miller (1971). By permission of USDA, Forest Service.]

detected in *Suillus cothurnatus, S. punctipes, Amanita rubescens,* and an unidentified species forming ectendomycorrhizae. However, the presence of cytokinin was not detected in the culture solutions of *Cenococcum graniforme* and *Telephora terrestris.* Similarly negative were assays of 21 nonmycorrhizal fungi (Miller, 1971). Gogala (1967, 1971) reported that fruiting bodies, mycelium, and also culture medium of *Boletus edulis* var. *pinicolus* contained: (1) three gibberellin-related compounds—gibberellic acid, a complex of compounds thought to be gibberellic esters, and an unidentified compound; and (2) one cytokinin, probably zeatin.

3. Vitamins

Shemakhanova (1957, 1958, 1962) provided evidence that ectomycorrhizal fungi synthesize and liberate vitamins. Culture solutions of these fungi, assayed with different species of yeast, showed (see Table I) that nicotinic acid, biotin, and pantothenic acid are produced by *Suillus (Boletus) luteus, Boletus scaber, S. (B.) bovinus, B. edulis* (the latter produced only a trace of pantothenic acid), and *Paxillus involutus.*

TABLE I

VITAMINS PRODUCED BY MYCORRHIZAL AND NONMYCORRHIZAL FUNGI IN PURE CULTURE. THE AMOUNT OF VITAMIN SYNTHESIZED EXPRESSED BY WEIGHT OF YEAST PRODUCED[a,b]

Fungus	Age of culture filtrate (days)	Amount of filtrate introduced (ml)	Weight of yeasts (mg)				
			Thiamine (B_1)	Nicotinic acid	Biotin (H)	Pantothenic acid	Pyridoxine (B_6)
Boletus luteus	14	2	1.13	4.55	14.5	7.65	—
Boletus scaber	38	2	1.68	13.65	9.3	7.1	—
Boletus bovinus	30	2	—	3.20	8.1	8.7	—
Boletus luridus	30	2	—	0	3.9	0	0
Boletus edulis	30	2	—	11.3	12.4	0.3	0.1
Paxillus involutus	30	2	—	1.4	9.5	4.5	1.0
Amanita muscaria	30	2	—	0	24.5	0	0
Amanita pantherina	30	2	—	0	0	0	35.7
Lepiota procera	30	2	—	1.2	2.2	6.3	0
Phallus impudicus	30	1	—	2.5	0.2	0	0
Cenococcum graniforme	14	2	—	—	—	3.22	0
Boletus subtomentosus	14	2	—	—	—	1.30	—

[a] From Shemakhanova (1962).

[b] When amounts less than 2.5 mg are produced of B_1 and B_6, Shemakhanova considers the results to be negative (personal communication).

Some of the symbiotic fungi tested liberated only one vitamin. Thus, *Amanita muscaria* and *B. luridus* synthesized only biotin, while *Xerocomus* (*Boletus*) *subtomentosus* and *Cenococcum graniforme* produced pantothenic acid. *Amanita pantherina* was the only species found to produce pyridoxine and was the only species tested which did not synthesize biotin. Calculation based on micrograms of amino acids present in 10-ml culture solutions showed that S. (*B.*) *bovinus* liberated 0.0014 mg of biotin, 0.09 mg of nicotinic acid, and 0.15 mg of pantothenic acid; and S. (*B.*) *luteus* liberated 0.0002 mg of biotin, 0.15 mg of nicotinic acid, and 0.57 mg of pantothenic acid. Vitamins were produced also by non-mycorrhizal fungi. Of the two species tested, *Lepiota procera* produced nicotinic acid, biotin, and pantothenic acid, but *Phallus impudicus* produced only nicotinic acid and biotin.

III. Effect of Symbiotic Fungus Exudates on Host Plant

A. MORPHOGENIC EFFECT ON SHORT ROOTS

The peculiar morphology of ectomycorrhizae of woody plants attracted the attention of scientists even before the symbiotic relationship was recognized. Gasparrini (1856) commented on the lack of root hairs and Janczewski (1874) described dichotomous branching of pine mycorrhizae. With the beginning of this century the morphology and anatomy of ectomycorrhizae of forest trees became a subject of intensive study in America (McDougall, 1914, 1922; McArdle, 1932), in Sweden (Melin, 1923), and in England (Laing, 1923, 1932; Aldrich-Blake, 1930).

Ectomycorrhizae are morphologically classified into three main types: simple (monopodial) mycorrhizae, dichotomously branched or coralloid mycorrhizae, and nodule (tubercle) mycorrhizae (Fig. 2). According to early views, mycorrhizal and nonmycorrhizal short roots differ structurally in many respects (cf. Aldrich-Blake, 1930; Laing, 1932). However, as a result of Hatch's intensive studies (1937; see also his data reported by Hatch and Doak, 1933) employing aseptic and sand cultures of pine, the number of these differences has been significantly reduced. According to him, both types of short roots are monarch; they do not have a root cap; the layers of their cortical cells are limited to four; and they do not exhibit secondary growth. As genuine structural characteristics of ectomycorrhizae Hatch acknowledged their swollen appearance, resulting from hypertrophy of cortical cells, delayed suberization of the cortex and endodermis, lack of root hairs, and, in pine mycorrhizae, a tendency to branch dichotomously in large numbers. According to Clowes (1951) and Chilvers and Pryor (1965), in mycorrhizal roots differentiation of meristem into stele, cortex, and endodermis begins nearer to the apex than in nonmycorrhizal roots. Chilvers and Pryor (1965) reported that eucalypt mycorrhizae exhibit also thickening of the radial and inner tangential walls of the inner cortex.

Melin (1923, 1925) used an ingenious method for mycorrhiza synthesis *in vitro* and proved that the structural characteristics of pine mycorrhizae developed only in the presence of the symbiotic fungus. Later, Slankis (1948) demonstrated that excised roots of *Pinus sylvestris* grown aseptically in nutrient solution also produced dichotomy and repeated dichotomy when inoculated with ectomycorrhizae-forming *Suillus* (*Boletus*) *luteus* and *S.* (*B.*) *variegatus* (Fig. 1). Dichotomy, resembling coralloid mycorrhizae, was also induced by adding cell-free culture filtrates from these fungi to the root cultures. Under the influence of the exudates, the newly formed apical regions were slightly swollen and devoid of root hairs (Fig. 1). Reports by other researchers show that also in pot cultures and under field conditions ectomycorrhizae-like root structures, including dichotomy, are induced by symbiotic fungi exudates. Inoculation of nursery soil at Wareham, Dorset, with *Rhizopogon luteolus*, stimulated lateral root initiation and extensive dichotomous branching of *Picea sitchensis* seedlings before the actual mycorrhizal infection was established (Rayner, 1934, 1939; Rayner and Neilson Jones, 1944; see also Rayner's data published by Levisohn, 1956). Levisohn (1952) allowed the growing roots of potted *Pinus sylvestris* to protrude through the drainage opening and reach the leached water in the earthenware below. Irrespective of the potting soil used, the short roots in the leached water developed abundant dichotomy when mycorrhizae developed in the pots. None of the roots grown in the water dish below showed any traces of mycorrhizal or other infections. Forking did not occur when the seedlings lacked mycorrhizal association. In Levisohn's view it suggests that the forking was induced by exudates of the symbiotic fungi.

Root structures bearing considerable resemblance to simple, coralloid, and tubercle mycorrhizae can be induced with synthetic auxins both in excised and attached roots of *Pinus sylvestris* (Slankis, 1949, 1950, 1951) and in attached roots of *Pinus strobus* seedlings (Slankis, 1958). Concentrations of 1.0–1.5 mg/liter of IAA induced dichotomy and inhibited root hair formation in excised pine roots. However, the branches of the developed forks were slender and only few had a slight increase in diameter. As IAA concentration was increased, the length of the newly developed dichotomous branches and the apical regions on the monopodial short roots decreased, but they became progressively more swollen.

Thus, concentrations of 5–10 mg/liter of IAA produced root structures that have a striking resemblance to simple and coralloid mycorrhizae, and concentrations of 10–20 mg/liter induced structural deviations that resemble tubercle mycorrhizae (Fig. 2). In all these mycorrhizae-like formations the swollen regions were devoid of root hairs, and cortical cells within these regions exhibited radial expansion (cf. Slankis, 1963). The strongest morphogenic effect was induced by NAA and, with the other auxins, decreased in the following order: IAA, IBA, IPA. Furthermore, it became apparent that the synthesized pine mycorrhizae-like structures were not permanent formations. To preserve them, a periodic supply of fungus exudates or synthetic auxin was required every second week or the swollen apices began to elongate and within less than 2 weeks the previous mycorrhiza-like characteristics disappeared. The elongated slender apical regions were densely covered with root hairs (Fig. 4). On the basis of these findings, Slankis (1951, 1958) concluded that the profound morphogenesis which nonmycorrhizal short roots undergo during conversion into mycorrhizae is induced by one or several metabolites exuded by the symbiotic fungus and that these metabolites are related to auxins.

Palmer (1954) has shown that mycorrhizae-like root structures, similar to those induced by synthetic auxins in roots of *Pinus sylvestris* (Slankis, 1949, 1950, 1951) and *Pinus strobus* (Slankis, 1958) can be induced by IAA also in the roots of *Pinus virginiana*. To improve aeration and provide more natural supporting medium for roots, the aseptic seedlings were grown in nutrient-soaked vermiculite. At the age of 1 month the root systems were subjected to IAA concentrations that varied from 1 to 40 mg/liter. A concentration of 2 mg/liter induced dichotomy in ~5% of the short roots, 10 mg/liter produced the maximum number (~30%) of dichotomously branched short roots, while for 20 and 40 mg/liter the dichotomy decreased to ~12 and ~7%, respectively. The swollen monopodial and dichotomously branched short roots, particularly at the higher IAA concentrations, conspicuously resembled simple and forked mycorrhizal roots of pine. In addition to the absence of root hairs, they had an apical meristem organization similar to that of ectomycorrhizae, and similar shape and arrangement of cortical cells. Concentrations of 20 mg/liter IAA induced prolific initiation of lateral roots. The results suggested to Palmer that in nature, forking and proliferation of new roots are induced by the symbiotic fungus extracellular IAA or substances with similar growth-promoting properties.

Gogala (1967, 1971) has shown that three indole derivatives and a cytokinin liberated in culture solution by *Boletus edulis* var. *pinicolus*

FIG. 4. Termination of auxin-induced mycorrhizae-like monopodial and dichotomized short roots caused by a discontinued supply of IAA. (A) and (B) depict

and added to aseptically grown root cultures of *Pinus sylvestris* inhibited elongation and root hair formation, and induced dichotomy. The tryptophan-related compound of the fungus induced distinctive dichotomous branching. Clowes (cited by Harley, 1969) found that IAA applied in lanolin to the root apices of *Fagus* seedlings caused decreased growth, and induced both local swelling and proliferation of lateral primordia. In the swellings, radial and tangential sizes of cortical cells were increased. Chilvers and Pryor (1965), experimenting with *Eucalyptus grandis* seedlings, reported that NAA concentrations of 10^{-5}–$10^{-6}\,M$ decreased root elongation and increased the diameter of growing apices by 70–100%. Although the volumes of cortical cells were enlarged four- to fivefold, the anatomical and histological changes only partially matched the effects of mycorrhizal infection.

In contrast to these findings are data of several investigators which indicate that ectomycorrhizae-like morphology can be induced by substances other than auxins and that auxin does not necessarily induce dichotomy. Barnes and Naylor (1959a,b) found that neither IAA nor NAA induced dichotomous branching in excised roots of *Pinus serotina* when they were supplied to root cultures at concentrations which appeared to be most effective in Slankis' (1951) experiments. On the other hand, in their experiments, dichotomy was induced by a great variety of chemically unrelated compounds, such as sulfanilamide, kinetin, amino acids, certain B vitamins, and antivitamins. Kinetin was the most effective inducer of dichotomy. Sulfanilamide induced dichotomous branching at concentrations of 1–10 ppm. Of the vitamins, thiamine, nicotinic acid, and folic acid induced branching at concentrations above 10 ppm. The amino acids citrulline and *o*-aminobutyric acid ($10^{-3}\,M$) produced dichotomy strongly resembling ectomycorrhizae. Arginine, ornithine, aspartic acid, and urea induced branching to a lesser degree, but the first three caused swellings on the lateral root apices. Glutamic acid and proline were ineffective.

Also, data published by Ulrich (1960b, 1963) about auxin effects on roots demonstrate that IAA concentrations from 10^{-11} to $10^{-4}\,M$ do not induce swellings or dichotomy in short roots of *Pinus lambertiana* seedlings. The effect was negative whether aseptic seedlings were grown in

fragments of two 1½-month-old subcultures of an excised, aseptic root system of *Pinus sylvestris* L., which for the last 2 months before being subcultured had grown in a nutrient solution containing 7.25 mg/liter IAA. After the development of swollen monopodial and dichotomously branched short roots the root system was divided into two subcultures. Subculture (A); addition of IAA (7.25 mg/liter) continued for 1½ months. Subculture (B); addition of IAA discontinued after subdivision (×7.5). [From V. Slankis (1951). By permission of *Symb. Bot. Upsal.*]

darkness totally submerged in Petri dishes containing nutrient solution (the appropriateness of this method may be questioned), or they were grown under greenhouse conditions with roots submerged in nonsterile nutrient solution. In both cases the only effect was that elongation of the main root was inhibited by 10^{-6} M concentration of IAA. The ineffectiveness of IAA may have resulted from the action of auxin-destroying enzymes present in *Pinus lambertiana* roots. Thus, after 2 weeks' growth, analysis of the nutrient from the Petri dishes revealed that IAA was still detectable only in the nutrient with 10^{-4} M IAA. However, in addition to the IAA spot, the chromatogram showed the presence of another compound. The R_f value of this compound and the color reaction with Salkovski reagent corresponded to that of indolealdehyde (IA). This spot was missing on the chromatogram of corresponding unused nutrient solutions. Also, in the greenhouse experiment, despite the large volume of nutrient used, no IAA could be detected after 4 days. Again, such a spontaneous auxin disappearance did not occur in the unused solutions. Naphthaleneacetic acid concentration of 10^{-5} M was toxic to roots, but 10^{-6} M inhibited elongation, induced swellings on root apices, and within 2 weeks, initiated many side roots. A supplement of 2,4-D at a concentration of 10^{-5} M not only induced swellings but also caused root mortality. Neither of these compounds produced dichotomy. The only auxin-related compound which induced both distinct swellings and limited dichotomy was indoleacetonitril (IAN) at 10^{-6} and 10^{-5} M concentrations. However, analyses of the used nutrient showed that IAN was rapidly converted into IAA and that the latter was further metabolized into compounds with higher R_f values. A test with 3.5 gm of intact root system, excised from a 7-month-old seedling, showed that IAA was formed from IAN within 24 hours and that the root system of such size could break down 0.8 mg of IAA in 40 ml of nutrient solution within 48 hours. Ulrich (1960b, 1963) did not ascribe the dichotomy induced by IAN to any direct auxin effect, since control seedlings in the experiments also branched dichotomously. In the Petri dishes such branching began after 2 months, and within 6 months all control seedlings had developed dichotomy. In the greenhouse experiment, the roots of control seedlings branched dichotomously in large numbers the following spring, if the seedlings had been subjected to low temperatures and short days during the winter months.

It is of interest that in Ulrich's (1960b, 1963) experiments inoculation of *Pinus lambertiana* seedlings with mycorrhizae-forming fungi did not cause changes in root morphology even when a Hartig net was formed. Thus, *Suillus (Boletus) variegatus* and *B. badius,* inoculated into flasks of seedlings grown aseptically in culture solutions, established a Hartig net and hyphal mantle, but only one dichotomy was induced with *S.* (*B.*) *variegatus.* Similarly, seedlings grown aseptically in nutrient-soaked terralite and inoculated with ten different species of mycorrhizal fungi produced only one dichotomy with *Amanita rubescens.*

Turner (1962) tested the effect of fungus exudates on excised roots of *Pinus sylvestris* by experimenting with 53 fungal species representing Phycomycetes, Fungi Imperfecti, and Basidiomycetes (mycorrhizal, non-mycorrhizal, and some of unknown state). The fungi were isolated from *Pinus sylvestris* root tips or from sporophores collected from a coniferous plantation. The effect of the seven mycorrhizal fungi tested on the roots differed greatly. Exudates of *Amanita muscaria* and *Suillus* (*Boletus*) *bovinus* stimulated the initial elongation of excised radicles, whereas exudates of *S.* (*B.*) *granulatus* and *Xerocomus* (*B.*) *subtomentosus* had no apparent effect. Lateral root formation was induced by exudates of *A. muscaria* and *S.* (*B.*) *variegatus,* and induction of root initials by the latter fungus was prolific. Exudates of only four of the mycorrhizae-formers induced dichotomous branching. Their efficiencies in inducing dichotomy were in the following order: *S.* (*B.*) *variegatus* > *S.* (*B.*) *badius* > *S.* (*B.*) *luteus* > *S.* (*B.*) *bovinus.* No dichotomous roots developed in cultures supplemented with culture filtrates of *A. muscaria,* *S.* (*B.*) *granulatus,* and *Xerocomus* (*B.*) *subtomentosus.* However, elongation was stimulated and dichotomous branching also induced by exudates of some Phycomycetes, Fungi Imperfecti, and nonmycorrhizal Basidiomycetes. Here, as was the case with mycorrhizae-formers, a marked variation in their efficiencies was observed. In contrast to mycorrhizal and nonmycorrhizal Basidiomycetes, the other groups of fungi did not stimulate lateral root initiation. Turner concluded that the ability to induce dichotomy was almost a characteristic of the Basidiomycetes as a whole, but not an exclusive property of the mycorrhizal symbionts.

Wargo (1966), experimenting with excised *Quercus alba* roots, inhibited root elongation and obtained an increase in dry weight (but not dichotomy) when IAA was added to culture solutions at concentrations known to induce mycorrhizae-like morphology in pine (cf. Slankis, 1951; Palmer, 1954; Shemakhanova, 1962). Exudates of mycorrhizal fungi (*Russula amygdaloides* and *Tricholoma personatum*) also inhibited root elongation, particularly during the first week of root growth, but did not induce dichotomy. Also, as already mentioned in Section I, Shemakhanova (1962) analyzed culture filtrates of five mycorrhizae-forming fungi and did not detect any auxin-like substances. Similarly, tests by Wargo (1966) for extracellular IAA of three mycorrhizae-formers produced negative results. Clowes (1951) reported that also with colchicine structural modifications resembling mycorrhizae can be induced. Thus,

0.01–0.02% aqueous solutions of colchicine reduced or arrested growth of *Fagus sylvatica* roots and caused formation of swelling. The swollen mycorrhizae-like roots had differentiated cells very close to the apex and all cells exhibited abnormally enlarged transversal expansion.

The contradictory experimental results obtained on the induction of mycorrhizae-like root structures of genus *Pinus* with synthetic auxins become less confusing if the following facts are taken into consideration:

1. Dichotomy is not a basic structural element for ectomycorrhizae of pine.

2. Since individual plants of the same species exhibit different genetic properties (Skinner, 1952; Brown, 1969), the degree and type of morphogenesis caused by auxins may also vary.

3. In any experiment the end result of any given treatment with growth hormones, whether it is auxin, cytokinin, or any other growth regulator, depends on both the general nutrient supply and the presence of endogenous hormones and growth regulators in the experimental plant.

For these reasons it is misleading to evaluate mycorrhizal development only by the frequency of dichotomously branched mycorrhizae and equally wrong to evaluate the effect of synthetic auxin on mycorrhizae-like morphogenesis on the basis of the dichotomy induced. It is well known that in most of the mycorrhizal root systems of pine grown under natural conditions simple and dichotomously branched mycorrhizae alternate on long roots and often the monopodial mycorrhizae are in the majority.

From studies with *Pinus ponderosa*, Goss (1960) concluded that in this species dichotomous branching is an inherent factor which becomes more pronounced as a result of the fungus association. It appears, however, that forking of mycorrhizae and the frequency of this type of mycorrhizal root are regulated not by the fungus auxin alone but result from an interaction between the auxin and some genetically controlled factors in the root system. In pine roots these factors may differ between root systems and even between different laterals of the same root system. It explains why different root systems of the same species and even individual roots within the same system of pine react differently to the same concentration of synthetic auxin. Thus, excised roots of *Pinus sylvestris* grown aseptically under similar experimental conditions, show great variation in the frequency of dichotomy when subjected to auxin. Generally, under the influence of synthetic auxin, short roots of such root systems develop apical swellings and many of these swellings branch dichotomously. However, in some root systems either very few short roots undergo dichotomy or all short roots may undergo a profuse branching (cf. Slankis, 1951, Figs. 2 and 9). Similar variations in dichotomy have been observed in the attached roots of *Pinus strobus* seedlings grown aseptically in specially designed long culture tubes with four root systems sealed in each tube (cf. Slankis, 1951, Fig. 39). Despite the identical nutrient, auxin concentration, and other growth conditions within a culture tube, the number of dichotomously branched roots per system was exceedingly variable and, in about 4% of the seedlings, all short roots dichotomized so profusely that the whole root system became a dense cluster of dichotomized and repeatedly dichotomized roots (V. Slankis, unpublished data). These observations convincingly demonstrate that in studies on the formation of the symbiotic relationship and on the role of fungus auxin in this formation process, conclusions cannot be based on the assumption that dichotomy is a basic structural element of ectomycorrhizae. Instead, the structural elements which should be used as criteria are the lack of root hairs, expansion and changed polarity of cortical cells (Hatch, 1937), and the distance of differentiated tissues from the root tip (Clowes, 1951; Chilvers and Pryor, 1965).

As mentioned above, several investigators have expressed the view that auxin is not necessary for the induction of mycorrhizae-like morphology. Reports show that colchicine induces mycorrhizae-like structures in roots of *Fagus* (Clowes, 1951) and *Eucalyptus* (Chilver and Pryor, 1965) and that dichotomy in pine roots is induced by cytokinin (Barnes and Naylor, 1959a; Gogala, 1971), some vitamins, antivitamins, and various amino acids (Barnes and Naylor, 1959a,b). Also data from studies with roots of herbaceous plants demonstrate that different chemical compounds may exert an auxin-like inhibitory effect on root elongation and stimulate lateral initiation. Clowes and Leyton (cited by Harley, 1969) observed that short roots dichotomize in aseptic cultures without the addition of auxin. Chilvers and Pryor (1965) have pointed out that factors inhibitory to root elongation also cause development of some anatomical modifications which are characteristic for the host plant tissues in mycorrhizae. In their view: "Aging roots, growing slowly through unfavorable media, and roots artificially restrained by overdoses of colchicine and naphthaleneacetic acid all exhibit the same phenomenon." Although this appears to be true, the question remains open about the root factors which under inhibited growth conditions induce the characteristic mycorrhizae-like structural deviations in the roots. Observations from experiments with pine roots have led the writer to speculate that endogenous auxin might be one of the factors responsible for the induced mycorrhizae-like structural deviations when inhibitory growth conditions for roots prevail. Due to the inhibition, the exogenous auxin might possibly accumulate in the root apices and reach concentrations above physiological levels. It is evident that excised roots of *Pinus sylvestris* and attached roots of *P. strobus* seedlings become totally inhibited and acquire a dark-brown color in sterile nutrient solution, containing only mineral salts and sucrose, if this solution is not changed for an extended period. A supply of fresh nutrient, however, brings about a sudden, renewed elongation within a few days. At first the elongating root tips develop miniature swellings and many of these swellings branch dichotomously. However, with continued elongation of the apices, the newly formed mycorrhizae-like structures disappear within 1 week or so (Slankis, unpublished data).

The reverse morphogenesis of the developed mycorrhizae-like structures, which occurs soon after growth is resumed by the roots, indicates that during the period of arrested growth certain metabolites must have accumulated in the roots and reached concentration levels able to induce such morphogenesis. However, after the elongation is resumed, these levels apparently cannot be maintained, and, therefore, the mycorrhizae-like structures revert to normal root morphology. Undoubtedly, there is a great similarity between the morphogenetic processes described above and those that are induced in pine roots by synthetic auxins and that disappear when the auxin supply is discontinued (cf. Section III,A). This similarity suggests that one of the metabolites accumulated during arrested root growth might be auxin. It is known that root tips are the main centers for auxin synthesis (cf. Thimann, 1972) and that auxin moves more acropetally than basipetally in roots (McCready, 1966; Wilkins and Scott, 1968 and others). Therefore, according to the "band theory" of Went and Thimann (1937), auxin may accumulate in the terminal regions of the roots, particularly under conditions when root elongation is inhibited.

B. Effect on Short- and Long-Root Development

According to views expressed by early investigators (Hartig, 1888, 1891; von Tubeuf, 1888; McDougall, 1914; and others) the fungal associate exerts a retarding influence on development of roots of the host by inhibiting elongation and destroying the infected roots. However, studies by Hatch (cf. Hatch and Doak, 1933; Hatch, 1937) proved otherwise. Employing aseptic and sand cultures, he showed that the final length of short roots of nonmycorrhizal pine seedlings was limited and thus verified the view expressed by Aldrich-Blake (1930) that the state of both short and long roots is physiologically predestined. Hatch also succeeded in demonstrating that mycorrhizal infections reactivated the dormant apical meristem even in old short roots of pine and that these laterals developed swellings and multiple branching (Fig. 5). It is true that mycorrhizal infections inhibit root hair formation. However, as pointed out by Hatch (1937), this reduction in absorbing surface is more than compensated for by the enormous and ramified hyphal network of the symbiotic fungus in the soil. In addition, the presence of the symbiotic fungus delays suberization of the endodermis and cortex (Aldrich-Blake, 1930; Laing, 1932; for more recent references, see Harley, 1969), thereby often keeping the cortex of mycorrhizal roots alive and active throughout the growing season. Conversely, in non-mycorrhizal short roots, the cortex soon dies due to the early deposition of suberin which cuts off the cortex from the conducting tissues.

As yet little is known about the effect of the symbiotic fungus on the rest of the root system, particularly its effect on long roots. Aldrich-Blake (1930), studying the evolution of diarch lateral primordia of Corsican pine, concluded that those which do not abort at an early stage may become infected by the mycorrhizal fungi at a short distance from the point of origin and dichotomize into two monarch roots,

Fɪɢ. 5. Profuse terminal branching of individual mycorrhizal laterals induced by the symbiotic fungus in roots of *Betula lutea* Michx. (A) and *Pinus strobus* L. (B).

which, in turn, may further dichotomize but never grow to an appreciable length. Alternately, these laterals may develop as mycorrhizac-bearing mother roots and eventually may be converted into pioneer roots capable of unlimited growth. By measuring the diameter of protoxylem of the dichotomized mycorrhizal and mother roots, he found that they fell into two significantly different size classes. The mean protoxylem diameter of the mother roots was 117 ± 6 μm while that of the basal, diarch portion of the dichotomized roots was only 73 ± 6 μm. The probability that the difference between these two measurements was significant was greater than 100 to 1. Aldrich-Blake (1930) suggested that dichotomy of lateral roots of Corsican pine represents the response of feeble roots to strong fungal infection. If that is true then to the same class belong those laterals that after infection establish as simple mycorrhizae. Studies by Wilcox (1968b) on *Pinus resinosa* root morphology showed that transition from short roots to long roots was of low probability, but transitions from one long-root category to another occurred readily. Only those emerging laterals, having over 50% of the diameter of the mother root, became long roots; only 10–21% attained this initial size.

It has become increasingly apparent that the hyphae of the symbiotic fungus also invade long roots. Such infections were already recorded by Noelle in 1910 and later verified by several investigators (Laing, 1923; Melin, 1925; Masui, 1926; Möller, 1947; Linnemann, 1955; Robertson, 1954; Wilcox, 1968b; and others). This phenomenon, which has gained little attention until recently, suggests, that long roots also are under direct influence of the symbiotic fungus. We may presume that a certain amount of the excess auxin in long roots originates from the established Hartig net in these roots. Also we may expect that an additional amount of the fungus auxin derives from mycorrhizae. In view of the fact that soil nutrients absorbed by the fungal hyphae are readily translocated from mycorrhizae into the root system of the host plant, it is rather difficult to envisage that the relatively high concentration of fungus auxin is confined to mycorrhizae until it becomes gradually inactivated or enzymatically destroyed. If a certain amount of this auxin is translocated into the long roots, then the total amount in the root system could reach a significant level, particularly in cases where root systems have numerous well-established mycorrhizae and where nutritional conditions are favorable for auxin production by the fungus. It is known that in the roots of herbaceous plants more auxin is translocated acropetally than basipetally (McCready, 1966; Wilkins and Scott, 1968), and pine roots seem to be no exception (see below). Also, it is known that roots are very sensitive to auxin and concentrations as low as 10^{-8} M are already inhibitory. In light of these findings it is possible that the acropetally moving fungus auxin could, at least in some cases, reach concentrations which both inhibit elongation of the weaker long roots (cf. Aldrich-Blake, 1930) and stimulate initiation of lateral roots. On the other hand, small dosages of this excess auxin could exert a stimulatory effect on elongation. Since IAA concentrations below 0.5–5 μg/liter stimulate elongation of excised roots of *Pinus sylvestris* (Slankis, 1951) and attached roots of *P. strobus* (Slankis, unpublished data) it indicates that in nonmycorrhizal pine roots the natural auxin is at suboptimal concentrations. It may, therefore, be expected that the excess auxin supplied by the fungal symbiont may significantly influence the host plant growth and development.

Fungus growth hormones seem to be deeply involved in conversion of a long root into a mother root. In nature these roots are frequently clavate-shaped and have a characteristic distribution of mycorrhizal laterals; their frequency and diameters increase toward the apical end of mother roots (Fig. 6). These structural features indicate a gradually increased inhibition of elongation as the mother root develops and must result from the hormonal effect of the symbiotic fungus, since structures similar to mother roots do not, as a rule, develop in uninfected root systems. They can, however, be induced with exogenous auxins if the concentration is increased gradually (Fig. 12,G). This suggests that mother roots are formed under the influence of growth hormones produced by the symbiotic fungus. According to Robertson (1954) and Wilcox (1968b), in mycorrhizal root systems of pine, the Hartig net also forms in the cortex of long roots. Therefore, it may be that, with the Hartig net advancing acropetally in long roots, the elongation of the physiologically weaker long roots (cf. Aldrich-Blake, 1930) becomes restricted or, during certain periods, even totally inhibited. As demonstrated experimentally with synthetic auxins, such inhibition can induce formation of swollen laterals in increased numbers (Slankis, 1951, 1958; Palmer,

Fig. 6. Mycorrhizae-bearing mother root of *Pinus resinosa* Ait. showing a gradual increase in diameter toward the apex. Note, that the frequency of mycorrhizae and the degree of their swelling also increase towards the apex of the mother root. ($\times$1.6).

1954). According to Wilcox (1968a), exceptionally swollen short laterals can be found on the distal regions of mother roots prior to establishment of a Hartig net in these regions. Yet these laterals had no mycorrhizal infections. This means that swollen mycorrhiza-like laterals can develop in a mycorrhizal root system even before their tissues are invaded by the symbiotic fungus hyphae. We may assume that in the cases reported by Wilcox the formation of swollen short roots was induced by translocated growth hormones of the fungus. Most likely these hormones are derived from both the Hartig net of the mother root and mycorrhizae alongside the mother root. As the number of mycorrhizae gradually increases so does the degree of structural change in the distal region of the mother root.

It is questionable whether experiments with potted seedlings and seedlings grown in axenic cultures can provide incontrovertible evidence on the effect of fungus auxin on root system development. Both methods, but particularly axenic cultures, allow root exudates as well as exudates of the symbiotic fungus to accumulate in the medium, thus creating an artificial environment which may considerably influence root system development and obscure the effect of the excess auxin of the fungus. In a natural habitat this situation does not arise because of dilution of the metabolites by diffusion, leaching, and assimilation by rhizosphere-inhabiting microorganisms. Similarly, natural conditions cannot be duplicated by experiments in which the whole root system is subjected to exogenous auxins. While in mycorrhizal root systems the fungus auxin is liberated from certain localized areas and by diffusion, its concentration gradually decreases with distance from the point of origin, there is no concentration gradient in an experiment in which the whole root system is submerged, and auxin concentrations which induce mycorrhizae-like morphology also strongly inhibit long-root elongation. Similarly, such experiments cannot provide necessary information on whether excess auxin, derived from an external source and absorbed by localized root areas, is translocated throughout the root system and how it affects root system development.

Recently, some information has been obtained on translocation of exogenous auxin by employing a method in which only one region of the root system was subjected to auxin (Slankis, 1958). This arrangement established a concentration gradient. The root systems of *Pinus strobus* seedlings were grown aseptically in horizontally placed and specially designed 80-cm-long (5-cm-diameter) glass tubes. The roots were grown in darkness, whereas the aerial parts of the seedlings protruded through protective closers above the tubes and were subjected to light. By means of thresholds these tubes were divided into two or three compartments, each having an inlet and outlet. This arrangement allowed periodic nutrient changes in any compartment and addition of auxin to a certain section of the root system. As the elongation of roots progressed in the first compartment, their tips crept over the thresholds and entered the adjacent compartment. When the long roots reached the far ends of the three-compartment culture tubes, IAA was added to the nutrient solution in one of the compartments. Addition of IAA in the middle compartment, at a concentration which caused formation of slight swellings in this compartment, considerably stimulated elongation in the two adjacent compartments. However, high IAA concentrations, strongly inhibitory in the middle compartment, induced a partial growth inhibition in roots of the other two compartments, indicating that the effect of auxin was transmitted in the root system of pine both acropetally and basipetally. However, the stronger inhibitory effect was observed in the distal part of the root system, suggesting that either more auxin was translocated acropetally, or that acropetally transferred auxin accumulated in the root apices (cf. Went and Thimann, 1937). Furthermore, the effect of exogenous IAA varied with distance from its source. When the basal region of the root systems in the first compartment was submerged in auxin solution, in the second compartment growth was partly inhibited, resulting in coarser long-root formation, whereas in the third compartment a considerable stimulation of elongation occurred. Of interest is the fact that in the third compartment even the most advanced long-root apices, although ~54 cm distant from the auxin source, were affected and their elongation stimulated. In another experiment the main root was cut off close to the base, leaving attached only two first-order laterals. These laterals were allowed to grow in separate compartments of a two-compartment culture tube, thus forming a divided root system. Addition of inhibitory concentrations of auxin to one compartment resulted in considerable stimulation of root elongation in the other compartment (Fig. 7). This indicated that the auxin effect was transferred across the basal part of the main root.

At present the generally accepted view is that auxins are transported as free acids (Thimann, 1972). Whether, in the above experiment, the growth and development of the distant regions of the root system were affected by translocated exogenous auxin or by some specific metabolites, formed in the root region submerged in the auxin-containing nutrient solution, remains to be demonstrated. Whatever the mode of action of exogenous IAA, the experiment demonstrated that exogenous auxin absorbed by one region of the root system for an extended period of time may affect relatively distant regions of the system. It is far from clear to what extent these findings with synthetic auxins reflect the

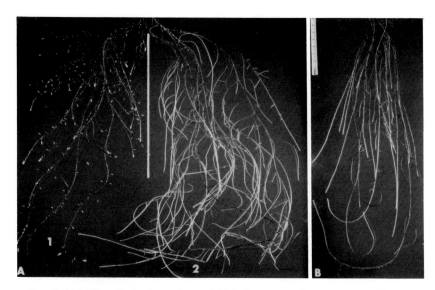

Fig. 7. (A) The effect of translocated IAA demonstrated in an aseptically grown, divided root system (A) of a *Pinus strobus* L. seedling: (1) IAA 3.5 mg/liter supplied directly to one-half of the root system; (2) IAA effect after being translocated into the other half of the root system. (B) Control (¼ natural size).

influence which the growth hormones produced by the symbiotic fungus exert on root system development in natural habitats. However, some structural phenomena of mycorrhizal root systems (see above) suggest that this hormonal effect is profound and reaches beyond the mycorrhizal roots.

C. Initiation of Laterals in Old Root Regions

Formation of mycorrhizal laterals and long roots within old regions of mycorrhizal root systems also seems to be induced by the symbiotic fungus growth hormones. This phenomenon, first briefly discussed by Laing (1932), is rather common, particularly in root systems of adult forest trees. The newly formed, simple and dichotomized mycorrhizae may develop individually or in comblike clusters, mostly besides old but still living mycorrhizae as well as moribund ones (Fig. 8). Less frequently, they initiate next to long roots or appear in the region between two long laterals. Occasionally, the apical meristems of very old mycorrhizae with dead and shrunken cortex reactivate and begin to elongate, and the new apical region establishes as mycorrhizal (Fig. 9).

Although the cause of this spontaneous reactivation of lateral and apical meristems in older root regions is still unknown, symbiotic fungus

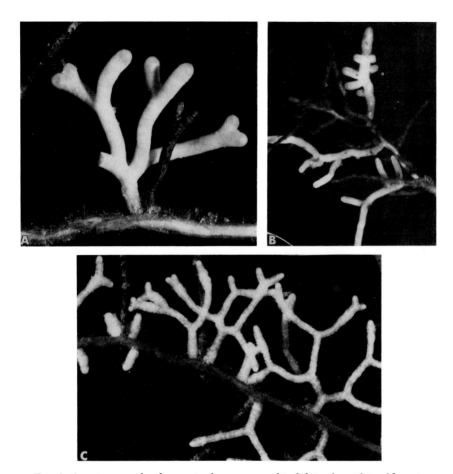

FIG. 8. Spontaneous development of new mycorrhizal laterals within old regions of root systems of *Pinus strobus* L. (A) and *Betula alleghaniensis Britton* (B). (C) A similar development of new mycorrhizae-like laterals induced within an older region of a long root of an excised root system of *P. sylvestris* L. by IBA 10 mg/liter. [(A) about ×7.4; (B) about ×12; (C) ×8]. [Fig. 8,C from V. Slankis (1951), by permission of *Symb. Bot. Upsal.*]

growth hormones may be involved. Laing (1932) observed that these new laterals carry the fungus web with them as they emerge from surfaces of the old long roots. Furthermore, formation of new laterals, resembling mycorrhizae, can be induced to initiate next to the old laterals by exogenous auxins (Fig. 8). Such formations occur in excised and attached pine root systems when these are subjected to auxins at concentrations above physiological levels (Slankis, 1951).

Melin (1925) reported short-root formation in dense groups of pine

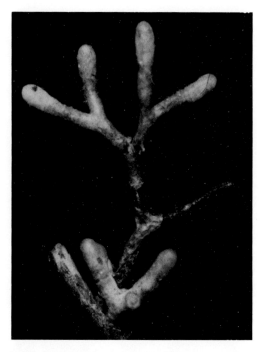

FIG. 9. A symbiotic relationship established within the newly developed apical region of one branch of a dichotomized, old mycorrhiza of *Pinus strobus* L. Note the shrunken, moribund cortex of the older region of the mycorrhizal root (×15).

and spruce seedlings grown in media containing nucleic acid as the sole nitrogen source, and with spruce grown in a nitrogen-deficient medium. Melin showed (1925, Fig. 23) that in such a group of short roots, consisting of two or more laterals, one lateral is longer. This suggests that the laterals did not initiate simultaneously but that the longer ones are older. Melin suggested the possibility that such formations result from nitrogen deficiency, since the nucleic acid which was the nitrogen source was assimilated with difficulty by roots. A supply of NH₄Cl and KCl did not induce the above phenomenon. It seems likely that the spontaneous reactivation of the lateral meristem in long root old regions is induced by an accumulation or a sudden increase in concentration of fungus growth hormones at some low nitrogen availability. Moser (1959) reported that production of auxin was influenced by nitrogen availability.

D. INFLUENCE ON SHORT- AND LONG-ROOT RATIO

According to Aldrich-Blake (1930), emerging laterals of mycorrhizal pine root systems are already predestined as short and long roots. In his view, the latter ones are sufficiently strong not to become arrested by the symbiotic fungus; they grow as mother roots with restricted elongation and eventually may become pioneer roots of unlimited growth. From detailed studies on root morphology of pine, Laing (1932) concluded that initiation of laterals is controlled by an internal mechanism. Wilcox (1968a,b), referring to the sparse branching of pioneer roots which results from abortion of primordia, considered the possibility that such losses of roots were caused by some internal hormonal correlative mechanism. Growth hormones, particularly auxins and cytokinins, play important roles in root formation and root growth processes (Fox, 1969; Thimann, 1969). It is known that roots of herbaceous plants are very sensitive to auxin, with a concentration of 10^{-8} M already inhibiting their elongation. Lower concentrations, on the other hand, are stimulatory (Thimann, 1969; Scott, 1972). Similarly, elongation of pine roots is considerably stimulated by low concentrations of exogenous auxins while higher concentrations are inhibitory and, as concentration increases, new laterals are initiated in increasing numbers (Slankis, 1951; Palmer, 1954; Barnes and Naylor, 1959a).

The long-root Hartig net effect on short-root initiation is apparent from data of Robertson (1954) and Wilcox (1968a,b). Robertson was the first to emphasize that the acropetal development of mycorrhizal short roots of *Pinus sylvestris* occurred with extension of the Hartig net in mother roots. He also found that in the proximal regions of long roots furnished with a Hartig net, short laterals became mycorrhizal and developed multiple dichotomy, whereas in distal regions, usually free from infection, the initiated laterals remained uninfected and monopodial or developed simple dichotomy. Those long roots, having most of the mycorrhizal short roots dichotomously branched, had a well-established Hartig net, and lateral initials became infected as they protruded through the Hartig net in the cortex. Robertson (1954) concluded that development of mycorrhizae on long roots depended on the presence of a Hartig net in the cortex. Studies of Wilcox (1968a,b) on the root morphology of *Pinus resinosa* verified that long roots have an established Hartig net and that the emerging laterals become infected as they pass through the Hartig net in the mother root. Only in cases where long roots elongate rapidly are the newly formed regions temporarily free of the symbiotic fungus as are the lateral initials within this region. It is of interest, that these laterals, before infection occurs, are already swollen.

A number of mycorrhizae-like short roots which form under the influence of high concentrations of exogenous auxins are potential long roots (Slankis, 1951). This fact became apparent when the periodic auxin addition, which was necessary to stabilize these structures, was discontinued. While the majority of short roots resembling mycorrhizae underwent an accelerated renewed elongation only for a limited period, the potential long roots, monopodial and dichotomized, continued to elongate and developed laterals (Fig. 4). These findings show that higher auxin concentrations arrest growth of both short roots and pre-

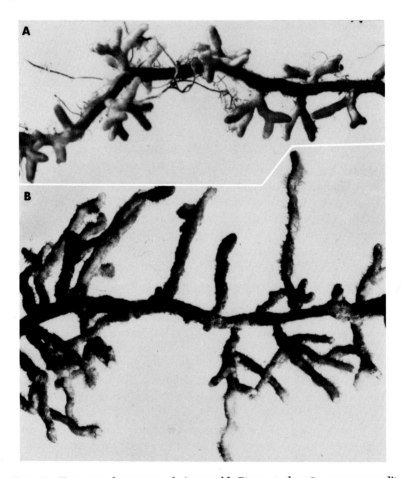

Fig. 10. Two root fragments of 4-year-old *Pinus strobus* L. nursery seedlings grown under different nitrogen concentrations for the last 7 months. Fragment (A): N = 2.5 mg/liter; depicts well-developed simple and corralloid mycorrhizae, enveloped by a thick hyphal mantle from which hyphal strings extend. Fragment (B): N = 159 mg/liter; depicts extensive renewed elongation of initially short, swollen mycorrhizal root tips as induced by an increase in nitrogen concentration. Note, the dense root hair development on the newly formed apical regions (×8.7). [From V. Slankis (1967).]

destined long roots. It is of interest that a similar phenomenon occurs in well-established mycorrhizal root systems of pine when seedlings are subjected to nitrogen levels that are known to inhibit formation of mycorrhizae (Slankis, 1967). At these increased nitrogen levels, the symbiotic relationship in mycorrhizae terminates, and their swollen apices begin to elongate (Fig. 10). While the majority of these apices undergo accelerated but limited elongation, a certain number of the simple and dichotomized mycorrhizae continue to elongate and become long roots. They show secondary growth and have laterals (Fig. 11). It appears likely that, contrary to the view of Aldrich-Blake (1930), a symbiotic relationship can be imposed also on predestined long roots at an early stage of their development and that under normal conditions they are bound to remain as mycorrhizal short roots. This may provide an explanation as to why characteristic mother roots often have only very few long laterals, or all of their laterals are entirely mycorrhizal short roots. We may presume that in such mother roots the fungus hormonal effect

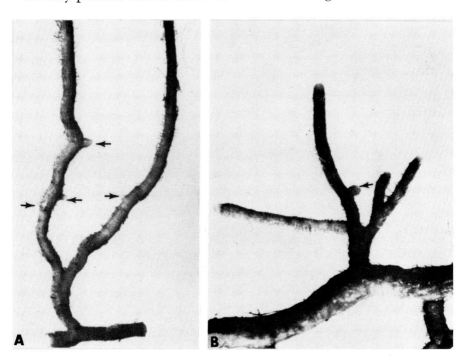

FIG. 11. Conversion of dichotomous mycorrhizal short roots of *Pinus strobus* L. into long roots during renewed elongation. The elongation was induced by an increased (159 mg/liter) nitrogen concentration. Long root properties in these elongating roots are indicated by the development of laterals (arrows). [(A) ×7.3; (B) ×13.] [From V. Slankis (1967).]

has reached the level at which also the potential long laterals are converted into mycorrhizae. As already mentioned, structural features resembling mother roots can be induced by gradually increasing concentrations of exogenous auxins (Fig. 2,G and I).

E. EFFECT ON HOST PLANT GROWTH AND DEVELOPMENT

Numerous investigators have verified that seedlings with mycorrhiza are, as a rule, better developed; they are taller, their dry weight is higher, and their foliage is darker green. Also it is known that mycorrhizal seedlings accumulate considerably more inorganic nutrients, particularly nitrogen, phosphorus, and potassium. For instance, Hatch (1937) found in mycorrhizal *Pinus strobus* seedlings 86% more nitrogen, 234% more phosphorus, and 75% more potassium than in nonmycorrhizal plants. He convincingly demonstrated that an enormously increased absorbing system results from the increased diameter of the swollen mycorrhizal roots and, particularly, from the immensely branched hyphal network of the fungal symbiont which connects the infected roots with the soil.

Despite the generally observed stimulatory effect of mycorrhizal symbiosis on host plant growth and development, the precise nature of the influence of the symbiotic fungus is little understood. It is apparent that this stimulation cannot be explained solely on the basis of the increased absorbing surface of mycorrhizae (cf. Hatch, 1937). Probably, the stimulated growth processes of the host plant are to a great extent influenced also by fungus-supplied metabolites. It has been shown that a single addition of 1–2 ml/liter culture filtrate of symbiotic fungi to the nutritional media of spruce seedlings considerably stimulated their growth during the 2-year experimental period (Lindquist, 1939). Shemakhanova (1962) obtained better shoot and root development of pine and oak seedlings by introducing under and onto the sown seeds culture filtrates from mycorrhizae-forming fungi in the amount of 7.5 ml/10 kg of sand. Also Shemakhanova found that the effect of culture filtrate of some fungi is long lasting. Thus, pine seeds immersed for 3 hours in 17-day-old culture filtrates of *Suillus* (*Boletus*) *luteus*, *S.* (*B.*) *bovinus*, *Amanita rubescens*, and of an unidentified fungus isolated from mycorrhizae, produced 2-year-old seedlings with higher dry weight than controls (Table II). The fact, that in the above experiments the stimulatory effect was still noticeable during the second year of growth, is of particular interest. This long-lasting effect cannot be caused by extracellular fungus auxins in the exudates, since the effect of a single addition of IAA at concentrations above physiological level to a medium rapidly decreases within 2 weeks (cf. Slankis, 1951). However, with increased IAA concentrations, better

TABLE II

EFFECT OF MYCORRHIZAL FUNGI CULTURE FILTRATES ON THE DEVELOPMENT OF
2-YEAR-OLD PINE SEEDLINGS[a,b]

	Dry Weight (mg)		
Fungus	Needles	Stem and roots	Total
Control	25	50	75
Suillus (Boletus) luteus	62	123	185
Suillus (B.) bovinus	56	147	203
Amanita rubescens	3	92	95
An unknown isolate from mycorrhizae	55	119	174

[a] From Shemakhanova (1962).

[b] Seeds prior to sowing immersed for 3 hours in the fungus 17-day-old culture filtrate.

developed stems and considerably longer needles have been obtained in aseptically grown *Pinus sylvestris* seedlings when the nutrient solution containing auxin was periodically renewed (Slankis, unpublished data). Whether the observed long-lasting growth stimulation of seedlings is induced by fungal hormones other than auxins, some growth regulators such as vitamins, or by some other as yet not identified metabolites in the fungus exudates, remains an open question.

F. CONCLUDING REMARKS

Even though our present knowledge pertaining to the symbiotic fungus growth hormones and their influence on the root system development is limited, certain conclusions can be drawn. It is evident that a number of ectomycorrhizae-forming fungi tested liberate growth hormones (Moser, 1959; Ulrich, 1960; Miller, 1967a, 1971; Gogala, 1967, 1971) and growth regulators (Shemakhanova, 1953, 1955, 1962; Turner, 1962) into culture media. Although the transfer of these hormones to the host plant has not yet been experimentally proven, the following data derived from experiments with synthetic auxins, mycorrhizal seedlings and general observations, strongly suggest this possibility:

1. The abundance of indole compounds in mycorrhizal roots (Mac-Dougal and Dufrenoy, 1944; Subba-Rao and Slankis, 1959), and in the swollen apical regions of mother roots (Subba-Rao and Slankis, 1959; partly published data).

2. The experimentally demonstrated transmission of auxin effect from

the region subjected to exogenous auxin throughout the root system (Slankis, 1958; see also Fig. 7).

3. The ability to induce with synthetic auxins morphological and histological root structures remarkably similar to those characteristic of mycorrhizae of pine (Slankis, 1951, 1958; Palmer, 1954; Shemakhanova, 1962; Gogala, 1971), *Fagus* (Clowes, 1950) and *Eucalyptus* (Chilver and Pryor, 1965).

4. The analogy between disappearance of the fungus-induced mycorrhizal morphology, when the symbiotic relationship terminates, and similar disappearance of auxin-induced, mycorrhizae-like morphology when the auxin supply is discontinued (Slankis, 1951, 1967; see also Figs. 4 and 10).

It has become apparent that the symbiotic fungus does not solely affect the predestined short roots by inducing morphological and histological deviations. Long roots are also affected and, hence the short- to long-root ratio. This change in mycorrhizal root systems is caused first, by the conversion of a number of predestined long root initials into mycorrhizae; second, by partial inhibition of elongation of a number of long roots, causing subsequent short-root formation in increased numbers (as is the case with mother roots); and third, by induction of new lateral formation in old regions of the root system.

Despite evidence that symbiotic fungus growth hormones influence development of roots of the host plant, the exact extent of this influence is still unknown. Reliable data are not available on the number of potential long roots converted into mycorrhizae and those which establish as mother roots or become pioneer roots. Also unknown is the percentage of mother roots which later become pioneer roots and those which remain as mother roots. Hatch (1937) stated that, because of induced swelling and branching, the absorbing surface of mycorrhizal roots is considerably greater than that of nonmycorrhizal roots. Yet the question still remains open about the total number of short roots in mycorrhizal root systems compared to those without mycorrhizae. Although the frequency of short roots per linear unit of long roots seems to be higher in mycorrhizal root systems than in nonmycorrhizal ones, it does not necessarily mean that their total number, compared to nonmycorrhizal root systems, is also higher. Probably, in mycorrhizal root systems, the frequency of short roots increases because the symbiotic fungus auxins inhibit the elongation of long roots. *In vitro* such a correlation was observed in the roots of herbaceous plants (Goldacre, 1959) and also in the roots of pine (Slankis, 1951, 1958; Palmer, 1954) at auxin concentrations which begin to inhibit long root elongation. It may well be that the same rule applies to mycorrhizal root systems growing in a natural habitat when the concentration of auxin liberated by the symbiotic fungus reaches levels that are inhibitory to long roots.

Similarly fragmentary is our knowledge about the total amount and initial concentration of growth hormones liberated into the root system by the fungal associate. Values obtained from pure cultures (Moser, 1959; Ulrich, 1960; Gogala, 1971) could be at great variance with those which exist in a natural habitat. Synthetic nutritional media are much simpler than those provided by the host plant and obtained as metabolic by-products from soil microflora. It has been generally confirmed that auxins move acropetally in roots (cf. Thimann, 1969; Scott, 1972). As fungus auxins and other growth hormones move acropetally, their concentration should gradually decrease with distance because of dilution and activity of auxin oxidases in the roots (Ray, 1958; Thimann, 1969). Therefore, it may be that in pioneer roots bearing mycorrhizae in limited numbers and having the greatest length, the concentrations of these hormones reaching their apical meristem might be very low. Such concentrations of auxins would stimulate elongation appreciably as has been shown with pine roots subjected to exogenous auxins (Slankis, 1951). In mother roots, on the other hand, having limited length and a higher frequency of mycorrhizae, the end concentration of the hormones should reach inhibitory levels. In fact, the structural features of these roots (Fig. 6) and excess indole compounds in their distal regions (Subba-Rao and Slankis, 1959) are indicative of inhibitory hormone levels.

To what extent the symbiotic fungus hormones affect the uptake and translocation of nutrients from the soil remains to be elucidated. It may well be that, besides the enlarged absorbing systems of mycorrhizae, the excess quantities of these hormones also play an important role in these processes.

In all probability, the presence of excess fungal hormones such as auxins and cytokinins profoundly changes the physiological and biochemical processes in mycorrhizal roots. Therefore, it would be of great importance to know the nature and extent of these changes and the effects they exert on both the host and the fungus. Data derived from research in general plant physiology show that cytokinins cause accumulation of amino acids, phosphates, and various other substances in the localized areas to which these hormones are applied (Möthes, 1960; Gunning and Barkley, 1963). Pozsár and Király (1966) have reported that rust infection and cytokinin treatment result in very similar changes in transport in the phloem, amino acid accumulation, and protein synthesis. Also extracellular auxins do induce nutrient mobilization; when applied to a stem or hypocotyl, the resulting swelling is accompanied by a marked increase in both dry weight and nitrogen content (Stuart, 1938; Mitchell and Stewart, 1939; Mitchell and Whitehead, 1941). Evidence continues to accumulate that nutrient translocation patterns are under hormonal control (Letham, 1967; Seth and Wareing, 1967).

Slankis (1958) suggested that at least part of the fungal auxin is not retained in the mycorrhizae, but is translocated into the mother root and then throughout the entire root system. If the fungal hormones move from mycorrhizae into the root system, and from there further into the shoots, the growth and development of the host plant would be affected to a much greater degree. Experiments with labeled auxins have shown that auxin is translocated both acropetally and basipetally in the roots, though the main flow is towards the tip (Pilet, 1964; Yeomans and Audus, 1964; Bonnett and Torrey, 1965). Also cytokinins, long assumed to be relatively immobile, are translocated in the root sap (Itai and Vaadia, 1965; Kende, 1965; Carr and Burrows, 1967). As discussed in Section III,B and shown in Fig. 7, exogenous auxin absorbed by one region of the root system of intact pine seedlings affects the whole root system.

These are only a few examples from the numerous known synergistic and antagonistic interactions of hormones and their influence on the biological and biochemical processes in the higher plant. It seems very probable that these effects may be even more pronounced at the increased hormone levels which are caused by growth hormones and growth regulators supplied by the symbiotic fungus. It may be expected, however, that the response of shoots to fungal hormones may vary as much as that of roots (see Section III). As for auxin, it has been found that increased concentrations produce better developed stems and considerably longer needles in *Pinus sylvestris* seedlings, but none of these effects is manifested in seedlings of *P. strobus* (Slankis, unpublished data).

IV. Factors Governing Formation of Mycorrhizae

Several theories have been proposed regarding factors which control ectomycorrhiza formation. According to Hatch (1937), mycorrhizal infection is conditioned by the internal nutrient state of the short roots. In Björkman's (1942, 1944) view, the prerequisite for mycorrhizae formation is a surplus of soluble sugars in the roots. According to him, the surplus develops when moderate availability of nitrogen and phosphorus is combined with high light intensity. Because of the limited supply of nitrogen and phosphorus, only a small amount of the photoassimilated carbohydrates becomes metabolized into nitrogenous compounds and a surplus of soluble sugars is accumulated. Slankis (1961) pointed out that the formation mechanism of the symbiosis is more complex than previously thought and that, in addition to nutritional factors, growth hormones of the symbiotic fungus are also involved. In his opinion, formation of the symbiotic relationship in ectomycorrhizae is based on a specific physiological state in the roots which develops from an interaction between fungus auxins and root metabolites of the host plant. Hacskaylo and Snow (1959) and Meyer (1962) suggested that the conditions which permit and promote formation of mycorrhizae are controlled by a complex of both internal and external factors.

A. CARBOHYDRATES AND NITROGEN

The paramount role ascribed by Björkman (1942) to excess sugars for formation of mycorrhizae has been recently challenged by several investigators. Warren-Wilson (1951, cited by Harley, 1969) in an analysis of sugar content in roots of experimental *Fagus* seedlings, did not find general correlation between sugar content of roots and mycorrhizal development. Meyer (1962) found a higher reducing sugar content in the roots and a greater abundance of mycorrhizae in *Fagus sylvatica* seedlings grown in nutritionally richer, eutrophic brown earth, than in those grown in podsol. The addition of nitrogenous and phosphatic fertilizers did not reduce, but, in fact, sometimes enhanced, mycorrhizae frequency. From this, Meyer concluded that an increase of soluble sugars in the roots was not the cause, but the effect of the symbiotic association. He believed that the higher reducing sugar level in mycorrhizal root systems was due to the effect of auxin released by the symbiotic fungus. Meyer (1962) based this conclusion on reports that exogenous auxins enhance the hydrolysis of starch into sugars (Borthwick *et al.*, 1937; Stuart, 1938; Alexander, 1938; Bausor, 1942). Schweers and Meyer (1970) arrived at the same conclusion in an experiment in which mycorrhizal and nonmycorrhizal seedlings of *Pinus sylvestris* were subjected to $^{14}CO_2$. After 4-hour photoassimilation of the labeled CO_2 and 4-hour exposure to normal air, mycorrhizal root systems liberated by respiration 40% of the labeled CO_2 assimilated by the needles while uninfected roots respirated less than 10%. After 6½-hour exposure to $^{14}CO_2$, roots of nonmycorrhizal seedlings contained 0.15–0.18% labeled carbon, whereas in root systems with 3 and 10% of short roots converted into mycorrhizae the labeled carbon content reached 0.38–0.45 and 6.99%, respectively. Also the ratio of activity between shoots and roots decreased with increasing mycorrhizae frequency. Richards and Wilson (1963), and Richards (1965) found no correlation between the percentage of mycorrhizae and reducing sugar levels in the roots of Pinus taeda and P. caribaea seedlings.

It has since become apparent that the contradictory data about the carbohydrate status in mycorrhizal roots have arisen from inappropriate analytical methods. As pointed out by Lewis (1963), the methods used by different investigators vary greatly and generally are not specific for carbohydrates, but instead measure the amount of reducing substances. According to Lewis, it is important to realize that the "easily soluble reducing substances" contain compounds which are not necessarily sugars but may be related to polysaccharide levels. Lewis and Harley (1965), employing more advanced methods, have found that in an 80% ethanol extract of beech mycorrhizae, only 50% of the reducing power was due to reducing sugars. Furthermore, as indicated by Harley (1969), mycorrhizae contain several nonreducing disaccharides such as sucrose, which might have been readily hydrolyzed during the preparation and estimated as reducing sugars. Recently, Lister et al. (1968), employing ^{14}C and analyzing the composition and amounts of individual sugars in roots by chromatography found that an excess of soluble sugars can be present in roots at nutritional conditions which, according to Björkman (1942) severely inhibit mycorrhizae formation. For this experiment they used 3-year-old nursery seedlings of Pinus strobus with abundant mycorrhizae and grew them for 13 weeks in several of the nutrient groups used by Björkman (1942). The seedlings were then given $^{14}CO_2$ as the carbon source for photoassimilation. The amount of translocated sugar in the root system was determined after an 8-hour photoassimilation period. As shown in Table III, 95–99.8% of the ^{14}C translocated to the roots was present

TABLE III

DISTRIBUTION OF ^{14}C AMONG THE ETHANOL-SOLUBLE SUGARS IN THE ROOTS OF Pinus strobus L. SEEDLINGS GROWN ON VARIOUS LEVELS OF NITROGEN AND PHOSPHORUS[a,b]

P (mg/liter):	0		173				692			
N (mg/liter):	53	265	0	2.5	53	265	0	2.5	53	265
Sugar	Percent of total ^{14}C in sugar fraction									
Sucrose	55.0	53.0	85.4	78.8	71.1	57.0	82.6	91.4	86.2	70.8
Glucose	14.2	10.9	4.8	7.4	12.1	17.2	4.9	2.9	5.7	8.6
Fructose	13.5	11.0	4.7	7.0	10.8	16.3	4.5	2.6	5.1	8.1
Raffinose	14.9	23.2	4.7	5.8	4.8	6.9	7.7	2.8	2.4	11.3
Unknown	2.4	1.9	0.4	1.0	1.2	2.6	0.3	0.3	0.6	1.2

[a] From Lister et al. (1968) by permission of Ann. Bot.
[b] Duration of $^{14}CO_2$ photoassimilation, 8 hours.

in the sugar fraction. The radioautograms showed that glucose, fructose, raffinose, and sucrose were present in the root systems of all nutritional groups. Except for the amounts of sucrose, which showed a tendency to decrease as the concentration of nitrogen was increased, the amounts of radioactive glucose, fructose, and raffinose were highest in the roots of those seedlings which had received either a moderate (173 mg/liter) or very high (692 mg/liter) phosphorus concentration in combination with the highest nitrogen concentration (265 mg/liter). Despite the presence of reducing sugars in the roots of high nitrogen seedlings, the mycorrhizal infections were inhibited and, as shown by Slankis (1967; cf. also Figs. 10 and 12) in mycorrhizae established prior to the experiment, the symbiotic relationship terminates.

Since the reports regarding soluble sugar content in the roots consider the sugar status of the whole root system, it may be questioned whether the results are applicable to the sugar economy of the short roots in which the actual symbiotic relationship becomes established. Meyer (1966), analyzing reducing sugar content in root regions of 2-year-old *Fagus sylvatica* seedlings with abundantly formed mycorrhizae, found a definite sugar concentration gradient within the root system. The highest sugar content was in the oldest, upper part of the tap root, the lowest in the younger parts of tender roots infected by symbiotic fungi. It has been reported (Slankis, 1967) that glucose, fructose, and sucrose are also present in mycorrhizal and uninfected short roots of *Pinus strobus* seedlings grown in Björkman's (1942) nutrient solutions containing a moderate (173 mg/ liter) amount of phosphorus and either very low (2.5 mg/liter), moderate (53 mg/liter), or very high (265 mg/liter) nitrogen levels. Employing Bidwell's (1962) chromatographic method, amounts as small as 7 mg by fresh weight, were analyzed. Mycorrhizal short roots from the two lower nitrogen levels and nonmycorrhizal short roots from the highest level both produced distinct and nearly similar density spots for glucose, fructose, and sucrose on the chromatograms.

Richards and Wilson (1963) and Richards (1965) stated that mycorrhizal infection in pine roots is more closely related to the soluble sugar–nitrogen ratio in roots than to the carbohydrate state alone. It is of interest that in the early 1940's carbohydrate–nitrogen balance was proposed as a regulating factor also for nodule formation by *Rhizobium* (Fred and Wilson, cited by Wilson, 1940). However, M. Raggio *et al.* (1965) presented evidence that is not compatible with the above theory. Studying the cause of the formation of various types of mycorrhizae, Kazakova (1968) concluded that an important condition in this formation is not the amount of carbohydrates in the roots but the presence of specific substances related to hydrocarbons and the degree of their exosmosis. Similarly, in contrast to the generally accepted view that fertile soils inhibit mycorrhizae formation, evidence continues to accumulate which shows that relatively excessive fertilization with properly balanced nitrogen and phosphorus, and particularly full (N, P, K) fertilizers, does not arrest mycorrhizae formation. Under laboratory (Meyer, 1962, 1966; Koberg, 1966) and field conditions (Scherbakov and Mishustin, 1950; Mishustin, 1951; Göbl and Platzer, 1967; Dumbroff, 1968), it has been demonstrated that such fertilization may produce large-size mycorrhizal seedlings.

It then appears that mycorrhizal infection and establishment of the symbiotic relationship are complex processes affected by a combination of factors rather than merely by soluble carbohydrates. This view has gained increasing support from several investigators (Melin, 1953; Hacskaylo and Snow, 1959; Levisohn, 1960; Slankis, 1961; Meyer, 1962; Harley, 1969; Harley and Lewis, 1969; Hacskaylo, 1971, and others). For additional interpretation of this problem the reader is referred to chapter 6 of this volume.

B. Fungus Growth Hormones

Although the structure of ectomycorrhizae in a natural habitat may show considerable variation, common features are a swollen appearance, lack of root hairs, and variable radial growth of cortical cells within the swollen region. Melin's (1925) mycorrhizae-synthesis experiments with *Pinus* and *Picea* seedlings *in vitro* provided evidence that these structural features develop in the presence of the symbiotic fungus. Later, Slankis demonstrated that similar morphological changes can be induced by the symbiotic fungus culture filtrates (1948) and by supraoptimal concentrations of synthetic auxins (1949, 1950, 1951, 1958). He suggested that mycorrhizae-forming fungi produce extracellular auxin(s), and that these auxins induce the characteristic morphology of ectomycorrhizae. The work of several investigators (Moser, 1959; Ulrich, 1960; Gogala, 1967, 1971) show that a large number of the symbiotic fungi tested liberate IAA and several other indole compounds into the culture medium. More recent experiments provide evidence that several mycorrhizal fungi may also produce cytokinins (Miller, 1967a, 1971; Gogala, 1967, 1971) and compounds related to gibberellins (Gogala, 1967, 1971). Reports by Turner (1962), Shemakhanova (1957, 1958, 1962), and Gogala (1971) confirm the fact that culture filtrates of some symbiotic fungi induce mycorrhizae-like morphological modifications in roots of *Pinus*. It has also been verified that similar structures are induced by synthetic auxins (Palmer, 1954; Shemakhanova, 1962; Chilvers and Pryor, 1965; Clowes cited by Harley, 1969) and cytokinins (Barnes and Naylor, 1959a; Gogala, 1971).

Proof that ectomycorrhizal structures are, in fact, induced by the symbiotic fungus becomes apparent from the observations that, with termination of the symbiotic relationship in mycorrhizae, their characteristic structural features also disappear, and the roots assume the appearance of nonmycorrhizal roots (Slankis, 1967). This has been demonstrated with mycorrhizal *Pinus strobus* seedlings which had been grown for 3 years in a forest nursery and were then transplanted to pots with granitic sand and provided with nutrient solutions in which the nitrogen concentration varied from 2.5 to 265 mg/liter. At nitrogen levels of 2.5 and 51 mg/liter, the newly formed short roots were mycorrhizal and displayed a well-established Hartig net and hyphal mantle (Figs. 10 and 12). In contrast, at nitrogen levels of 159 and 265 mg/liter, the newly formed short roots were uninfected. At the two higher nitrogen levels significant structural changes were displayed by the older mycorrhizae which had already developed in the nursery prior to the experiment. The previously swollen apices had elongated, and the newly formed regions were slender and densely covered with root hairs (Fig. 10). Sections of previously mycorrhizal regions of these roots showed a partly sloughed-off primary cortex with remnants of the Hartig net. In the newly formed tissues, however, no traces of fungal hyphae were found. These newly formed apical regions, resulting from renewed elongation, had normally shaped and oriented cortical cells. At the two lower nitrogen levels, the older mycorrhizae retained their mycorrhizal appearance and their symbiotic state and, when sectioned, displayed a Hartig net and external hyphal mantle (Fig. 12).

Evidence that structural modifications of mycorrhizae are induced by the fungus growth hormones and disappears when the hormonal influence ceases, is also indicated by the fact that mycorrhizae-like root structures, induced either by fungal exudates or synthetic auxins, are not permanent formations, but persist only as long as the exudates or auxins are periodically added to the nutrient solution (Slankis, 1948, 1951). A delay of more than 2 weeks in the addition of the exudates or auxins causes renewed elongation. When such additives are left out altogether, the elongating mycorrhizae-like short roots revert to a nonmycorrhizal appearance. As was the case with genuine mycorrhizae at high nitrogen concentrations, the elongating mycorrhizae-like monopodial and dichotomized roots acquire a slender shape and become densely covered with root hairs (Fig. 4). The remarkable similarity between the two phenom-

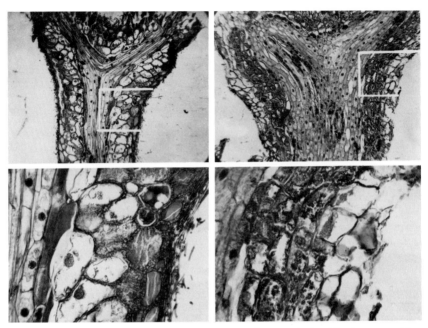

Fig. 12. Nitrogen-induced termination of symbiotic relationship in mycorrhizae of *Pinus strobus* L. nursery seedlings. Above: longitudinal sections of two dichotomized and originally mycorrhizal roots after 7 months of controlled cultivation of seedlings under different N concentrations. Left: N = 2.5 mg/liter; the previously established symbiotic relationship is still present as indicated by the well-developed Hartig net surrounding the hypertrophied cortical cells. Right: N = 265 mg/liter; the previously established symbiotic relationship has terminated as indicated by the partly sloughed-off cortical cells and the hyphae-free secondary tissues. ×80; below, higher magnification (×290) of the outlined areas. From V. Slankis (1967).

ena, one induced by increased availability of nitrogen and the other by witholding auxin, is emphasized by the fact that in both cases a certain number of the elongating mycorrhizae and the mycorrhizae-resembling short roots develop as long roots (cf. Figs. 4 and 11).

These findings suggest that the structural morphogenesis undergone by nonmycorrhizal roots during their conversion into mycorrhizae is induced by fungal growth hormones. Whether these structural and physiological changes in mycorrhizal roots are induced by the fungus-liberated auxins and/or cytokinins remains to be elucidated. Barnes and Naylor (1959a) and Gogala (1971) have reported that cytokinins induce dichotomy in pine roots. However, the dichotomous branching of pine roots induced by cytokinins (Gogala, 1971) only slightly resembles the morphology referred to by Melin as A and B type ectomycorrhizae (cf. Melin, 1927). Svensson (1972) has shown that in the roots of maize and wheat, changes in polarity could be established only in the swellings caused by auxin, both in connection with cell division and cell expansion. In contrast, kinetin-induced swellings were found to be the result of a "balloon effect," i.e., an increase in width of cylindric cells while retaining the cylindrical form. Therefore, detailed data on morphological and histological structure of mycorrhizae-like roots induced by auxins, cytokinins and other substances are required in evaluating their role in formation of the symbiotic relationship.

Although the actual transfer of indole compounds or other hormones by the symbiotic fungus to roots has not yet been demonstrated, an abundance of indole compounds has been found in mycorrhizal roots of pine (MacDougal and Dufrenoy, 1944; Subba-Rao and Slankis, 1959) and in the swollen apical regions of mother roots (Subba-Rao and Slankis, 1959). It has also been demonstrated that exogenous auxin supplied to one region of the root system is translocated throughout the whole system (Fig. 7). The structural features of ectomycorrhizae, such as lack of root hairs and the transverse expansion of cortical cells, indicate the presence of excess auxin in these symbiotic organs. Such morphogenic and histogenic deviations are known to occur in roots of herbaceous plants supplied with exogenous auxin at concentrations above physiological levels (cf. Thimann, 1972). Taking into consideration the fact that roots are very sensitive to auxin, and that auxins take part in many physiological and metabolic processes, it can be expected that the presence of excess auxin in mycorrhizae would profoundly affect their physiology and metabolism. This leads to the conclusion that ectomycorrhizal morphology should not be considered as merely a structural deviation per se from uninfected roots. Rather, these structural features, as suggested by Slankis (1961), reflect a specific physiological and metabolic state, which is necessary for the proper functioning of the symbiosis and which prevails in mycorrhizae as long as the symbiotic fungus is able to synthesize auxin in amounts necessary to stabilize this specific condition.

The fact that increased nitrogen availability terminates the symbiotic relationship in ectomycorrhizae suggests several possibilities. Under such nutritional conditions in roots, the excess auxin could be converted into inactive compounds (Siegel and Galston, 1953; Andreae and Good, 1955; Andreae and Ysselstein, 1956; and others). Alternately, the auxin might be destroyed by the host plant auxin oxidases. It is known that roots normally possess very efficient auxin oxidases (Ray, 1958). However, according to Ritter (1968), symbiotic fungi in pure cultures liberate compounds which are strongly inhibitory to auxin oxidases in the roots. In his view, the hyperauxiny in mycorrhizal roots is more likely to result from the host plant's endogenous auxins than from the fungus auxin. A third pos-

sibility is that high nitrogen concentrations may seriously inhibit production of fungus auxin. It is apparent from a study by Moser (1959) that
mycorrhizae-forming fungi in pure cultures cease extracellular IAA production from tryptophan as precursor if the medium contains large
amounts of amino acids, which stimulate mycelial growth and can be
assimilated more readily than tryptophan. On the basis of experiments
with *Pinus strobus* seedlings, Slankis (1971) has also suggested that high
nitrogen concentrations inhibit the synthesis of fungus auxin. The root
systems of the seedlings were aseptically sealed in large culture tubes
while aerial parts of the seedlings protruded above the protective clo-
sures. The concentration of nitrogen, as ammonium nitrate, in the nutrient
solution was 25 mg/liter. After 10 months, half of the seedlings were subjected to a nitrogen at a concentration of 5 mg/liter and half to 159
mg/liter. Three weeks later, IAA was added to the nutrient solution,
initially at a concentration of 2.5 mg/liter, and then gradually increased
to 5.0 mg/liter. In both groups the exogenous auxin induced apical swellings on old laterals as well as the initiation of many swollen new laterals.
However, the auxin-induced swellings were more pronounced at the
higher nitrogen concentration (Fig. 13). This indicated that the exogenous auxin is not inactivated or destroyed at high nitrogen concentra-

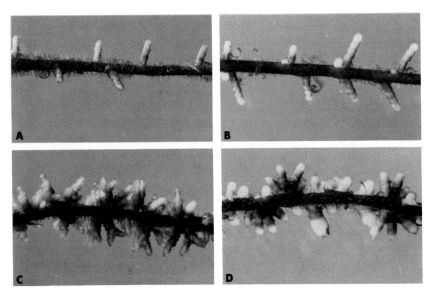

Fig. 13. The degree of mycorrhizae-like swelling induced by IAA (5 mg/liter)
under different nitrogen concentrations in roots of *Pinus strobus* L. seedlings, grown
aseptically in nutrient solutions. (A) 5 mg N/liter, control; (B) 159 mg N/liter,
control; (C) 5 mg N/liter and IAA, (D) 159 mg N/liter and IAA. (×6).

tions but even exerts a stronger effect than at lower nitrogen levels. Therefore, the disappearance of mycorrhizal structures and termination of the
symbiotic relationship in mycorrhizae at high nitrogen concentrations
(see Figs. 10 and 12) cannot be explained on the basis of fungus auxin
increased destruction or inactivation.

In view of the data discussed above, the logical explanation for both
the inhibition of mycorrhizal infection at high nitrogen concentrations
(cf. Hatch, 1937; Björkman, 1942) and the termination of well-established
symbiotic relationships in mycorrhizae subjected to high nitrogen levels
(Slankis, 1967) is that high nitrogen concentrations inhibit the fungus
auxin synthesis. This is supported by data reported by Moser (1959),
that in pure culture the amount of extracellular auxin produced by isolates of symbiotic fungi decreases with increased supply of readily
assimilated nitrogenous compounds. The fact that IAA induces more
pronounced mycorrhizae-like swellings in the roots of *Pinus strobus*
seedlings grown at higher nitrogen concentrations (Slankis, 1971) pre-
cludes speculation about an extensive fungus auxin inactivation or destruction in the host plant roots at high nitrogen levels. Also, as demonstrated by Slankis (1951, 1967), there is a striking similarity between the
reverse morphogenesis undergone by auxin-induced, mycorrhizae-like
root structures when periodic auxin supply is discontinued, and that of
ectomycorrhizae when mycorrhizal seedlings are subjected to increased
nitrogen concentrations. In both cases, these roots revert to nonmycorrhizal root structures (cf. Figs. 4 and 10) and in previously mycorrhizal
roots the symbiotic relationship also terminates (cf. Fig. 12).

Although the above experimental data tend to indicate that the observed inhibition of symbiosis formation at high nitrogen concentrations (Hatch, 1937; Björkman, 1942; and others) may be explained on
the basis of inhibited auxin production by the fungus, the mode of action
of high nitrogen levels on the mechanism of mycorrhizal formation is
far from fully understood. M. Raggio *et al.* (1956) have shown that
rhizobial nodulation in excised bean roots receiving organic nutrient via
the basal end of the roots was inhibited when nitrate was added to the inorganic salts surrounding the roots. However, nodulation was stimulated
if nitrate was supplied with the organic moiety via the basal end of the
root (N. Raggio *et al.*, 1959). In further studies on nodule formation in relation to nitrate and carbohydrate interaction, M. Raggio *et al.* (1965)
concluded that there are two apparently separate effects of nitrate which
depend on the place of application to the root. Thus, nodulation of roots
which were receiving 2% sucrose via their base was inhibited when nitrate,
at a concentration as low as 27 ppm, was added to the medium surrounding the roots, while 814 ppm of nitrate supplied to the base of the roots

did not inhibit nodulation. That the nitrate supplied to the base of the roots had been absorbed and translocated acropetally was indicated by enhanced primary root growth and increase in the number and length of lateral roots. The authors suggested that the inhibition by nitrate was exerted on the Rhizobia themselves, while the stimulatory effect appeared to be exerted on the host or, at least, to be mediated by the host.

The inhibitory effect of high concentrations of inorganic nitrogen on synthesis of symbiotic fungus auxin also might result from its direct influence on the fungus hyphal network in the soil, and/or it might be caused by the inorganic nitrogen metabolized in the roots and exuded by them into the rhizosphere. By exuding of nitrogenous metabolites, roots of the host plant may affect both the fungus virulence and production of auxin even before a symbiosis is established. Depending on the composition of exudates, the mycorrhizal infection can be stimulated or inhibited. Reports show that pine roots exude amino acids (Bowen, 1969), amides (Slankis et al., 1964; Bowen, 1969), organic acids (Slankis et al., 1964), and unidentified metabolites which stimulate the fungus (Melin, 1963). The effect of nutrition on root exudation has been convincingly demonstrated by Bowen (1969). Phosphate deficiency increased exudation of amino acids and amides from roots of *Pinus radiata* seedlings, while nitrate deficiency greatly reduced the amounts. Apparently, the same factors which influence mycorrhizae formation also influence root exudation (Rovira, 1969). Reports by several investigators show that some amino acids, either singly or in certain combinations, can stimulate as well as strongly inhibit mycelial growth of ectomycorrhizal fungi in pure culture (see references in Melin, 1953; Harley, 1969). Moser (1959) found that increased concentrations of both inorganic and organic nitrogen inhibited extracellular auxin production by the fungus in pure culture.

As shown by Lister et al. (1968), an increase in nitrogen concentration changed the amino acid content in roots. After an 8-hour photoassimilation of $^{14}CO_2$, radioactivity in the amino acid fraction from the whole root system of *Pinus strobus* seedlings increased with an increase in nitrogen concentration from 2.5 to 265 mg/liter. α-Alanine, aspartic acid, glutamine, glutamic acid + serine, glycine, and tyrosine and/or β-aminobutyric acid were the main recipients of ^{14}C. High nitrogen levels increased recovery of ^{14}C in some compounds, particularly in glycine, but decreased in others, such as aspartic, glutamic-serine and/or β-aminobutyric acid. Analytical data of Slankis (1971) about amino acid composition in mycorrhizae of *Pinus strobus* seedlings grown under experimental conditions similar to those of Lister et al. (1968) also showed that with an increase in nitrogen concentration, the number of amino acids present, and particularly the amount of some amino acids, increased. In mycorrhizal roots grown at 2.5 mg/liter nitrogen, paper chromatography revealed the presence of aspartic and glumatic acids, glutamine, alanine, and serine. Mycorrhizae grown at 53 mg/liter nitrogen contained larger amounts of glutamic acid and glutamine, and showed as additional amino acids, arginine, lysine, asparagine, and one unidentified compound. As shown by Durzan (1964), high nitrogen availability also appreciably increased the total amount of free alcohol-soluble organic nitrogenous compounds in needles of *Pinus banksiana* and *Picea glauca*. The composition of the compounds also changed qualitatively and quantitatively. The high nitrogen needles of *Picea glauca* produced 19 alcohol-soluble, free nitrogenous metabolites, while the low-nitrogen needles contained only 11. The numbers of such compounds in *P. banksiana* needles were 19 and 15, respectively.

Available data about the mode of action of nitrogen in the complex mycorrhizae-formation process are too scanty to permit any definite conclusions. It seems, however, that the inhibitory effect of high nitrogen concentrations on symbiosis formation cannot be related solely to Björkman's (1942, 1944) postulated deficiency of soluble sugars in the roots. Without denying the general significance of carbohydrates of the host plant as sources of energy and cell development of the symbiotic fungus, it is apparent from the above discussed data, that soluble sugars are present in the root systems and short roots over a wide range of nitrogen concentrations. More likely, the formation of the symbiotic relationship is influenced to a great extent also by the organic nitrogen derivatives in the host plant which, depending on their concentration and composition, may stimulate or inhibit the hormone synthesis by the fungal associate.

C. Light and Auxin

Gast (1937) suggested that the considerable reduction in mycorrhizae frequency of pine at low light intensities is caused by a decrease in carbohydrate production. Subsequently, Björkman (1942) showed that with a decrease in light intensity below 25% of full daylight, mycorrhizae frequency decreased considerably in pine seedlings, and at 6% light intensity no infection occurred. Analyses of the root systems showed that the reducing sugar content of the roots decreased with a decrease in light intensity. Björkman, therefore, concluded that in addition to nitrogen and phosphorous (cf. Hatch, 1937; Björkman, 1942) light intensity also influences the amount of free soluble sugars in the roots, and thus regulates the development of mycorrhizal infection.

Further experiments on the correlation between light intensity and mycorrhizae frequency have brought conflicting results. Some investigators working with different species of *Pinus* obtained results which supported this correlation (Wenger, 1955; Hacskaylo and Snow, 1959; Shemakhanova, 1962). Essentially similar results were obtained with *Fagus* by Harley and Waid (1955). Boullard (1961) found that increasing day length from 6 to 16 hours/day or longer, enhanced root system development, short lateral initiation, and mycorrhizal infection in *Pinus sylvestris, P. pinaster, Cedrus atlantica,* and other tree species. However, in other cases, light intensity and mycorrhizae frequency are not correlated. Mikola (1948) reported that mycorrhizae frequency was higher at 10% than 39% of daylight in *Betula pubescens* inoculated with *Cenococcum graniforme*. Harley and Waid (1955) observed a similar phenomenon in *Fagus,* infected with *C. graniforme;* the highest frequency of mycorrhizae was at the lowest light intensity. Extensive studies by Warren-Wilson (1951, cited by Harley, 1969) on *Fagus* mycorrhiza in woodland soils produced variable results and, in general, did not support the hypothesis outlined by Björkman (1942, 1944).

The correlation between light intensity and soluble sugar content in roots as postulated by Björkman (1942) has not been confirmed either. Handley and Sanders (1962) repeating Björkman's experiment grew uninoculated *Pinus sylvestris* seedlings aseptically in quartz sand, using the same nutrient solution which in Björkman's experiment produced the highest frequency of mycorrhizae. After 4½ months the amount of reducing substances in the root system was determined in plants grown at 50, 25, or 12% of full daylight. Despite considerable differences in the dry matter produced per seedling at various light intensities, there was no apparent decrease in the amounts of readily reducing substances in the roots with decreasing light intensity. In their view, the increased amounts of easily soluble reducing substances found by Björkman in roots with mycorrhizae may have been derived from sugars accumulated in the fungal mycelium.

There is evidence that active photoassimilation is necessary for mycorrhizal development as indicated by the observation that seedlings of various species become mycorrhizal only after their primary leaves have developed (Huberman, 1940; Harley, 1948; Warren-Wilson, 1951; Robertson, 1954; Boullard, 1960, 1961; Laiho and Mikola, 1964). Generally, this phenomenon has been attributed to an increased carbohydrate supply to the symbiotic fungus resulting from active photoassimilation. Yet, there is a report by Mikhalevskaya (1952, cited by Shemakhanova, 1962) about mycorrhizal development in unsprouted *Quercus* germinants. She suggested that photosynthesis was not a prerequisite for formation of a symbiotic relationship. Shemakhanova (1962) comments that in this particular case, mycorrhizae developed at the expense of the carbohydrates stored in the cotyledons. It seems more likely that, in addition to carbohydrates, these relatively large-size cotyledons may also supply other organic metabolites in amounts necessary for mycorrhizal development before photoassimilation begins. Thus, Fortin (1966) demonstrated that well-developed mycorrhizae formed in pine explants (see below) when sucrose, thiamine, and choline chloride were supplied through the attached hypocotyl.

For a better understanding of the mechanism of mycorrhizae formation it is necessary to realize that the influence of light on the host plant is more complex than earlier anticipated. In addition to photoassimilation, light is involved in other biosynthetic processes as well. Some reports show that light enhances the synthesis of phenolic compounds (Zucker, 1963; Engelsma and Meijer, 1965; Jaffe and Isenberg, 1969) and of carotenoids (Virgin, 1967; Valadon and Mummery, 1969). In addition, it has been suggested that abscisic acid is metabolized from carotenoids (Taylor and Smith, 1967) and that growth inhibitors may be formed from certain xanthophylls (Taylor, 1968) by the action of light. Light is also known to increase the amounts of growth inhibitors in plants (Masuda, 1962; Wright, 1968), and Eliasson and Palèn (1972) reported that light enhances the 2,4-D inhibitory effect in roots of pea seedlings. Fortin (1966) showed that specific stem metabolites may participate in development of the root system of pine explants. When the organic nutrient moiety was supplied to the base of excised radicles of 15-day-old *Pinus sylvestris* seedlings, radicles ceased growing after 10 days. However, sustained growth and normal root system development were obtained when a 5-mm-long basal portion of the hypocotyl was left attached to the radicle and the organic constituents of the nutrient supplied through its cut surface.

Studies on factors causing mycorrhizae morphology have revealed that at low light intensity, the fungus auxin cannot induce the specific physiological and metabolic changes in the roots which are required for establishment of the symbiotic relationship. It has been demonstrated (Slankis, 1963) that at a light intensity of 500 ft-c (approximately 5% of full daylight) IAA does not induce mycorrhizae-like swellings in roots of aseptically grown intact *Pinus strobus* seedlings, whereas at 2500 ft-c such swellings are readily formed (Fig. 14). At the lower light intensity, the long roots of control seedlings continued to elongate, though to a much lesser degree than at the higher light intensity, and new laterals were initiated. This indicated that under the lower light intensity, photoassimilation still continued, and the assimilates formed were translocated

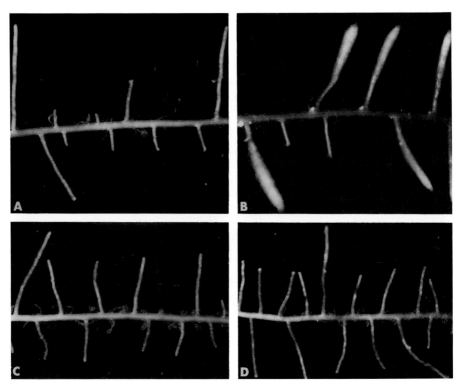

Fig. 14. Induction of mycorrhizae-like root swellings of aseptically grown *Pinus strobus* L. seedlings by IAA in relation to light intensity and supply of sugar. (A) 2500 ft-c, control; (B) 2500 ft-c and IAA; (C) 500 ft-c and IAA; (D) 500 ft-c, IAA and glucose 5 gm/liter. The IAA concentration during a 1½ month period was gradually increased from 0.1 mg to 3.2 mg/liter. (×3.5).

to the roots. The lack of swellings at the low light intensity can hardly be ascribed to a deficiency of soluble sugars in the roots, as an addition of glucose, fructose, or sucrose to the nutrient solution did not bring about any change (Fig. 14). This means that at low light intensities neither auxin alone nor auxin together with soluble sugars induces the profound morphological modifications characteristic for ectotrophic mycorrhizae.[2] It should be noted that swellings are readily induced by exogenous auxins in excised pine roots grown in darkness (Slankis, 1951). One has to agree with the view expressed by Harley (1969) that light-

[2] The writer would like to point out that in some recent publications (Harley and Lewis, 1969; Kozlowski, 1971) these results are erroneously interpreted, giving the impression that also at low light intensity synthetic auxin induces mycorrhizae-like morphology if sugar concentration in the nutrient solution is adequately high.

induced changes in the constitution, growth, or excretion of the roots are not solely or directly due to carbohydrate supply.

D. Concluding Remarks

Despite considerable progress in our knowledge about beneficial effects of ectomycorrhizal symbiosis on the host plant, and the role of the symbiotic fungus in this relationship, the mechanism of formation of the symbiotic association is still poorly understood. It is curious that the study of this process, which is fundamentally important both from a purely academic point of view as well as for applied science, has made so little progress. In contrast, research on *Rhizobium* symbiosis has delved much more deeply into the physiological and metabolic processes involved in the symbiotic relationship. This may be due to the fact that the concept of hormonal involvement in the formation of legume nodules has been accepted more readily than for ectomycorrhizae. Although hormonal involvement in ectomycorrhizal symbiosis was demonstrated more than two decades ago (Slankis, 1949, 1951) this concept has gained recognition only recently.

In the light of information accumulated during recent years on the state of soluble sugars in roots under different nutritional and light conditions, it is apparent that the postulated correlation between carbohydrates of the roots and mycorrhizae formation can no longer be accepted as a satisfactory explanation for mycorrhizal development. To avoid further confusion, it is absolutely necessary to distinguish between the fungus requirement for soluble sugars and conditions under which these sugars become available from the host plant. Since it is true that, except for few species, ectomycorrhizal fungi are unable to utilize complex carbohydrates, there is no reason to minimize the significance of host plant soluble sugars as nutritional source for fungus development. However, data obtained with radioactive carbon tracers and with chromatography, show that soluble sugars are present in roots even at nitrogen levels which inhibit formation of the symbiotic relationship (Lister *et al.*, 1968) and terminate the already established symbiosis (Slankis, 1967). It follows, therefore, that mycorrhizal formation is a complex process which is not conditioned by soluble sugars alone. Even an established symbiotic relationship within the host plant tissues does not necessarily secure permanently the beneficial gains to the fungus. An increase in nitrogen availability can terminate the symbiotic relationship for the fungus despite the fact that soluble carbohydrates are still available in the short roots (see above).

The data discussed in Section IV,A,B strongly suggest that for the formation of the symbiotic relationship in the roots, certain amounts of growth hormones must be provided by the fungal associate. It has been demonstrated that a continuous supply of one of these hormones, namely auxin, is needed to induce and to stabilize mycorrhizae-like morphology in excised roots of *Pinus sylvestris* (Slankis, 1951) and attached roots of *P. strobus* (Slankis, unpublished data). It is apparent that synthesis of auxin is influenced by the concentration of available nitrogen. Data by Moser (1959) show that an ample supply of easily available nitrogenous compounds inhibits extracellular auxin production by the fungus in pure culture.

The possibility that high nitrogen concentrations extensively enhance the fungus auxin inactivation or destruction in the host plant roots seems to be slight. Several investigators have reported that increased availability of nitrogen to nonwoody plants results in a considerable increase in endogenous auxin content. Avery *et al.* (1936) found that under field conditions, the auxin content was proportional, within limits, to the nitrate supplied. Gustafson (1946) working with tomato plants, concluded that poor mineral nutritive conditions generally lowered the auxin content of plants, and Gorter (1954) suggested close correlation between nitrate availability and auxin content in higher plants. Also, as shown by Slankis (1971), IAA induces more pronounced mycorrhizae-like swellings in the roots of *Pinus strobus* seedlings grown at higher nitrogen concentrations (cf. Fig. 13). These data preclude speculation about the fungus auxin being extensively inactivated or destroyed at high nitrogen levels. The only feasible explanation is that ample availability of nitrogen to the host plant inhibits synthesis of fungus auxin. This may explain the phenomenon that, with an increase in nitrogen concentration, mycorrhizal root apices undergo accelerated renewed elongation, root hairs are formed on the newly developed root regions, and concurrently, the symbiotic relationship terminates (Slankis, 1967).

Although it is apparent that the nitrogen concentration affects the fungus auxin synthesis, the mode of action of nitrogen in this metabolic process is not yet understood. Moser (1959) and Ulrich (1960) have shown that IAA and other indole compounds are produced by mycorrhizal fungi in pure culture at moderate nitrogen concentration if tryptophan is available as precursor. Ritter (1968) expressed doubts that the fungus is capable of supplying the required increase in auxin concentrations required to induce the characteristic mycorrhizae morphology which, according to Slankis (1961), reflects a specific physiological state in mycorrhizal roots. Ritter argued that, despite the provision of relatively large amounts of tryptophan in Moser's (1959) experiments, the amounts of auxin obtained in pure cultures were too small to induce the characteristic morphogenic changes. In his view, the concentrations of tryptophan used by Moser do not normally exist either in the host plant or in the soil. Ritter found that mycorrhizae-forming fungi liberate substances inhibitory to auxin oxidases in the roots. On this basis, he postulated that hyperauxiny in mycorrhizal roots originates mainly from the host plant endogenous auxin being protected from destruction by the symbiotic fungus metabolites and that the fungus auxin plays a relatively minor role in this respect.

It is true that in higher plants, tryptophan is rapidly converted into endogenous auxin and is, therefore, seldom available in a free state. However, higher plants contain other compounds which can be readily metabolized into auxins and auxin-related substances (Thimann, 1972). Moreover, Moser (1959), Ulrich (1960a), and Horak (1963, 1964) have shown that some symbiotic fungi are able to synthesize indole compounds from substances other than tryptophan. Gogala (1971) has detected an indole compound, probably tryptophan, in the fruiting bodies and in the culture medium of *Boletus pinicola*. This compound caused a considerable increase in the endogenous auxin content in *Pinus sylvestris* seedlings when supplied to the roots. It seems, therefore, that under certain nutritional conditions, some of the auxin precursors might originate even in the symbiotic fungus itself.

Auxin precursors do not appear to be the only metabolites influencing the extracellular auxin production by the fungus. Rather, the amount of auxin produced is regulated by a complex interaction of several factors. According to Moser (1959), the symbiotic fungus in pure culture does not assimilate tryptophan in the presence of those amino acids which can be more readily assimilated, and, consequently, under such nutritional conditions extracellular auxin is not produced. Lister *et al.* (1968), experimenting with *Pinus strobus* seedlings, showed that the amino acid content in their roots changed with the different nitrogen concentrations in the nutrient. Particularly informative in this connection are data about amino acid content obtained from short roots of *P. strobus* seedlings in which the symbiotic relationship had terminated due to increased nitrogen concentration in the nutrient (Slankis, 1971). In these previously mycorrhizal short roots, compared to mycorrhizal short roots of seedlings still growing at low nitrogen levels, the number of amino acids and especially the amounts of some amino acids had increased. According to Durzan (1964), high nitrogen concentrations increase the number of amino acids in the needles of *Picea glauca* and *Pinus banksiana* seedlings. As shown by Moser (1959), certain amino acids impair auxin synthesis by the symbiotic fungus. These fragmentary data lead to the speculation that amino acids might be factors which affect fungal auxin synthesis.

Furthermore, it is probable that the synthesis of this auxin is to a great extent hormonally controlled. In higher plants, for instance, cytokinin regulates the endogenous auxin content (for references, see Hemberg and Larsson, 1972). It is known that in addition to auxins, mycelia of symbiotic fungi may produce cytokinins (Miller, 1967a, 1971; Gogala, 1971) and gibberellins (Gogala, 1967, 1971). It is possible that under certain conditions these hormones may enhance auxin production in the hyphae to a degree that is autoinhibitory to hyphal development and consequently to auxin synthesis. Fortin (1967) inhibited mycelial growth of mycorrhizal fungi by subjecting pure cultures to increased concentration of synthetic auxins. Gogala (1970) demonstrated that natural auxins, cytokinins, and gibberellins from *Pinus sylvestris* root exudates exert similar inhibitory effects. Probably the lower symbiont's virulence might be decreased by increased auxin levels in the host plant. An increased nitrogen supply considerably increases the endogenous auxin content in higher plants (Avery *et al.*, 1936; Gustafson, 1946; Gorter, 1954). On the other hand, as shown by Gogala (1971), endogenous auxin production in higher plants is also stimulated by metabolites released by symbiotic fungi. Thus a metabolite exuded by the mycelium of mycorrhizal fungus *Boletus edulis* var. *pinicolus* considerably increased the auxin content in pine seedlings. This metabolite was found to be an analogue of Harada's (1962, cited by Gogala, 1971) substance E. A combination of several metabolites of hormonal nature liberated by the same fungus also increased cytokinin levels in the seedlings.

In view of the data discussed above, it is apparent that auxin synthesis by ectomycorrhizae-forming fungi is regulated also by factors other than auxin precursors. Furthermore, the results also suggest that the amounts of extracellular auxins obtained in pure cultures may not necessarily correspond to the amounts produced by the fungus when in association with the host plant.

It is probable that at low nitrogen concentrations, the auxin concentration required for induction of the specific physiological state and, consequently, of the mycorrhizal morphology may arise from a synergistic interaction of several factors. Fortin (1967) has shown that in pure culture, mycelial growth of several symbiotic fungi is inhibited by increased exogenous auxin concentrations. In another experiment, Fortin (1970) observed that the auxin concentration in the nutrient solution rapidly decreased in the presence of excised pine roots. This decrease was directly proportional to the number of root systems per culture dish. Since exudates of the roots did not affect the exogenous auxin concentration, Fortin suggested that the exogenous auxin was actively absorbed by the roots. In his opinion, this protects the symbiotic fungus from autoinhibition by the auxin, and thus secure continuous auxin production required for establishment of the symbiotic relationship. Ritter (1968) found that mycorrhizae-forming fungi of the genera *Amanita, Paxillus, Boletus, Leccinum, Suillus,* and *Xerocomus* liberated several substances which strongly inhibited auxin oxidases in the roots of pine, birch, and bean plants. It is known that increased nitrogen concentration considerably increases the endogenous auxin content in higher plants (Avery *et al.,* 1936; Gustafson, 1946; Gorter, 1954). It is probable, therefore, that at high nitrogen levels the endogenous auxin of the host plant may also inhibit the symbiotic fungus to some extent.

Very little is known about the mechanism that controls mycorrhizal infection. Apparently, the entrance of the symbiotic fungus into the root tissues is affected by several factors both of host and of fungus origin. Palmer (1954) reported that mycorrhizal infection was stimulated when roots of *Pinus virginiana* seedlings were simultaneously inoculated with a suspension of *Amanita rubescens* and subjected to 1:100,000 dilution of IAA. To what extent the fungus auxin in the rhizosphere stimulates mycorrhizal infection is not yet known. Also the host plant root exudates presumably play a significant role in the infection process. Of the exuded metabolites it is probably the amino acids which significantly influence the virulence of the symbiotic fungus and auxin production. Several investigators have demonstrated that some amino acids stimulate ectomycorrhizae-forming fungi, while others strongly inhibit them (for references, see Melin, 1953). According to Moser (1959), the readily assimilated amino acids inhibited fungus auxin synthesis. Since a change in nitrogen concentration also changes the composition of the exuded amino acids (see above) it is possible that at low nitrogen concentrations the amino acids exuded by the host plant into the rhizosphere stimulate mycorrhizal infection while those exuded at high nitrogen concentrations are inhibitory.

Melin (1963) found that roots of *Pinus sylvestris* exude a metabolite (M factor) which significantly stimulates mycelial growth of symbiotic fungi in pure cultures. Nilsson (cited by Melin, 1963) obtained evidence that the stimulatory effect of the M factor can be replaced by diphosphopyridine nucleotide. Gogala (1970, 1971), however, maintains that the M factor is related to cytokinins. She found that *Pinus sylvestris* seedlings contained large amounts of cytokinins and that at certain concentrations the response of the symbiotic *Boletus edulis* var. *pinicolus* was similar to that obtained by Melin (1963) with the M factor on other symbiotic fungi. The cytokinins extracted from pine seedlings stimulated the fungus mycelial growth at low concentrations (up to 100 μg/liter) while at higher concentrations they were strongly inhibitory. Based on these results, Gogala (1971) postulated that the fungal vigor, both before infection and after the symbiosis is established, is controlled by the host plant cytokinins.

Melin (1963) has suggested that the susceptibility to mycorrhizal infection is determined by diffusible inhibitory substances in the roots. Recently, Krupa and Fries (1971) reported that roots of *Pinus sylvestris* seedlings contained volatile metabolites (primarily terpenes and sesquiterpenes) with fungistatic properties. Compared with uninoculated seedlings, inoculation with *Boletus variegatus* resulted in a two- to eightfold increase in the concentration of the volatile compounds. That monoterpenes and sesquiterpenes are inhibitory to symbiotic fungi was demonstrated by Melin and Krupa (1971). They found that the growth of *Suillus* (*Boletus*) *variegatus* and *Rhizopogon roseolus* mycelia in pure culture was inhibited 55–86% when their mycelia were exposed for 5 days to vapors generated by 20 μl of these compounds.

The suggested control of mycorrhizal infection by growth hormones and by specific growth stimulators and inhibitors of either host plant or fungal origin requires further elucidation. It is apparent that the production of one of these participating hormones, namely, the fungal auxin, is controlled by the concentration of available nitrogen. Whether the nitrogen affects the synthesis of other participating hormones and specific metabolites as well, remains to be answered. A deeper insight into the nitrogen-induced physiological and biochemical processes in both the host and the fungus would widen our understanding of the mechanisms by which the mycorrhizal infection, the establishment of the symbiotic relationship, and the termination of already established symbiosis is controlled.

The recently accepted hormonal involvement in the ectomycorrhizal formation mechanism is complex and further studies are required to fully understand it. It is logical to assume that growth hormones and growth regulators supplied by the symbiotic fungus raise the hormonal levels in infected roots. Particularly in cases where the fungus supplied hormone is identical to the host plant's endogenous hormone, the total concentration of this hormone in the roots could reach relatively high levels. Thimann (1936) was the first to demonstrate that structures resembling nodules can be induced in roots of legumes by auxin concentrations above physiological levels. Data discussed in Section III,A show that also ectomycorrhizae-like root structures are induced in pine by increased concentrations of auxin. To what extent the fungus cytokinins are responsible for these structural deviations remains to be elucidated. Recently, several investigators have shown that cytokinins, supplied at concentrations which are too high for formation of new laterals, induced nodule-like structures in tobacco roots (Arora *et al.*, 1959), peas (Torrey, 1962), and *Solanum tuberosum* (Palmer and Smith, 1969). However, according to Svensson (1972), only auxin-induced swelling in the cortex of excised and attached roots of maize, wheat, and pea displayed a polarity change in cell division and cell expansion. Under the influence of cytokinins, only the width of the cylindrical cells increased while the cylindrical form was retained. The polarity change of cortical cells is a characteristic phenomenon in ectomycorrhizae and is particularly expressed in the mycorrhizae of *Eucalyptus* (Chilvers and Pryor, 1965; see also Fig. 12). It has been demonstrated with pine roots (Palmer, 1954; Slankis, 1963) that both expansion and changed polarity of cortical cells can be induced with synthetic auxins alone.

The pronounced deviation in structural features of ectomycorrhizae, caused by symbiotic fungus growth hormones, strongly suggests that the physiological state in mycorrhizal roots is profoundly different from that in uninfected roots. In all probability, the presence of excess hormones in mycorrhizae affects to a great degree the physiological and biochemical processes in these roots. Therefore, it would be of great importance to know the nature and extent of these changes and the resultant effect on the host plant. Auxins and cytokinins deserve the highest priority in these studies because data derived from research in general physiology show that both hormones are involved in many biological and biochemical processes. Auxins and cytokinins have been shown to regulate division, expansion, and differentiation of cells, RNA and protein synthesis, and many other physiological and biochemical processes (cf. Thimann, 1969, 1972; Cleland, 1969; Skoog and Armstrong, 1970; Skoog and Schmitz, 1972). These processes are apparently regulated by a delicate balance of the hormones; therefore, a slight shift in this balance may produce considerably different results. For instance, experiments with tissue cultures show that a high cytokinin–auxin ratio induces formation of shoots in large numbers while only roots are formed when this ratio is appropriately lowered (Skoog and Miller, 1957; Skoog and Schmitz, 1972). Similar observations have been reported for pea roots (Torrey, 1962). In higher plants, cytokinin regulates both the endogenous IAA formation and activity (Hemberg and Larsson, 1972) and auxin transport (for references, see Hemberg, 1972). At low concentrations, cytokinin activates auxin oxidases, while at higher concentrations it inhibits their activity (Lee, 1971).

Björkman (1970) argued that the recently published data on sugar economy in mycorrhizae and on fungal hormonal involvement in the formation of the symbiotic relationship are derived from experiments carried out under very artificial conditions rather than from *in situ* studies where ecologically more relevant results might be obtained. Although these objections have a certain validity, it is also true that our knowledge and understanding of the equally complex *Rhizobium* symbiosis with legume roots has progressed much further as a result of both field studies and carefully planned experiments performed under laboratory conditions *in vitro*.

In conclusion, one must admit that the information presently on hand provides only a few clues about the process through which the symbiotic relationship of ectomycorrhiza becomes established. Much work still must be done before this complex mechnism is fully understood. Even less is known about the processes by which the growth and development of the host plant are stimulated by the symbiotic fungus. With the discovery that the fungus supplies growth hormones and growth regulating substances it has become apparent that the beneficial effect on the host plant cannot be ascribed solely to the increased efficiency of mycorrhizal roots in nutrient absorbtion. The hormone-induced changes in the physiology and metabolism of the higher symbiont must also play a significant role in this stimulation. The exact nature and extent of the fungus hormonal effect on the host plant is not yet known and needs to be further studied. It is a research problem of great complexity because it deals with growth hormones at concentrations above physiological levels and also because of the hereditary variations in the physiology and metabolism of species of both the higher and lower symbiotic associates and even among strains of the same species.

Acknowledgments

The valuable support and helpful cordiality provided by the personnel of the Research Branch, Ontario Ministry of Natural Resources have made possible the writing of this chapter.

In particular, I would like to express my sincere gratitude to Dr. W. R. Henson Director of the Research Branch (presently Director of the Policy Branch), Ontario Ministry of Natural Resources; who, after my retirement, provided me with office space, with facilities for preserving biological research material, and with access to library facilities. I am deeply indebted to Mr. D. H. Burton, Supervisor of the Forestry Section (presently Director of the Forest Research Branch), for his readiness to provide administrative assistance and advice on many technical problems. Also, I wish to express my thanks to Mrs. P. Marks for valuable assistance with library references; and to Miss J. F. Robinson for skillful preparation of illustrative material.

References

Agnihotri, V. P., and Vaartaja, O. (1969). Stimulation of *Witea circinata* by root exudates of *Pinus cembroides. Can. J. Microbiol.* 15, 1319.

Aldrich-Blake, R. N. (1930). The plasticity of the root system of Corsican pine in early life. *Oxford Forest. Mem.* 12, 1.

Alexander, T. R. (1938). Carbohydrates of bean plants after treatment with indoleacetic acid. *Plant Physiol.* 13, 845.

Andreae, W. A., and Good, N. E. (1955). The formation of indoleacetylaspartic acid in pea seedlings. *Plant Physiol.* 30, 380.

Andreae, W. A., and van Ysselstein, M. W. H. (1956). Studies on 3-indoleacetic acid metabolism. III. The uptake of 3-indoleacetic acid by pea epicotyls and its conversion to 3-indoleacetylaspartic acid. *Plant Physiol.* 31, 235.

Arora, N., Skoog, F., and Allen, O. N. (1959). Kinetin-induced pseudonodules on tobacco roots. *Amer. J. Bot.* 46, 610.

Avery, G. S., Jr., Burkholder, P. R., and Creighton, H. B. (1936). Plant hormones and mineral nutrition. *Proc. Nat. Acad. Sci. U. S.* 22, 673.

Barnes, R. L., and Naylor, A. W. (1959a). *In vitro* culture of pine roots and the use of *Pinus serotina* roots in metabolic studies. *Forest. Sci.* 5, 158.

Barnes, R. L., and Naylor, A. W. (1959b). Effect of various nitrogen sources on the growth of isolated roots of *Pinus serotina. Physiol. Plant.* 12, 82.

Bausor, S. S. (1942). Effect of growth substances on reserve starch. *Bot. Gaz.* (*Chicago*) 104, 115.

Bidwell, R. G. S. (1962). Direct paper chromatography of soluble compounds in small samples of tissue adhering to the paper. *Can. J. Biochem. Physiol.* 40, 758.

Björkman, E. (1942). Über die Bedingungen der Mykorrhizabildung bei Kiefer und Fichte. *Symb. Bot. Upsal.* 6, 1.

Björkman, E. (1944). The effect of strangulation on the formation of mycorrhiza in pine. *Sv. Bot. Tidskr.* 38, 1.

Björkman, E. (1970). Forest tree mycorrhiza—The conditions for its formation and the significance for tree growth and afforestation. *Plant Soil* 32, 589.

Bonnett, H. T., Jr., and Torrey, J. G. (1965). Auxin transport in *Convolvulus* roots cultured *in vitro. Plant Physiol.* 40, 813.

Borthwick, H. A., Hammer, K. C., and Parker, M. W. (1937). Histological and microchemical studies of the relations of tomato plants to indoleacetic acid. *Bot. Gaz.* (*Chicago*) 98, 491.

Boullard, B. (1960). La lumière et les mycorrhizes. *Annee Biol.* 36, 231.

Boullard, B. (1961). Influence du photopériodisme sur le mycorrhization de jeunes conifères. *Bull. Soc. Linn. Normandie* [10] 2, 30.

Brown, J. H. (1969). Variation in roots of greenhouse grown seedlings of different scotch pine provenances. *Silv. Genet.* 18, 111.

Bowen, G. D. (1969). Nutrient status effects on loss of amides and amino acids from pine roots. *Plant Soil* 30, 139.

Brushnell, W. R., and Allen, P. J. (1962). Induction of disease symptoms in barley by powdery mildew. *Plant Physiol.* 37, 50.

Carr, D. J., and Burrows, W. J. (1967). Studies on leaflet abscission of blue lupin leaves. I. Interaction of leaf age, kinetin and light. *Planta* 73, 357.

Chilvers, G. A., and Pryor, L. D. (1965). The structure of eucalypt mycorrhizas. *Aust. J. Bot.* 13, 245.

Cleland, R. E. (1969). The gibberellins. In "Physiology of Plant Growth and Development" (M. B. Wilkins, ed.), pp. 49–81. McGraw-Hill, New York.

Clowes, F. A. L. (1951). The structure of mycorrhizal roots of *Fagus sylvatica*. *New Phytol.* **50**, 1.

Dumbroff, E. B. (1968). Some observations on the effects of nutrient supply on mycorrhizal development in pine. *Plant Soil* **28**, 463.

Durzan, D. J. (1964). Nitrogen metabolism of *Picea glauca* (Moench) Voss and *Pinus banksiana* L. with special reference to nutrition and environment. Ph.D. Thesis, Cornell University, Ithaca, New York.

Eliasson, L. (1971). Growth regulators in *Populus tremula* III. Variation of auxin and inhibitor level in roots in relation to root sucker formation. *Physiol. Plant.* **25**, 118.

Eliasson, L., and Palèn, K. (1972). Effect of light on the response of pea seedling roots to 2,4-dichlorophenoxyacetic acid. *Physiol. Plant.* **26**, 206.

Engelsma, G., and Meijer, G. (1965). The influence of light of different spectral regions on the synthesis of phenolic compounds in gherkin seedlings in relation to photomorphogenesis. I. Biosynthesis of phenolic compounds. *Acta Bot. Neer.* **14**, 54.

Fortin, J. A. (1966). Synthesis of mycorrhiza on explants of the root hypocotyl of *Pinus sylvestris* L. *Can. J. Bot.* **44**, 1087.

Fortin, J. A. (1967). Action inhibitrice de l'acide 3-indolyl-acétique sur la croissance de quelques Basidiomycètes mycorrhizateurs. *Physiol. Plant.* **20**, 528.

Fortin, J. A. (1970). Interaction entre Basidiomycètes mycorrhizateurs et racines de Pin en presence d'acide indol-3yl-acétique. *Physiol. Plant.* **23**, 365.

Fox, J. E. (1969). The cytokinins. In "The Physiology of Plant Growth and Development" (M. B. Wilkins, ed.), pp. 85–114. McGraw-Hill, New York.

Gasparrini, G. (1856). "Ricerche sulla natura dei succlatori la escrezione delle radici," p. 1–113. Naples.

Gast, P. R. (1937). Studies on the development of conifers in raw humus. III. The growth of Scots pine (*Pinus sylvestris* L.) seedlings in pot cultures of different soils under varied radiation intensities. *Medd. Skogsförsöksanst. Stockholm* **29**, 587.

Göbl, F., and Platzer, H. (1967). Düngung und Mykorrhizabildung bei Zirbenjungpflanzen. *Mitt. Forstl. Versuchsanst., Wien* **74**, 1.

Gogala, N. (1967). Die Wuchsstoffe des Pilzes *Boletus edulis* var. *pinicolus* Vitt. und ihre Wirkung auf die keimenden Samen der Kiefer *Pinus silvestris* L. *Biol. Vestn.* (*Lublin*) **15**, 29.

Gogala, N. (1970). Einfluss der natürlichen Cytokinine von *Pinus silvestris* L. und anderer Wuchsstoffe auf das Mycelwachstum von *Boletus edulis* var. *pinicolus* Vitt. *Oesterr. Bot. Z.* **118**, 321.

Gogala, N. (1971). Growth substances in mycorrhiza of the fungus *Boletus pinicola* Vitt. and the pine-tree *Pinus sylvestris* L. Dissertation, Cl. IV, (XIV/5). Akad. Sci. Art, Slovenica.

Goldacre, P. L. (1959). Potentiation of lateral root induction by root initials in isolated flax-roots. *Aust. J. Biol. Sci.* **12**, 388.

Gorter, C. J. (1954). Nutrition and auxin-production in seedlings of *Raphanus sativus* and *Zea Mais. Proc. Kon. Ned. Akad. Wetensch., Ser. C* **57**, 617.

Goss, R. W. (1960). Mycorrhizae of ponderosa pine in Nebraska grassland soils. *Univ. Nebraska Coll. of Agric. Res. Sta. Bull.* **192**, 1.

Gruen, H. H. (1959). Auxins and fungi. *Annu. Rev. Plant Physiol.* **10**, 405.

Gunning, B., and Barkley, W. (1963). Kinin-induced directed transport and senescence in detached oat leaves. *Nature (London)* 199, 262.

Gustafson, F. G. (1946). Influence of external and internal factors on growth hormone in green plants. *Plant Physiol.* 21, 49.

Hacskaylo, E. (1971). Metabolite exchanges in ectomycorrhizae. *In* "Mycorrhizae" (E. Hacskaylo, ed.), USDA, Forest. Serv. Misc. Publ. No. 1189, pp. 175–82. US Govt. Printing Office, Washington, D. C.

Hacskaylo, E., and Snow, A. G., Jr. (1959). Relation of soil nutrients and light to prevalence of mycorrhizae on pine seedlings. *U. S. Forest Serv., Northeast. Forest Exp. Sta., Pap.* 125, 1.

Handley, W. R. C., and Sanders, C. J. (1962). The concentration of easily soluble reducing substances in roots and the formation of ectotrophic mycorrhizal associations. A re-examination of Björkman's hypothesis. *Plant Soil* 16, 42.

Harada, H. (1962). Etude des substances naturelles de croissance en relation avec la floraison. *Rev. Gén. Bot.* 69, 201.

Harley, J. L. (1948). Mycorrhiza and soil ecology. *Biol. Rev.* 23, 127.

Harley, J. L. (1969). "The Biology of Mycorrhiza." Leonard Hill, London.

Harley, J. L., and Lewis, D. H. (1969). The physiology of ectotrophic mycorrhizas. *Advan. Microbiol. Physiol.* 3, 53.

Harley, J. L., and Waid, J. S. (1955). The effect of light upon the roots of beech and its surface population. *Plant Soil* 7, 96.

Hartig, R. (1888). Die pflanzlichen Wurzelparasiten. *Allg. Forst-Jagdztg.* 64, 118.

Hartig, R. (1891). "Lehrbuch der Anatomie und Physiologie der Pflanzen unter besonderer Berücksichtigung der Forstgewächse." Springer Verlag, Berlin.

Hatch, A. B. (1937). The physical basis of mycotrophy in *Pinus. Black Rock Forest Bull.* 6, 1.

Hatch, A. B., and Doak, K. D. (1933). Mycorrhizal and other features of the root systems of *Pinus. J. Arnold Arboretum, Harvard Univ.* 14, 85.

Hemberg, T. (1972). The effect of kinetin on the occurrence of acid auxin in *Coleus blumei. Physiol. Plant.* 26, 98.

Hemberg, T., and Larsson, U. (1972). Interaction of kinetin and indoleacetic acid in the *Avena* straight-growth test. *Physiol. Plant.* 26, 104.

Horak, E. (1963). Untersuchungen zur Wuchsstoffsynthese der Mykorrhizapilze. *In* "Mykorrhiza" (W. Rawald and H. Lyr, eds.), pp. 147–163. Fischer Verlag, Jena.

Horak, E. (1964). Die Bildung von IES-Derivation durch ectotrophe Mykorrhizapilze (*Phlegmacium* spp.) von *Picea abies* Karsten. *Phytopathol. Z.* 51, 491.

Huberman, M. A. (1940). Normal growth and development of southern pine seedlings in the nursery. *Ecology* 21, 323.

Itai, C., and Vaadia, Y. (1965). Kinetin-like activity in root exudates of water-stressed sunflower plants. *Physiol. Plant.* 18, 941.

Jaffe, M. J., and Isenberg, F. M. R. (1969). Red light photoenhancement of the synthesis of phenolic compounds and lignin in potato tuber tissue. *Phyton* 26, 51.

Janczewski, E. (1874). Das Spitzenwachstum der Phanerogamenwurzeln. *Bot. Ztg.* 32, 113.

Kazakova, G. I. (1968). The causes of the formation of various types of mycorrhizae. *Uch. Zap. Perm. Gos. Pedagog. Inst.* 64, 228; translation in *Biol. Abstr.* 51, No. 99649 (1970).

Kende, H. (1965). Kinetin-like factors in the root exudate of sunflowers. *Proc. Nat. Acad. Sci. U. S.* 53, 1302.

Koberg, H. (1966). Düngung und Mykorrhiza. Ein Gefässversuch mit Kiefern. *Forstwiss. Centralbl.* **85**, 371.

Kozlowski, T. T. (1971). "Growth and Development of Trees." Vol. II. Academic Press, New York and London.

Krupa, S., and Fries, N. (1971). Studies on ectomycorrhizae of pine. I. Production of volatile organic compounds. *Can. J. Bot.* **49**, 1425.

Laiho, O., and Mikola, P. (1964). Studies on the effect of some eradicants on mycorrhizal development in forest nurseries. *Acta Forest. Fenn.* **77**, 1.

Laing, E. V. (1923). Tree roots. *Trans. Roy. Scot. Arbor. Soc.* **37**, 6.

Laing, E. V. (1932). Studies on tree roots. *Bull. Forest Comm., London* **13**, 1.

Lee, T. T. (1971). Cytokinin-controlled indoleacetic acid oxidase isoenzymes in tobacco callus cultures. *Plant Physiol.* **47**, 181.

Letham, D. S. (1967). Chemistry and physiology of kinetin-like compounds. *Annu. Rev. Plant Physiol.* **18**, 349.

Letham, D. S., Shannon, J. S., and McDonald, I. R. C. (1967). Regulators of cell division in plant tissues. III. The identity of zeatin. *Tetrahedron* **23**, 479.

Levisohn, I. (1952). Forking in pine roots. *Nature (London)* **169**, 715.

Levisohn, I. (1953). Growth response of tree seedlings to mycorrhizal mycelia in the absence of a mycorrhizal association. *Nature (London)* **172**, 316.

Levisohn, I. (1956). Growth stimulation of forest tree seedlings by the activity of free-living mycorrhizal mycelia. *Forestry* **29**, 53.

Levisohn, I. (1960). Physiological and ecological factors influencing the effect of mycorrhizal inoculation. *New Phytol.* **59**, 42.

Lewis, D. H. (1963). Uptake and utilization of substances by beech mycorrhiza. Ph.D. Thesis, Oxford University.

Lewis, D. H., and Harley, J. L. (1965). Carbohydrate physiology of mycorrhizal roots of beech. I. Identity of endogenous sugars and utilization of exogenous sugars. *New Phytol.* **64**, 224.

Lindquist, B. (1939). Die Fichtenmykorrhiza im Lichte der modernen Wuchstoff-forschung. *Bot. Notis.* **1939**, 315.

Link, G. K. K., and Eggers, V. (1941). Hyperauxiny in crown-gall of tomato. *Bot. Gaz. (Chicago)* **103**, 87.

Linnemann, G. (1955). Untersuchungen über die Mykorrhiza von *Pseudotsuga taxifolia* Britt. *Zentralbl. Bakteriol., Parasitenk., Infektionskr. Hyg.* **108**, 398.

Lister, G. R., Slankis, V., Krotkov, G., and Nelson, C. D. (1968). The growth and physiology of *Pinus strobus* L. seedlings as affected by various nutritional levels of nitrogen and phosphorus. *Ann. Bot. (London)* [N.S.] **32**, 33.

McArdle, R. E. (1932). The relation of mycorrhizae to conifer seedlings. *J. Agr. Res.* **44**, 287.

McCready, C. C. (1966). Translocation of growth regulators. *Annu. Rev. Plant Physiol.* **17**, 283.

MacDougal, D. T., and Dufrenoy, J. (1944). Mycorrhizal symbiosis in *Aplectrum, Corallorhiza,* and *Pinus*. *Plant Physiol.* **19**, 440.

McDougall, W. B. (1914). On the mycorrhizas of forest trees. *Amer. J. Bot.* **1**, 51.

McDougall, W. B. (1922). Mycorrhizas of coniferous trees. *J. Forest.* **20**, 255.

Masuda, Y. (1962). Effect of light on a growth inhibitor in wheat roots. *Physiol. Plant.* **15**, 780.

Masui, K. (1926). A study of the mycorrhiza of *Abies firma*, S. et Z. with special reference to its mycorrhizal fungus, *Cantharellus floccosus* Schw. *Mem. Coll. Sci., Kyoto Imp. Univ.* **2**, 15.

Melin, E. (1923). Experimentelle Untersuchungen über die Konstitution und Ökologie der Mykorrhizen von *Pinus sylvestris* L. and *Picea abies* (L.) Karst. *Mykol. Unters.* **2**, 73.

Melin, E. (1925). "Untersuchungen über die Bedeutung der Baummykorrhiza. Eine ökologisch-physiologische Studie." Fischer, Jena.

Melin, E. (1927). Studier över Barrträdsplantans utveckling i råhumus. *Medd. Skogsforsoksanst. Stockholm* **23**, 433.

Melin, E. (1953). Physiology of mycorrhizal relations in plants. *Annu. Rev. Plant Physiol.* **4**, 325.

Melin, E. (1963). Some effect of forest tree roots on mycorrhizal Basidiomycetes. *In* "Symbiotic Associations" (P. S. Nutman and B. Mosse, eds.), pp. 125–45. Cambridge Univ. Press, London and New York.

Melin, E., and Krupa, S. (1971). Studies on ectomycorrhizae of pine. II. Growth inhibition of mycorrhizal fungi by volatile organic constituents of *Pinus silvestris* (Scots pine) roots. *Physiol. Plant* **25**, 337.

Meyer, F. H. (1962). Die Buchen- und Fichtenmykorrhiza in verschiedenen Bodentypen, ihre Beeinflussung durch Mineraldüngung sowie für die Mykorrhizabildung wichtige Faktoren. *Mitt. Bundesforschungsanst. Forst-Holzwirt.* **54**, 1.

Meyer, F. H. (1966). Mycorrhiza and other plant symbiosis. *In* "Symbiosis" (S. M. Henry, ed.), Vol. 1, pp. 171–255. Academic Press, New York.

Mikhalevskaya, O. B. (1952). Biological characteristics of the initial stages of development and growth of oak (*Quercus robur* L.). Ph.D. Thesis, Moscow State University, Moscow (in Russian).

Mikola, P. (1948). On the physiology and ecology of *Cenococcum graniforme* especially as a mycorrhizal fungus of birch. *Commun. Inst. Forest. Fenn.* **36**, 1.

Miller, C. O. (1967a). Zeatin and zeatin riboside from a mycorrhizal fungus. *Science* **157**, 1055.

Miller, C. O. (1967b). Cytokinins in *Zea mays. Ann. N. Y. Acad. Sci.* **144**, 251.

Miller, C. O. (1971). Cytokinin production by Mycorrhizal fungi. *In* "Mycorrhizae" (E. Hacskaylo, ed.), USDA, Forest. Serv. Misc. Publ. No. 1189, pp. 168–174. US Govt. Printing Office, Washington, D. C.

Mishustin, E. N. (1951). Soil microflora and mycorrhiza formation with oak. *Agrobiologiya* **2**, 27.

Mitchell, G. W., and Stewart, W. S. (1939). Comparison of growth responses induced in plants by naphthalene acetamide and naphthalene acetic acid. *Bot. Gaz.* (*Chicago*) **101**, 410.

Mitchell, G. W., and Whitehead, M. R. (1941). Responses of vegetative parts of plants following application of extract of pollen from *Zea mays. Bot. Gaz.* (*Chicago*) **102**, 770.

Möller, C. (1947). Mycorrhizae and nitrogen assimilation. *Forstl. Forsöksv. Danm.* **19**, 105.

Moser, M. (1959). Beiträge zur Kenntnis der Wuchsstoffbeziehungen im Bereich ectotropher Mycorrhizen. *Arch. Mikrobiol.* **34**, 251.

Möthes, K. (1960). Über des Altern der Blätter und Möglichkeit ihrer Wiederverjüngung. *Naturwissenschaften* **47**, 337.

Nielsen, N. (1930). Untersuchungen über einen neuen wachstumsregierenden Stoff: Rhizopin. *Jahrb. Wiss. Bot.* **73**, 125.

Nielsen, N. (1932). Über das Vorkommen von Wuchsstoff bei *Boletus edulis. Biochem. Z.* **249**, 196.

Noelle, W. (1910). Studien zur vergleichenden Anatomie und Morphologie der Koniferwurzeln mit Rücksicht auf die Systematik. *Bot. Z.* **68**, 169.

Palmer, C. E., and Smith, O. E. (1969). Cytokinins and tuber initiation in the potato *Solanum tuberosum* L. *Nature (London)* **221**, 279.

Palmer, J. G. (1954). Mycorrhizal development in *Pinus virginiana* as influenced by growth regulators. Ph.D. Thesis, George Washington University, Washington, D. C.

Pilet, P. E. (1964). Auxin transport in roots. *Lens culinaris. Nature (London)* **204**, 561.

Pozsár, B. L., and Király, Z. (1966). Phloem transport in rust infected plants and the cytokinin-directed long-distance movement of nutrients. *Phytopathol. Z.* **56**, 297.

Raggio, M., Raggio, N., and Torrey, J. G. (1956). The nodulation of isolated leguminous roots. *Amer. J. Bot.* **44**, 325.

Raggio, M., Raggio, N., and Torrey, J. G. (1965). The interaction of nitrate and carbohydrates in rhizobial root nodule formation. *Plant Physiol.* **40**, 601.

Raggio, N., Raggio, M., and Burris, R. H. (1959). Enhancement by inositol of the nodulation of isolated bean roots. *Science* **129**, 211.

Ray, P. M. (1958). Destruction of auxin. *Annu. Rev. Plant Physiol.* **9**, 81.

Rayner, M. C. (1934). The mycorrhiza of conifers: A review. *J. Ecol.* **22**, 308.

Rayner, M. C. (1939). The mycorrhizal habit in relation to forestry. III. Organic composts and the growth of young trees. *Forestry* **13**, 19.

Rayner, M. C., and Neilson-Jones, W. (1944). "Problems in Tree Nutrition." Faber & Faber, London.

Richards, B. N. (1965). Mycorrhiza development of loblolly pine seedlings in relation to soil reaction and the supply of nitrate. *Plant Soil* **22**, 187.

Richards, B. N., and Wilson, G. L. (1963). Nutrient supply and mycorrhiza development in Caribbean pine. *Forest Sci.* **9**, 405.

Ritter, G. (1968). Auxin relations between mycorrhizal fungi and their partner trees. *Acta Mycol., Warsaw* **4**, 421.

Robertson, N. F. (1954). Studies on the mycorrhiza of *Pinus sylvestris* L. *New Phytol.* **53**, 253.

Rovira, A. D. (1969). Plant root exudates. *Bot. Rev.* **35**, 35.

Scherbakov, A. P., and Mishustin, E. N. (1950). Nutritional conditions as means of enhancing growth of oak seedlings and development of mycorrhiza with their roots (in Russian). *Agrobiologiya* **5**, 121.

Schweers, W., and Meyer, F. H. (1970). Einfluss der Mykorrhiza auf den Transport von Assimilaten in die Wurzel. *Ber. Dtsch. Bot. Ges.* **83**, 109.

Scott, T. K. (1972). Auxins and roots. *Annu. Rev. Plant Physiol.* **23**, 235.

Seth, A. K., and Wareing, P. F. (1967). Hormone-directed transport of metabolites and its possible role in plant senescence. *J. Exp. Bot.* **18**, 65.

Shemakhanova, N. M. (1957). Role of mycorrhiza-forming fungi in the nutrition of forest trees (in Russian). *Izv. Akad. Nauk SSSR, Ser. Biol.* **3**, 317.

Shemakhanova, N. M. (1958). Contributions to the nature of mycotrophy in trees (in Russian). *Tr. Inst. Mikrobiol., Akad. Nauk Latvi. SSR* **2**, 7.

Shemakhanova, N. M. (1962). "Mycotrophy of Woody Plants." *Akad. Nauk SSSR,* Moscow (transl., Isr. Program Sci. Transl., Jerusalem, 1967).

Siegel, S. M., and Galston, A. W. (1953). Experimental coupling of indoleacetic acid to pea root protein. *Proc. Nat. Acad. Sci. U. S.* **39**, 1111.

Skinner, J. C. (1952). Genetical variation in excised root cultures of (*Senecio vulgaris* L.). *Heredity (Washington, D. C.)* **43**, 299.

Skoog, F., and Armstrong, D. J. (1970). Cytokinins. *Annu. Rev. Plant Physiol.* **21**, 359.

Skoog, F., and Miller, C. O., (1957). Chemical regulation of growth and organ formation in plant tissues cultured *in vitro. Symp. Soc. Exp. Biol.* **11**, 118.

Skoog, F., and Schmitz, R. Y. (1972). The natural plant hormones. Sect. IX. Cytokinins. *In* "Plant Physiology" (F. C. Steward, ed.), Vol. 6B, pp. 181–213. Academic Press, New York.

Slankis, V. (1948). Einfluss von Exudaten von *Boletus variegatus* auf die dichotomische Verzweigung isolierter Kiefernwurzeln. *Physiol. Plant.* **1**, 390.

Slankis, V. (1949). Wirkung von β-Indolylessigsäure auf die dichotomische Verzweigung isolierter Wurzeln von *Pinus sylvestris. Sv. Bot. Tidskr.* **43**, 603.

Slankis, V. (1950). Effect of α-naphthalene acetic acid on dichotomous branching of isolated roots of *Pinus sylvestris. Physiol. Plant.* **3**, 40.

Slankis, V. (1951). Über den Einfluss von β-Indolylessigsäure und anderen Wuchsstoffen auf das Wachstum von Kiefernwurzeln. I. *Symb. Bot. Upsal.* **11**(3), 1.

Slankis, V. (1958). The role of auxin and other exudates in mycorrhizal symbiosis of forest trees. *In* "Physiology of Forest Trees" (K. V. Thimann, ed.), pp. 427–43. Ronald Press, New York.

Slankis, V. (1961). On the factors determining the establishment of ectotrophic mycorrhiza of forest trees. *In* "Recent Advances in Botany." pp. 1738–1742. University of Toronto Press, Toronto.

Slankis, V. (1963). Der gegenwärtige Stand unseres Wissens von der Bildungder ektotrophen Mycorrhiza bei Waldbäumen. *In* "Mykorrhiza" (W. Rawald and H. Lyr, eds.), pp. 175–183. Fischer Verlag, Jena.

Slankis, V. (1967). Renewed growth of ectotrophic mycorrhizae as an indication of an unstable symbiotic relationship. *Proc. 14th Int. Union Forest Res. Organ. Congr., 1967* Sect. 24, Part V, p. 84.

Slankis, V. (1971). Formation of ectomycorrhizae of forest trees in relation to light, carbohydrates, and auxins. *In* "Mycorrhizae" (E. Hacskaylo, ed.). USDA, Forest. Serv. Misc. Publ. No. 1189, pp. 151–167. US Govt. Printing Office, Washington, D. C.

Slankis, V., Runeckles, V. C., and Krotkov, G. (1964). Metabolites liberated by roots of white pine (*Pinus strobus* L.) seedlings. *Physiol. Plant.* **17**, 301.

Smith, W. H. (1969). Release of microbial nutrients from the roots of selected tree seedlings. *Forest Sci.* **15**, 138.

Stuart, N. W. (1938). Nitrogen and carbohydrate metabolism of Kidney bean cuttings subsequent to rooting with IAA. *Bot. Gaz.* (*Chicago*) **100**, 298.

Subba-Rao, N. S., and Slankis, V. (1959). Indole compounds in pine mycorrhiza. *Proc. Int. Bot. Congr. 9th, 1959* Vol. 2, p. 386.

Svensson, S. B. (1972). A comparative study of the changes in root growth, induced by coumarin, auxin, ethylene, kinetin and gibberellic acid. *Physiol. Plant.* **26**, 115.

Taylor, H. F. (1968). Carotenoids as possible precursors of abscisic acid in plants. *SCI* (*Soc. Chem. Ind., London*) *Monogr.* **31**, 22.

Taylor, H. F., and Smith, T. A. (1967). Production of plant growth inhibitors from xantophylls: A possible source of dormin. *Nature* (*London*) **215**, 1513.

Thimann, K. V. (1935). On the plant growth hormone produced by *Rhizopus suinus. J. Biol. Chem.* **109**, 279.

Thimann, K. V. (1936). On the physiology of the formation of nodules of legume roots. *Proc. Nat. Acad. Sci. U. S.* **22**, 511.

Thimann, K. V. (1969). The auxins. *In* "Physiology of Plant Growth and Development" (M. B. Wilkins, ed.), pp. 1–37. McGraw-Hill, New York.

Thimann, K. V. (1972). The natural plant hormones. *In* "Plant Physiology" (F. C. Steward, ed.), Vol. 6B, pp. 1–365. Academic Press, New York.

Torrey, J. G. (1962). Auxin and purine interactions in lateral root initiation in isolated pea root segments. *Physiol. Plant.* **15**, 177.

Turner, P. D. (1962). Morphological influence of exudates of mycorrhizal and nonmycorrhizal fungi on excised root cultures of *Pinus sylvestris* L. *Nature (London)* **194**, 551.

Ulrich, J. M. (1960a). Auxin production by mycorrhizal fungi. *Physiol. Plant.* **13**, 429.

Ulrich, J. M. (1960b). Effect of mycorrhizal fungi and auxins on root development of sugar pine seedlings (*Pinus lambertiana*, Dougl.). *Physiol. Plant* **13**, 493.

Ulrich, J. M. (1963). Wurzelmorphogenese der Sämlinge von *Pinus lambertiana* (Dougl.) bei Anwesenheit von Mykorrhizapilzen oder Auxinen. *In* "Mykorrhiza" (W. Rawald and H. Lyr, eds.), pp. 165–73. Fischer Verlag, Jena.

Valadon, L. R. G., and Mummery, R. S., (1969). The effect of light on carotenoids of etiolated mung bean seedlings. *J. Exp. Bot.* **20**, 732.

Virgin, H. I. (1967). Carotenoid synthesis in leaves of etiolated wheat seedlings after varying light and dark treatments. *Physiol. Plant.* **20**, 314.

von Tubeuff, K. F. (1888). "Beiträge zur Kenntniss der Baumkrankheiten." Berlin.

Wargo, P. M. (1966). Influence of fungal exudates on growth of excised roots of white oak (*Quercus alba*). Ph.D. Thesis, Iowa State University, Ames.

Warren-Wilson, J. (1951). Micro-organisms in the rhizosphere of beech. Ph.D. Thesis, Oxford University.

Wenger, K. F. (1955). Light and mycorrhiza development. *Ecology* **36**, 518.

Went, F. W., and Thimann, K. V. (1937). "Phytohormones." Macmillan, New York.

Wilcox, H. E. (1968a). Morphological studies of the root of red pine, *Pinus resinosa.* I. Growth characteristics and patterns of branching. *Amer. J. Bot.* **55**, 247.

Wilcox, H. E. (1968b). Morphological studies of the roots of red pine, *Pinus resinosa.* II. Fungal colonization of roots and the development of mycorrhizae. *Amer. J. Bot.* **55**, 688.

Wilkins, M. B., and Scott, T. K. (1968). Auxin transport in roots. III. Dependence of the polar flux of IAA in *Zea* roots upon metabolism. *Planta* **83**, 335.

Wilson, P. W. (1940). "The Biochemistry of Symbiotic Nitrogen Fixation." Univ. of Wisconsin Press, Madison.

Wright, S. T. C. (1968). Multiple and sequential roles of plant growth regulators. *In* "Biochemistry and Physiology of Plant Growth Substances" (F. Wightman and G. Setterfield, eds.), pp. 521–542. Runge Press, Ottawa.

Yeomans, L. M., and Audus, L. J. (1964). Auxin transport in roots. *Vicia faba. Nature (London)* **204**, 559.

Zucker, M. (1963). The influence of light on synthesis of protein and chlorogenic acid in potato tuber tissue. *Plant Physiol.* **38**, 575.

CHAPTER 8

The Rhizosphere of Mycorrhizae

ANGELO RAMBELLI

I. Introduction

A. CONCEPT OF THE RHIZOSPHERE

1. The Problem in Perspective

In 1904, Hiltner introduced the term "rhizosphere" to mean the portion of the soil subject to the influence of the root systems of plants, and he noted that in this area there was greater microbial activity than in portions of the soil further away from the roots.

Since 1904, there have been numerous researches on the subject. Starkey (1929a,b) interpreted the influence of higher plants on the soil in a broader sense. According to him, this influence manifested itself by a diminution in the soil of certain mineral nutrients owing to root absorption, with the partial drying of the soil owing to removal of water by the roots, by an increase in the carbonates produced by roots secreting carbon dioxide, and by the growth of all microorganisms which feed on portions of disintegrating roots or their secretions. Some of these effects directly concern the soil population, others concern the soil population only indirectly by modifying the physical composition of the soil.

Studies have also been conducted on the influence exerted by rhizospheric organisms on plants. This influence is exerted through production of nutrients useful to plants, through decomposition of organic matter in the soil, and through the solvent action of carbon dioxide and other acids produced by those organisms, with transformation of nitrogen and sulfur compounds into readily available forms for which the higher plants compete. A variety of nutrients, minerals, and vitamins are found in the rhizosphere, leading to complex forms of competition and interaction.

From the very earliest investigations, the area of influence of the root has been one of the most important aspects in the study of rhizospheric microorganisms.

Gräf (1930) and Poschenrieder (1930) distinguished between the rhizosphere of the soil adhering to the roots and the rhizosphere of the surface of the roots themselves. The location of the microorganisms on the roots and around the root hairs was demonstrated by the contact slide method (Hulpoi, 1936; Isakova, 1938; Starkey, 1938; Linford, 1939).

To estimate the "rhizospheric effect," based on the number of microorganisms present, it is necessary to make a comparison with the number of microorganisms present in a portion of soil further from the roots. Comparisons with test soils are particularly important in showing a selective action of the roots on defined types or groups of microorganisms.

The number of microorganisms in the root zone is naturally influenced by many factors, such as the nature and age of the plant, soil type, moisture, soil reaction, and agricultural operations, and it is not always possible to determine which of these is responsible for the observed effects. Soil operations, for instance, may stimulate the growth of the plant, which, in turn, might cause an increase in the number of microorganisms on the root surface.

Numerous quantitative comparisons of rhizospheric microorganisms in different plants, grown under identical environmental conditions, have been conducted. However, these data cannot always be compared due to the fact that, as a rule, they are obtained from plants of the same age but assuredly not at the same stage of development, because of their different rates of growth.

Starkey (1929c) noted that the roots of alfalfa (*Medicago sativa*) stimulated fungi only slightly, whereas eggplant (*Solanum molongena*) had a more appreciable effect leading to an increase in the number of actinomycetes and bacteria, while beetroot and corn were less active. These phenomena are even more evident in forests. In fact, in a young forest of relatively undeveloped specimens, the influence of the trees on the microbial flora of the soil is limited, since there is a modest production of root exudates, owing to the scanty root development. These conditions change visibly, however, with tree growth, accompanied by a progressive increase in the root systems which gradually invade the available soil, intertwining and influencing practically all the microbial flora of the soil environment.

In general, the majority of the plants do not exert any appreciable effect during the first stages of growth, and usually the maximum effect appears after the plant has reached a moderate stage of development, at the time of flowering or at the start of the senescing process. The rhizospheric effect, therefore, varies during the plant's growth. The germination of the seed also coincides with the appearance of new soil microorganisms, i.e., those which constitute the spermosphere. The phenomenon also involves liberation of rapidly degradable substances which form the sheath of the seed and stimulate microbial multiplication. Some investigators consider that the microorganisms of the spermosphere take an active part in colonizing the roots (Tardieux *et al.*, 1961), while others make a contrary claim that root colonization is solely the work of soil microorganisms (Parkinson *et al.*, 1963).

In many cases the greatest rhizospheric effect corresponds to flowering or the period preceding this stage (Vančura and Hovadik, 1965). Views of the subject are, however, still at variance, and, according to some investigators, the greatest rhizospheric effect occurs at other stages in the plant's growth (Rivière, 1959). When roots age, the rhizospheric effect is gradually masked by proliferation of microorganisms which intervene in the decomposition of the dead root tissues.

The rhizospheric effect also varies greatly from plant to plant. In forest trees it proceeds regularly and in the majority of cases in proportion to growth. In herbaceous plants, on the other hand, the greatest effect can be characterized by a particular stage of development. Potatoes (*Solanum tuberosum*) and oats (*Avena sativa*) display a rapid increase in the number of rhizospheric microorganisms after the first stage of growth and a progressive decline after maturity and death. Biennial plants show a marked effect towards the later stages of their development.

Gräf (1930), besides noting a distinction between the rhizosphere of the root surface and that of the soil adhering to the root, also noted through weekly checks that a considerable increase in the number of microorganisms on the root surfaces of cereals during the growth period was followed by a slight decrease during the plant's maturation period, and a considerable increase after harvesting. He also noted that the increase in the total number of microorganisms was small in the case of wheat (*Triticum vulgare*) and much greater with legumes and turnips (*Brassica*).

In a number of annual plants Thom and Humfeld (1932) found greater stimulation in the number of bacteria rather than the fungi and the actinomycetes. However, they indicated that this phenomenon appeared to be linked primarily to the nature of the soil. In 1934, Krasilnikov noted that during growth the rhizosphere of groundnuts (*Arachis hypogaea*) hosted 10 to 100 times the number of microorganisms than that encountered in soil distant from the roots. This condition continued until the flowering stage. Toward maturity the number of bacteria decreased, whereas the number of fungi and actinomycetes started to increase.

Obraztzova (1935) pointed out a remarkable rhizospheric effect in the roots of tea (*Camellia theifera*) that was linked to the age of the plant. Twice the number of microorganisms were observed in 2-year-old plants than those present in the rhizosphere of an old plant. In many leguminous plants, the number and activity of the bacterial population of the rhizosphere varies considerably with species and stage of plant development.

Sabinin and Minina (1932) observed an extraordinary example of the influence of root systems on terrestrial microflora in desert soil. Microbial life was present only in the root systems of plants. Clark (1940a), studying the influence of fertilizers on rhizospheric populations, noted a greater number of rhizospheric microorganisms in a treated soil sample than in an untreated soil sample. This difference was also evident in relation to the nonrhizospheric microflora present in a soil that had been given massive quantities of fertilizer. In further studies, he compared the rhizosphere of cotton (*Gossypium*) with that of herbaceous plants and found a greater number of bacteria in the former.

Adati (1939) performed extensive work on the number of bacteria, actinomycetes, and fungi in the rhizospheres of different plants growing in sandy, claylike soils of various types and in humus soils. He made extensive comparisons with soil taken from around roots and some distance away from the roots. Many organisms were found in the soil directly adjacent to the roots, with the number inversely proportional to the distance from the roots. The bacteria were markedly stimulated, the fungi less stimulated, and, in general, the actinomycetes even less stimulated.

Timonin (1940), comparing the rhizospheres of cotton (*Gossypium*), oats, alfalfa, and peas (*Pisum sativum*), found that the bacteria and actinomycetes together were from 7 to 71 times more numerous and the fungi from 0.75 to 3.1 times more numerous in the rhizosphere than in the control soil. The second group of organisms was considerably more abundant in the rhizosphere of oats and barley (*Hordeum vulgare*), while the first was more abundant in that of alfalfa. In general, an increase in the rhizospheric effect was noted during the growth of the plants.

Zukovskaya (1941), studying the rhizospheres of potato, flax (*Linum usitatissimum*), and clover (*Trifolium pratense*), showed that each plant had the capacity to stimulate individually a specific microflora. Berezova (1941) identified different types of organisms on flax plants, associated with the plant at different growth stages. From both the quantitative and the qualitative standpoint, he distinguished the rhizosphere of the soil next to the roots, populated by aerobic bacteria, fungi, and actinomycetes, and the rhizosphere of the root surface with the denitrifying bacteria of the *Agrobacterium radiobacter* group and *Bacillus macerans*.

The large number of microorganisms normally present around plant roots presupposes intensive metabolic activity in this region. This activity has been measured in terms of CO_2 production and naturally is much greater in soils covered by vegetation than in plantless soils. The production of CO_2 stems mainly from two sources: the plant, as a product of respiration of the root cells, and the microbial activity near the roots, which causes decomposition of root secretions. Which of these two sources provides a more abundant quantity of CO_2 depends upon the general environmental conditions, the stage of growth, and the nature of the plant. According to Waksman (1931), 45% of this gas can be produced by microbial activity.

Starkey (1929c) observed that small amounts of CO_2 were produced in the first stages of plant growth, followed by the greatest production at the stage of fastest growth, and finally a decrease during the mature stage. In further studies, Starkey (1931) noted greater amounts of CO_2 in soils taken from the regions of major root development than in soils taken from regions more distant from the roots, thereby again showing a correlation between intensive CO_2 production and abundance of microbes in the soil.

Stille (1938) compared production of CO_2 by mustard roots (*Brassica nigra*) grown in sterile sand and in sand inoculated with a soil suspension. He found greatest CO_2 production in the inoculated sand. Barker and Broyer (1942), checking the production of CO_2 by pumpkin (*Cucurbita maxima*) roots with the addition of manganese, obtained 0.279 μl in a sterile aerobic culture and 0.549 μl in a nonsterile culture, i.e., about 50% of the CO_2 was produced by the roots. Under anaerobic conditions, the sterile system produced 0.108 μl and the nonsterile system 0.289 μl, i.e., about 37% was produced by the roots.

2. Qualitative Aspects of the Rhizosphere

Numerous studies have been made of the qualitative influence exerted by higher plants on the microflora of the soil, especially on bacteria. Investigations have been conducted along two main lines: the study of bacterial groups using selective media and the study of individual bacterial types from morphological and physiological viewpoints. An attempt will now be made to provide an idea of the groups of microorganisms living in the rhizosphere according to a qualitative criterion which stresses the specific function of each group.

Many writers, rather than dwelling on quantitative investigations of a sterile root environment, have considered it important to understand the interactions between plant roots and soil microflora in order to estimate the physiological activity and nutritional needs of the organisms encountered on the root surfaces. The production of CO_2 by the soil has often been interpreted as an index of microbial activity, in view of the greater production of this gas in the soil adjacent to the roots. This activity is related to the abundance of organisms in the soil adjacent to the roots and to the availability of organic substances to be broken down. Thom (1935) observed that many of the organisms associated with the rhizosphere belong to the species active in the decomposition of organic matter. Clark (1940b) noted that the *Pseudomonas* bacteria present in the soil were different from those associated with roots. The majority of those isolated from the soil were unable to utilize saccharose or to hydrolyze amides, whereas those isolated from the rhizosphere fermented the more complex sugars.

By using slides and isolating material on nonselective agar, Lochhead (1940) also found greater microbial activity in the rhizosphere than in the soil distant from the roots. He found a higher percentage of forms in the rhizosphere which grew well on nutrient agar and dissolved the substrate and forms which cause acid or alkaline reactions in the presence of dextrose. He also observed that a large percentage of the bacteria belonged to the motile, chromogenous type. Relatively abundant chromogenous forms were found in the rhizosphere of tobacco (*Nicotiana tabacum*) plants. Lochhead also observed a physiologically more active microflora in the rhizosphere of certain varieties of flax and tobacco.

A qualitative study of soil bacteria, using the technique of Taylor and Lochhead (1938), was carried out by West and Lochhead (1940). They investigated the nutritional needs of rhizospheric organisms isolated with the use of a nonselective cultural medium (soil extract) by adding

TABLE I

COMPARISON BETWEEN THE NUMBERS OF CULTURES WITH POSITIVE GROWTH IN EACH MEDIUM IN *Pinus radiata*, *Pinus pinea*, *Populus euramericana*, AND *Eucalyptus camaldulensis*

Nutritional groups	Rhizosphere (%)				Soil control (%)			
	P. radiata	P. pinea	Populus	Eucalyptus	P. radiata	P. pinea	Populus	Eucalyptus
Good growth in basal medium	11	18.6	0	2	7	12.5	0	2
Require amino acids	17	31.8	21	18	2	10.0	20	3
Require growth factors	3	4.3	14	8	2	3.7	3	2
Require amino acids + growth factors	22	8.7	9	14	0	12.5	11	9
Require yeast extract	19	31.8	19	16	24	16.2	14	67
Require soil extract	3	1.1	0	4	9	20.0	2	0
Require yeast + soil extract	11	2.1	11	9	20	23.7	27	8
No growth	14	1.1	27	29	36	1.2	23	9

amino acids, growth factors, yeast extract, and soil extract to an inorganic medium, and testing the development of individual types in these media. They identified the nutritional requirements of a representative group of rhizosphere organisms. It was thus possible to ascertain that the roots of a number of plants favored the development of bacteria which needed thiamin, biotin, and amino acids for growth and microorganisms capable of synthesizing the essential growth substances on the simple basal medium only. This microbial distribution was typical for a number of plants.

West and Lochhead (1940) adopted the bacterial balance index (BBI) to identify the rhizospheric effect. This index, calculated by assigning a negative value to the percentage of bacteria growing abundantly in a simple inorganic medium and a positive value to organisms which needed amino acids and growth factors, provided interesting data on distribution of the rhizospheric flora.

A more extensive differentiation of nutritional groups of bacteria was proposed by Lochhead and Chase in 1943. The method consists of determining the development that can be noted in seven cultural media of increasing complexity. Using this method, Lochhead and Thexton (1947) and others ascertained the presence of bacteria able to synthesize growth factors and use amino acids in the rhizosphere of a number of plants. Similar results were later obtained by Rambelli (1962, 1963, 1965, 1966), who studied the rhizosphere of the mycorrhizae of *Eucalyptus camaldulensis, Populus euramericana, Pinus pinea,* and *P. radiata.* He obtained good utilization of amino acids in the mycorrhizosphere of the trees studied. With respect to synthesis of growth factors, however, positive results were noted in the rhizosphere of *Populus* and *Eucalyptus,* while the phenomenon appeared uncertain in that of *P. radiata* and *P. pinea.* These results, which indicate some degree of similarity in the physiological activity of the rhizosphere of *P. radiata* and *P. pinea,* are also borne out by checking the behavior of the bacterial types with respect to amino acid and growth factor requirements (Tables I–III). This phenomenon should cause no surprise because, whereas *Populus* and *Eucalyptus* formed mycorrhizae with systematically distant symbionts, *P. radiata* and *P. pinea* both formed mycorrhizae with *Boletus granulatus.* Considering that the rhizospheric microflora studied was isolated from the surface of the fungoid cover, it is logical to assume that this microflora undergoes similar stimuli in the last two plants.

In 1943, Lochhead and Chase also observed that filtered bacterial cultures, grown in a simple inorganic medium, gave a greater stimulus to the organisms which needed amino acids, whereas they exerted an antagonistic action on bacteria which required growth factors. This induced

TABLE II

AMINO ACIDS IN THE NUTRITION OF RHIZOSPHERE AND CONTROL BACTERIA. COMPARISON OF *Pinus radiata*, *Pinus pinea*, *Populus euramericana*, AND *Eucalyptus camaldulensis*

Amino acids	Rhizosphere (%)				Soil control (%)			
	P. radiata	*P. pinea*	*Populus*	*Eucalyptus*	*P. radiata*	*P. pinea*	*Populus*	*Eucalyptus*
No growth or submaximum growth in basal medium								
Maximum growth with amino acids	17	24	19	18	2	8	20	4
Maximum growth with cysteine	0	4	7	11	2	0	2	2
No growth in basal medium								
Growth (maximum or submaximum) with amino acids	19	10	15	14	13	34	23	8
Growth (maximum or submaximum) with cysteine	8	4	9	11	13	10	17	4
Submaximum growth in basal medium								
Stimulated by amino acids	19	45	32	35	13	18	35	40
Stimulated by cysteine	3	14	21	28	4	5	15	20
Submaximum growth in growth factor medium[a]								
Stimulated by amino acids	50	44	28	33	16	30	38	37

[a] See Lockhead and Texton (1947) for nomenclature.

TABLE III

GROWTH FACTORS IN THE NUTRITION OF RHIZOSPHERE AND CONTROL BACTERIA. COMPARISON OF *Pinus radiata*, *Pinus pinea*, *Populus euramericana*, AND *Eucalyptus camaldulensis*

Growth factors	Rhizosphere (%)				Soil control (%)			
	P. radiata	*P. pinea*	*Populus*	*Eucalyptus*	*P. radiata*	*P. pinea*	*Populus*	*Eucalyptus*
No growth or submaximum growth in basal medium or A.[a]								
Maximum growth with growth factors	3	2	14	28	2	1	3	4
Maximum growth, growth factors + amino acids	25	8	16	64	2	11	13	15
No growth in basal medium or A.[a]								
Growth (maximum or submaximum) with growth factors	11	0	12	44	2	0	2	7
Growth (maximum or submaximum) with growth factors + amino acids	11	0	16	60	2	1	5	8
Submaximum growth in basal medium Stimulated by growth factors	6	16	35	41	16	11	17	29
Submaximum growth on A.[a] Stimulated by growth factors	22	16	22	15	7	25	5	29

[a] See Lockhead and Thexton (1947) for nomenclature.

interactions in the rhizospheric microflora, which accompanied the influence exerted by the root exudates and decomposition of the root fragments.

Another important character of the rhizosphere (Wallace, 1947) concerns partial suppression of bacteria dependent on more complex substances contained in the soil extract. The investigations, carried out on herbaceous plants, were confirmed by Rambelli (1962, 1963, 1965, 1966) studying the rhizosphere of *Eucalyptus, Pinus pinea, Populus*, and *Pinus radiata*.

Wallace and Lochhead (1950), while studying the amino acid requirements of rhizospheric bacteria of a number of herbaceous plants, checked the development of strains in a medium containing 23 amino acids. The amino acid group containing sulfur (cysteine, methionine, and taurine) proved of special significance because it was indispensable for development of some microbial groups. Many studies, on organisms able to utilize this group of amino acids, have brought to light the importance of methionine in microbial metabolism.

Krasilnikov and Koreniako (1946) observed that certain species of *Pseudomonas* and *Achromobacter*, present in the rhizosphere of clover (*Trifolium pratense*), strongly influenced development of bacterial root nodules. While a number of species stimulate growth of the bacterial nodules, others inhibit it, and still others show no effect on it whatever.

It is generally admitted that legumes exert a very marked stimulatory effect on *Rhizobium*. By means of slide cultures, the existence was ascertained of an abundant microbial population in the rhizosphere of this plant (Nutman, 1957). In the rhizosphere of the legume *Astragalus* sp. the quantitative rhizosphere soil ratio varies between 3000 and 1000 for *Rhizobium*, but drops to between 100 and 400 for the overall microflora. There is therefore a preferential stimulation of *Rhizobium* which represents 1–10% of the total rhizospheric micropopulation.

Vančura *et al.* (1965) studied the rhizospheric effect on *Azotobacter* in Egyptian soils in which this microorganism is very abundant. The rhizospheric effect does not vary essentially from the description of this phenomenon in scientific literature, although the number of *Azotobacter* colonies in the rhizosphere of a number of plants may increase up to 10^7/gm of soil.

Investigations have also been carried out on the distribution of microorganisms belonging to the genus *Azotobacter* in the rhizosphere of other plants. Nevertheless, despite the interest which these microorganisms hold from a theoretical and a practical point of view, there is surprisingly little information of a quantitative nature concerning their natural participation in the rhizosphere of plants. Beijerinck, as long ago as 1888, observed a greater number of *Azotobacter* organisms in a legume root soil than in a nonlegume root soil. Kostychev *et al.* (1926) found an accumulation of *Azotobacter* in the vicinity of tobacco roots. Comparing rhizospheric soil and nonrhizospheric soil, Gräf (1930) concluded that the most intensive nitrogen fixation by *Azotobacter* and by *Bacillus amylobacter* occurred in the rhizosphere.

In a study on the microflora of the root systems of 31 families of plants, Rokitzkaya (1932) observed *Azotobacter* in all except members of the *Pinaceae, Betulaceae,* and *Fagaceae*. The relative intensity of development of *Azotobacter* was at its maximum in members of the *Polygonaceae, Rosaceae, Pomaceae,* and others.

Undoubtedly many plants (*Pinaceae, Betulaceae, Fagaceae*, etc.) either have no effect on *Azotobacter,* or have no significant effect (Katznelson and Strzelczyk, 1961). Others, such as clover and vetch (*Vicia sativa*), however, seem capable of stimulating growth of *Azotobacter* (Vanćura *et al.,* 1965); while others, such as *Chelidonium* sp. and *Ephedra distachya* (Daste, 1956), inhibit it. Hence, extensive research shows that a single plant could have, according to prevailing ecological conditions, an unfavorable or favorable effect, or no effect on *Azotobacter* development.

According to many investigators, *Pseudomonas* is frequently stimulated by *Arthrobacter*. This can be explained by the capacity of the former to produce substances inhibiting the development of other microorganisms. Chan and Katznelson (1961) found that different species of the genus *Pseudomonas* were not stimulated to the same degree in the different plants.

Using suitable media and microscopic observation, it was shown that the filamentous forms are abundant on the root surface. These are bacteria (*Corynebacterium*) and actinomycetes (*Mycobacterium* and *Nocardia*). *Clostridium* is frequently found in the rhizospheres of lupin (*Lupinus albus*), peas, poppies (*Papaver somniferum*), and radishes (*Raphanus sativus*) (Strzelczyk, 1958). The examples quoted show that it is not possible to detect a generalized behavior pattern in composition of the rhizospheric microflora.

In a broad sense, in the rhizosphere various spore-forming and non-spore-forming bacteria, such as *Bacillus mycoides and B. fluorescens,* play an active part in breaking down protein molecules and the simplest nitrogen compounds to liberate ammonia. Gräf (1930) noted that decomposition of peptone and the production of ammonia occurred abundantly in specimens taken from the root surface, increasing in intensity from May through July and dropping slightly in August. Rokitzskaya (1932) demonstrated the presence of an active ammonifying microflora in the root zone of 178 different plants. Krasilnikov *et al.* (1936) established that the predominant forms in the rhizosphere were usually non-spore-forming bacteria with pronounced ammonifying activity. Lochhead (1940) observed forms that dissolved gelatin, and Timonin (1946) noted that forms which hydrolyzed casein were more numerous in the rhizosphere of various plants than in control soils.

Regarding ammonification, Rambelli (1966), while studying the development of bacteria isolated from the rhizosphere of *P. radiata* mycorrhizae in three different geographic areas of Italy, found a decidedly greater need for amino acids than those present in the control.

Very little research has been carried out on nitrifying bacteria in the rhizosphere. Many comparisons of rhizospheric and nonrhizospheric soil samples have shown the positive influence of plants in the first stage of growth and, in contrast, microbial inhibition in subsequent stages.

Starkey (1929c) compared the accumulation of nitrates in cultivated and noncultivated soil, and, in general, observed an increase in nitrates in the latter. Starkey (1931) showed that nitrification is favored in soils characterized by intense root development. He thus considered it reasonable that the soil constituents required by nitrifying bacteria were present in relative abundance in the vicinity of the roots. Truffaut and Vladykov (1930) isolated strains of *Nitrosomonas* from the rhizosphere of wheat, and Rokitzkaya (1932) observed nitrification of the root zone in all the plants examined. Rovira (1956) studied the effect of root exudates on terrestrial microflora by adding them to the soil. He obtained an increase in the number of gram-negative microorganisms, whereas nitrification and oxygen absorption only increased if glucose or peptone was added. Molina and Rovira (1964) confirmed the stimulation of *Nitrosomonas* and *Nitrobacter* by corn (*Zea mays*) and alfalfa roots. According to Rambelli (1970), nitrite- and nitrate-forming flora are stimulated in the mycorrhizae of *P. radiata*.

Investigations on the denitrifying group are more numerous. Chalvignac (1958), Gräff (1930), and others observed considerable denitrification of the root surface of certain herbaceous plants. Denitrifying organisms have been found in great abundance on the roots of wheat in fertilized soil, but not on alfalfa roots in fertilized soil. Investigations into the presence of denitrifying organisms in the rhizosphere of *P. radiata* (Rambelli, 1970) have shown a reduced number of denitrifiers in the rhizosphere of adult plants, but a number in excess of the control in seedlings grown in plastic pots.

Several investigators have shown that cellulose-decomposing bacteria are present in the rhizosphere of many plants. Krasilnikov (1936) described a marked increase of cellulose-decomposing bacteria in the rhizosphere of wheat, maize, sunflower (*Helianthus annuus*), and soybeans

(*Glycine soja*). He observed that as the plants grew there was an increase in the total number of anaerobic bacteria and organisms which decompose cellulose in the rhizosphere of sunflowers and soybeans. Some investigators claim that these plants take part in the decomposition of dead root fragments and that the products of the process are immediately metabolized by other microorganisms. Katznelson (1946) found numerous aerobic cellulose-decomposing bacteria in the rhizosphere of mangels (*Beta vulgaris* v. *macrorhiza*) grown in fertilized soil, but they occurred far more abundantly in that of mangels grown in unfertilized soil.

Investigations have also been conducted on the rhizospheric microflora capable of oxidizing manganese. This group of organisms has been studied especially in relation to plant diseases caused by manganese deficiency. Timonin (1946) observed a large number of microorganisms capable of oxidizing manganese in the rhizosphere of a variety of oats susceptible to diseases. Such a microbial population, on the other hand, is relatively scanty in the rhizosphere of resistant varieties.

Anaerobic microorganisms have been surprisingly neglected by many investigators working on the rhizosphere. In 1903, Velich found a great abundance of *Clostridium gelatinosum* in the soil close to the roots and the root surface of beet (*Beta vulgaris*), mangels, barley, rye, oats, and wheat. Rokitzkaya (1932) demonstrated the presence of *Clostridium* in the root zone of 31 plant families. The intensity of growth of this microorganism was greatest in the *Chenopodiaceae, Solonaceae,* and *Valerianaceae* and least in the *Pinaceae, Betulaceae,* and *Fagaceae* (the same families that do not possess *Azotobacter* in the rhizosphere). Katznelson (1946) observed a remarkable increase in the number of anaerobic forms on the root surface of mangels grown in both fertilized and nonfertilized soils. Here, too, an anaerobic microbial population developed in the rhizosphere.

3. Nitrogen Fixation[1]

In 1967, Foster and Marks, with the help of an electron microscope, identified bacterial cells in the fungal cover of mycorrhizae of *P. radiata* and suggested that these may be associated with nitrogen fixation. Rambelli (1970), studying the rhizosphere of mycorrhizae of *P. radiata*, isolated, in a pure culture, a bacterial strain that developed well in a liquid mineral medium without any nitrogen and in an environment of air filtered through a 20% sulfuric acid solution. This microorganism seems able to fix small quantities of atmospheric nitrogen. It probably obtains

[1] See Chapter 5.

its nutrition from the fungal symbiont (apart from the possible fixation of atmospheric nitrogen). This bacterium showed rapid and vigorous growth when cultivated in a nitrogen-free Hagem liquid medium (pH 7.3) in an atmosphere devoid of ammoniated nitrogen vapors and with 0.01% of the metabolites extracted from the fungal symbiont of *P. radiata.* This observation probably has some significance, especially in the light of the supposed fixation of atmospheric nitrogen by bacteria within the fungal cover. The phenomenon is an extension of the mycorrhizal concept since this microorganism would be adjacent, for symbiotic purposes, to two other elements characteristic of mycorrhizal symbiosis, the plant and the fungus.

Subsequently, Rambelli (1970) investigated the distribution of the nitrogen-fixing microflora of mycorrhizal and nonmycorrhizal seedlings of *Pinus radiata.* He carried out the following experiment in the nursery of the Rome Agriculture and Forestry Experimentation Center.

Carefully washed siliceous river sand was placed in two wooden boxes $1 \times 1 \times 0.60$ m in size. The sand was washed primarily to remove as much inorganic matter as possible. *Pinus radiata* seeds were sown in one of the boxes. After 8 months growth the seedlings were very small, chlorotic, and lacking in vigor. At that time a certain number of *P. radiata* seedlings of the same age, i.e., 8 months old, grown in plastic pots with soil rich in mycorrhizal organisms, were taken from the Agriculture and Forestry Experimentation Center nursery. The root systems of the latter seedlings were well furnished with mycorrhizae of the chestnut-brown form. These seedlings were considerably better developed. They were given a careful preliminary washing to free their root systems from adhering soil fragments and were then subjected to prolonged washing in running water, after brushing away all traces of mineral fragments or organic substances adhering to the root system. The root portions which could not be washed carefully were eliminated. After this treatment, the seedlings were planted in a second box similar to the first containing washed sand. No fertilizer was added to either box. They were watered periodically in line with ordinary cultural practices.

One month after planting, the seedlings in the second box had taken root and growth had started, although it was not as vigorous as that of plants in plastic pots or in the open field. The seedlings in the first box, on the other hand, were growing slowly, and they exhibited distinct symptoms of chlorosis. After another 3 months, i.e., 12 months after sowing, the differences in size were even more obvious. It then seemed of interest to examine the mycorrhizal condition of both boxes. A few seedlings from each box were sampled and their roots were washed very carefully. The plants from the first box showed a completely nonmycorrhizal root system, the plants in the other box had abundant, well developed, chestnut-brown mycorrhizae. The rhizospheric microflora of the root tips of both boxes was examined as before, by washing a certain number of root tips six times in sterile distilled water. The water from the sixth washing was used for isolation. Twenty-seven bacterial strains were isolated from the nonmycorrhizal tips, and 40 strains from the mycorrhizal tips of the second box. In the first batch, 24 of the strains were able to develop in a mineral medium without nitrogen and in the presence of air filtered through a solution of sulfuric acid. All 40 bacterial strains from mycorrhizal root tips in the second box were able to develop in a liquid mineral medium without nitrogen and in the presence of air filtered through a sulfuric acid solution (Table IV).

4. Succession of Rhizospheric Microflora

Direct microscopic examination has revealed the presence of a considerable population of actinomycetes on the root surface where these microorganisms are most likely to play a prominent role in the relationships of the rhizosphere. Quantitative evaluation techniques generally show a positive rhizospheric stimulation for fungi, but, except for some exceptions, this aspect is less obvious for bacteria. Rovira (1965b) pointed out in this respect that if the rhizospheric effect is considered in terms of the cellular mass and not the number of cells and propagules, the relationship is probably as distinct for bacteria as it is for fungi.

Harley and Waid's techniques (1955) have led to much progress in the study of the fungal rhizosphere. Their method makes it possible to show a clear succession of fungi on the roots in relation to root development (Parkinson *et al.*, 1963). The initial disorderly colonization effected by a fairly large number of fungal species is superseded by stable

TABLE IV

GROWTH OF BACTERIAL STRAINS (ISOLATED FROM MYCORRHIZAL AND NONMYCORRHIZAL TIPS OF *P. radiata* GROWN IN WASHED SAND) IN A NITROGEN-FIXING MEDIUM AND IN AIR FILTERED THROUGH 20% SULFURIC ACID SOLUTION

Bacterial strains isolated from					
Mycorrhizal tips			Nonmycorrhizal tips		
No. of strains	N-fixing agar	N-fixing solution	No. of strains	N-fixing agar	N-fixing solution
40	+38 −2	+40 −0	27	+27 −0	+24 −3

colonization of specialized forms closely dependent on both the plant and soil. As the roots age, the cortical tissue is colonized more and more deeply by the fungi.

Parkinson and Thomas (1969), working with dwarf bean (*Phaseolus vulgaris*) plants, studied changes in the fungal flora during colonization of roots at different stages of growth of the plant. In general, there was weak microbial stimulation in the first stages of growth which increased in parallel with vegetative growth of the plant and lessened considerably during senescence. Respiration studies on samples of rhizospheric and nonrhizospheric soil indicated greater activity in the former. The degree of maximum microbial stimulation appears to correspond to the plant's maximum vegetative activity.

Timonin (1964) studied the rhizosphere of healthy and diseased seedlings of *Pinus contorta*, and showed a distinct difference in rhizospheric organisms present. Isolations from the rhizosphere of plants and seeds of both healthy and diseased plants have shown that *Aspergillus, Phoma, Pythium, Rhizoctonia,* and *Rhizopus* occurred only in the former, and *Alternaria, Cephalosporium, Metarrhizium, Spicaria,* and *Tilachlidium* in the latter. Rambelli (1962), studying the rhizosphere of the mycorrhizae of *Eucalyptus camaldulensis*, identified the following fungi that were absent in the control: *Penicillium implicatum, P. sclerotiorum, P. cyclopium, P. decumbens, P. fellutanum, Aspergillus flavipes, Tieghemella spinosa,* and *Trichoderma koningi*. Rambelli (1965) subsequently observed a smaller number of fungi in the rhizosphere of *Populus euramericana* than in that of the control, and he considered that rhizospheric stimulation is able to select the fungal microflora, enabling only given microbial forms to develop. These findings however do not agree with those made by studying the fungal rhizosphere of mycorrhizae of *Pinus radiata* in three geographically different stations in Italy. In the latter investigations, Rambelli (1966) constantly showed a distinct dominance of the fungal flora in the rhizosphere.

Investigations on the qualitative distribution of the fungal population, however, are relatively scarce. Thom and Humfeld (1932) refer to the colonization of wheat roots in acid soils by *Trichoderma*, while the roots in the alkaline soils appeared associated, above all, with *Penicillium* of the biverticillata series (*P. luteum*). According to Timonin (1940), there is a significant difference in the rhizospheric and nonrhizospheric fungal population of many plants. He provided information in 1941 on the isolation of ten fungi present only in the rhizosphere of flax. Certain genera such as *Alternaria, Aspergillus, Cephalosporium, Fusarium, Helminthosporium,* and *Verticillium* were numerically more abundant in the rhizosphere of varieties of flax sensitive to root rot, while others, such as *Mucor, Cladosporium, Hymenula, Penicillium, Scolecobasidium,* and *Trichoderma,* were more abundant in the rhizosphere of resistant varieties.

Katznelson and Richardson (1943), in a study on the qualitative differences of the fungi of the rhizosphere of the tomato, identified a selective action by the roots, especially regarding an unidentified component of the fungal population. The rhizosphere of plants in soil treated with acids was characterized by forms of *Penicillium* (26%), *Verticillium* (28%), and *Cladosporium* (4%). The control soil, on the other hand, contained *Aspergillus* (6%), *Trichoderma* (15%), *Oothecium* (16%), an unidentified fungus (24%), and other forms such as *Mucor* and *Fusarium.* Their later researches yielded results which indicate a selective action by roots on soil fungi. The remaining groups of microorganisms of the soil, such as protozoa and algae, have not been examined from the qualitative standpoint.

5. Interrelationships of Rhizosphere Microorganisms

Within the rhizosphere, the phenomena of synergism, satellitism, and antagonism are particularly intense. Since these interactions play an important role in the balance between saprophytes and parasitic microorganisms, their study is open to noteworthy applications in biological control of infectious plant diseases caused by bacteria, fungi, and actinomycetes.

Numerous investigations have shown rhizospheric stimulation of a number of microbial associations, generally microorganisms which synthesize and use vitamins and various growth factors. These associations are of some use to the plant, if they are likely to favor its growth directly or indirectly. Synergistic associations can also be formed among phytopathogenic microorganisms which thereby increase their virulence toward the host, e.g., root rot (caused by *Thielaviopsis basicola*) of a number of varieties of tobacco is assisted by rhizospheric bacteria which permit development of the fungal pathogen. In contrast, in species resistant to root rot, there is a rhizospheric stimulation of bacteria which inhibit the development of the parasitic fungus (Strzelczyk, 1966).

Antagonism can also develop among rhizospheric microorganisms. Such phenomena become manifest, with differing intensity, in rhizospheric and nonrhizospheric soils. According to Chan and Katznelson (1961), *Pseudomonas* sp., a typical microorganism of the rhizosphere, and *Arthrobacter globiformis,* a typical microorganism of nonrhizospheric soil, present much more accentuated antagonism when cultivated together on soil extract medium than on root extract. The inhibition of *Arthrobacter globiformis* is said to be due to acidification of the culture medium by *Pseudomonas* sp., and perhaps also to production of antibiotics, favored by their culture medium. The same investigators simultaneously inoculated a root extract medium with *Agrobacterium radiobacter, Arthrobacter citreus, Azotobacter, Bacillus cereus,* and *Pseudomonas* and found a clear-cut inhibition of *Bacillus*, a less marked inhibition of *Azotobacter chroococcum*, and, in contrast, a predominance of *Pseudomonas*, a result similar to the observed distribution in the natural rhizosphere (Chan *et al.*, 1963).

Antibiosis is a particularly spectacular form of antagonism, observed *in vivo*, but its importance in the rhizosphere ranks alongside others, as that stemming from the competition for nutrients or space and the antagonism between parasitic organisms. The latter types of antagonism result in roots being protected from pathogenic microorganisms. Competition among rhizospheric microorganisms suggests that in this environment the sources of nutrition and energy are limited. The frequency with which *Fusarium* is isolated in the rhizosphere is due to the great competitive power of this fungus.

Chalvignac (1966) noted that the selective effect of root exudates on the microflora is the result of competition which occurs among the slow- and rapid-growing species and that the latter are particularly favored in the rhizosphere.

Competition plays an important role in protecting plants attacked by pathogens. In this respect, Katznelson (1965) showed that rhizospheric and soil bacteria can limit the development of a pathogenic fungus of alfalfa roots by competing with the fungus for thiamine and mineral nutrients.

The microbial formation of antibiotics is often greater in the rhizosphere than in nonrhizosphere soil. These microorganisms are essentially actinomycetes whose antagonism is exerted toward bacteria (e.g., *Azotobacter* or *Rhizobium*) or toward fungi. In the rhizospheric environment, production of antibiotics by microorganisms can be shown quite easily, but it is more difficult to determine the actual role of these microorganisms. Production of antibiotics by rhizospheric microorganisms today cannot be denied, even if their presence sometimes seems exceptional. Strzelczyk (1966) found that resistance of certain varieties of flax (*Linum usitatissimum*) to *Fusarium oxysporum, F. lini,* and a number of varieties of tobacco to *Thielaviopsis basicola* is due to the presence in the rhizosphere of a large number of bacteria having great competitive power against fungi and not to the presence of bacteria which synthesize antibiotics.

6. Nutrition of Rhizospheric Microflora

The quantitative and qualitative differences between rhizospheric and nonrhizospheric microorganisms are due, above all, to their different sources of nutrition. Whereas the microorganisms which live in soil are not influenced by root systems because they feed on organic residues in varying stages of decomposition, the rhizospheric microorganisms receive their nutrients from root exudates. These substances, although not always readily utilized by microorganisms, are of great importance to the population of the plant's root surface and, under all circumstances, are the factors which determine the rhizospheric effect (Vančura, 1964).

The process of root exudation is still little understood, since plant physiologists have taken greater interest in root absorption than the reverse process. One of the most interesting methods for demonstrating root exudation consists of applying radioactive substances to the foliage and identifying the radioactive element at the root level. In this way Rakhtcenko (1958) found that after injecting ^{32}P in the leaves of forest trees this element was present in the soil at a distance of 0.25–2 m from the stem. Indirect proof of root exudation was observed in situations where the application of certain substances to leaves led to a spectacular change in the rhizospheric microflora.

According to Katznelson *et al.* (1955), alternating wet and dry conditions in the soil foster exudation of amino acids. High light intensity and temperatures also activate exudation (Vančura, 1967).

Studies on the nature of root exudates are numerous, but it is not possible to give any general rules about the qualitative composition of exudates. This varies with each variety (which clearly explains the great specificity of the rhizospheric effect) and with the state of development of the plant. Melin (1962) demonstrated the presence of a factor strongly stimulating development of mycorrhiza-forming mycelia in pine root culture.

Slankis *et al.* (1964) investigated the composition of root exudates of seedlings of white pine (*Pinus strobus*), grown for 9 months in a controlled environment. Using $^{14}CO_2$ they identified 35 components including two sugars (glucose and arabinose), two amides (glutamic acid and asparagine), and six organic acids (glycolic, malonic, malic, shikimic, oxalic, and *cis*-aconitic acid). Vančura (1964) studied the root exudates of barley and wheat plants in their initial growth stages, and showed through chromatographic separation the following sugars in wheat: maltose, glucose, arabinose, xylose, ribose, and rhamnose; and the following sugars in barley: maltose, galactose, glucose, arabinose, xylose, ribose, deoxyribose, and another unidentified sugar. Vančura and Hovadik (1965), studying the composition of the root exudates found fourteen sugars in watermelon (*Cucumis sativus*),three sugars in turnip cabbage (*Brassica oleracea* v. *gongylodes*), eleven sugars in tomatoes (*Lycopersicon esculentum*), and nine sugars in red peppers (*Capsicum annuum*). Greater quantities of arabinose were found in watermelons, and greater quantities of fructose in tomatoes and red peppers. In turnip cabbage only two unidentified oligosaccharides and xylose were found. These sugars diminish in quantity as the various plants approach flowering stage. Rovira (1956), found that when green peas and oats were grown aseptically in sterile sand, fructose and glucose were exuded by the roots of both plants only during the first 10 days of growth.

Various writers have used the electron microscope to demonstrate a layer of mucilaginous substances or mucigel around the root tips of many plants. Such layers, which are several microns in thickness, apparently are composed of polymers of pectic substances (Jenny and Grossenbacher, 1963; Dart and Mercer, 1964; Rovira, 1965a,b). This mucigel is not only a source of carbohydrates for the rhizospheric microflora, but it also changes the rhizospheric environment, fixing enzymes of plant or microbial origin and extending the ion exchange area around roots.

Root exudates contain many amino acids, such as leucine, isoleucine, valine, glutamine, α-alanine, β-alanine, asparagine, serine, glutamic acid, aspartic acid, glycine, phenylalanine, threonine, tyrosine, proline, lysine, methionine, cystathionine, γ-aminobutyric acid, and tryptophan. The list of these compounds varies considerably according to growing conditions and the plant varieties considered. Scheffer *et al.* (1964) identified homoserine, threonine, α-alanine, glutamine, asparagine, and serine in root exudates of peas, and serine, lysine, and glycine in those of oats.

Quantitative data on production of amino acids are rare. However, an investigation by Paul and Schmidt (1961) is noteworthy because it showed that the rhizospheric soil of soybean contained from 2 to 4 μg of amino acids per gm of soil.

According to Bowen (1969), exudation is directly related to the nutritional state of the plants. He detected greater release of amides and amino acids in *Pinus radiata* seedlings that had developed in nutrient solution lacking phosphates than in plants in complete nutrient solutions. This was ten times greater in plants that had grown in a solution lacking nitrogen.

Vančura (1964) identified fourteen and eighteen amino acids, respectively, in root exudates of barley and wheat. The common amino acids were asparagine, aspartic acid, serine, glycine, glutamic acid, threonine, α-alanine, proline, tyrosine, phenylalanine, isoleucine, and leucine. Those unique to barley were cysteic acid and α-aminoadipic acid, and those unique to wheat were cysteine, cystathionine, glutamine, β-alanine, and γ-aminobutyric acid. Vančura and Hovadik (1965) observed that in watermelon the chief amino acid was β-pyrazolylalanine, which was also present in the free state in the seeds. They observed that turnip cabbage exudates contained eighteen amino acids, with larger quantities of glycine, α-alanine, leucine, and methionine, and that tomato roots in their initial growth stage contained fifteen amino acids, with large quantities of glutamic acid, threonine, and α-alanine, while citrulline, ornithine, and proline were present only as traces. At the flowering stage, tomato root exudates contained tyrosine and three unidentified substances in addition to those found earlier. At this stage the quantities were generally smaller. This means that during the vegetative period of growth quantitative changes occur in the distribution of the individual components of root exudates. They also observed that the root exudates of red peppers contained eighteen amino acids in the initial growth stage, with larger quantities of serine, glycine, threonine, and glutamic acid, while in the flowering stage, tyrosine and six further unidentified substances were found with a relatively greater quantity of asparagine.

Rovira (1956), studying the root exudates of pea and oat plants grown aseptically in quartz sand, noted that the pea plants contained amino acids in larger proportions than in oat plants during the first 21 days of growth. In later studies Rovira (1959) found that the root exudates of tomato (*Lycopersicum esculentum*), subterranean clover (*Trifolium subterraneum*), and *Phalaris tuberosa,* produced in nutrient solutions under different environmental conditions, differed in the amount of amino acids exuded in the three types of plants. It was larger in *Phalaris tuberosa* than in tomato. Exudation appeared more abundant during the first 2 weeks of growth than in the following 2 weeks. Light intensity proved to have a strong influence on root exudation of amino acids in clover. In the exudates of tomato, he found a diminution of aspartic acid, glutamic acid, phenylalanine, and leucine as the light diminished, while serine and asparagine appeared to increase. Finally, high temperatures increased the quantities of amino acids exuded by the roots of tomato and subterranean clover. Furthermore, a number of vitamins were isolated from the root exudates; these included biotin, thiamin, calcium pantothenate, niacin, and riboflavin. Biotin is frequently found in sufficiently high concentrations to be biologically active.

Rovira subsequently (1969) reviewed the role of root exudates and provided interesting data about their nature, the factors which influence exudation, the mechanism of exudation, and the influence of exudates on terrestrial microorganisms.

Sulochana (1962), while examining the root exudates of diploid and amphidiploid strains of cotton in soils with and without *Fusarium*, found choline, pyridoxine, *p*-aminobenzoic acid, and traces of biotin and inositol in the infected soil. In the noninfected soil he found pyridoxine, choline, thiamine, *p*-aminobenzoic acid, and traces of biotin and inositol. Thiamine, biotin, pyridoxine, and *p*-aminobenzoic acid were found in greater quantities in the root exudates of diploid cotton strains grown in soils that were not infected by *Fusarium*. The amphidiploid strains, with very few exceptions, show comparatively smaller quantities of vitamins in the root exudates. It is probable that the vitamin needs of terrestrial microorganisms are satisfied not only by production of root exudates, but also by other rhizospheric microorganisms with which they live in association, and that they are capable of synthesizing vitamins.

Organic acids have also been identified in root exudates. These include formic, acetic, propionic, butyric, valeric, glycolic, oxalic, succinic, fumaric, malic, tartaric, nitric, pyruvic, and oxalacetic. The acids can be exuded in relatively large quantities. Thus a single wheat plant can, up to tillering, viz., in 6 weeks, free 13 mg of acetic acid, 3.5 mg of propionic acid, 2 mg of butyric acid, and 1.5 mg of valeric acid (Rivière, 1959, 1960). Therefore, it is likely that the organic acids from the exudates play an important role in the rhizosphere, as energy substrates for microorganisms and as acidifying agents or in chelating metal ions. Vančura (1964) identified the following organic acids in root exudates of barley and wheat: two keto acids (pyruvic and oxalacetic) and a number of hydroxy acids and di- and tricarboxylic acids (oxalic, malic, glycolic, succinic, and fumaric).

Vančura and Hovadik (1965) identified lactic and oxalic acids in root exudates of watermelon. In the exudates of turnip cabbage they found, in addition to the above-mentioned acids, malic and glycolic acids, while the root exudates of tomato and red peppers, in the initial growth stage, have the same organic acids (oxalic, citric, malic, glycolic, succinic, and fumaric). These same organic acids were identified in root exudates of tomato during the flowering stage, but in distinctly smaller quantities. Oxalic, citric, and fumaric acids were identified in root exudates of red pepper during flowering, but the other organic acids produced during the initial stage of growth were not noted.

7. Other Factors Influencing Composition of the Rhizosphere

Rhizospheric soil is generally richer in enzymes than is nonrhizospheric soil. The enzymes are of both plant and microbial origin. Enzymes of plant origin may be exuded or arise from deteriorating root cells. The latter case, however, is not one of exudation in the strict sense. Considerable work has been carried out on root exudation of enzymes. Knudson (1920) found glucose and fructose present in a saccharose solution in which plants developed aseptically. He claimed that this indicated that the root exudes invertase. Subsequently, however, Knudson (1920) showed that the saccharose is first absorbed by the roots and then hydrolyzed to glucose and fructose, and small amounts of these sugars are then exuded. Krasilnikov (1952) identified invertase, amylase, and protease enzymes in wheat, maize, and pea plants. The enzyme polygalacturonase was identified in exudates of alfalfa and clover. This enzyme manifests itself only in the presence of *Rhizobium* sp. Krasilnikov and Kotelev (1959) established that the enzyme phosphatase is exuded by maize. When the atmosphere is saturated with moisture, plants are able to exude appreciable amounts of water (Schippers *et al.*, 1967).

Apart from the substances listed, products derived from degradation of nucleic acids, flavones, glucosides, etc., have recently been discovered in root exudates. The exudates are not released uniformly along the whole root system, instead they are produced in greatest concentrations in the root elongation zone and at the root hairs, i.e., where metabolic activity is more intense and permeability is greatest.

Quantitative data on the amount of root exudates are relatively rare and, except for some exceptions (Harmsen and Jager, 1963), they have been established on the basis of hydroponic cultures which certainly do not reflect the course of the phenomenon in nature.

According to Demidenko (quoted by Macura, 1966), the weight of the root exudates accounts for 27% of the plant mass (wheat and tobacco). Vanćura (1961), however, proposed considerably lower values, e.g., 7–10%. These assessments are almost of the same order or slightly higher than Meschkov's (quoted by Macura, 1966); according to the latter investigator, the exudates of maize, expressed as carbon, represent 1.0–1.7% of the plant mass. It is therefore probable that production of exudates in soil is at least as great as in hydroponic cultures, since they have undergone biodegradation by the rhizospheric microflora or have been absorbed by soil colloids.

While research on chemical composition of root exudates has increased rapidly in the past 10 years, information on the effect of these exudates on the different sectors of the pathogenic or saprophytic microflora is still incomplete. One of the major difficulties is the fact that the action of the exuded substances varies according to their concentration in the rhizosphere (Daste, 1956) and that the concentration of these substances, their area of diffusion, their importance, and the speed of their inactivation through biodegradation or through absorption of soil colloids are not known. The variations in concentration of biologically active substances in root exudates could explain the sequence of the phenomena of stimulation and depression frequently observed in certain microorganisms, such as *Azotobacter* sp. and nitrite-forming bacteria (Molina and Rovira, 1964). However, there are cases in which the stimulating or inhibiting effect is well established in particular experimental conditions. These quantitative and qualitative indications are also incomplete because in the majority of studies the fraction or fractions derived from metabolism of the symbiont fungi, normally found on the root tips in symbiotic association, are not considered. Above all, in plants characterized by ectotrophic mycorrhizal symbiosis, the contribution of these metabolites is undeniable and significant enough to modify the composition of the exudates which are liberated in the adjacent soil.

According to Rubenchik (1960), *Beijerinckia indica* is stimulated by root exudates of sugar cane (*Saccharum officinarum*), and *Azotobacter chroococcum* is stimulated by exudates of *Canna indica*. Cátská (1965) ascertained that fungi (*Rhizopus arrhizus, Mucor racemosus, Fusarium oxysporum,* and *Penicillium chermesinum*) isolated from the root surface of wheat fructify better when they are close to germinating plants than in the soil.

Experiments with pathogenic fungi (*Fusarium oxysporum, F. pisi,* and *Plasmodiophora brassicae*) show that plants resistant to these parasites exude substances which inhibit spore production, whereas susceptible plants stimulate this process (Buxton, 1957; Bochow, 1965). However, the stimulatory effect of root exudates on spore production in pathogenic fungi does not always manifest itself. Schroth and Hendrix (1962) observed that formation of chlamydospores in *Fusarium solani f. phaseoli* took place in the presence of exudates of susceptible and resistant plants.

According to Samtsevich (1965), during the first stages of growth, roots produce considerable quantities of toxic substances, limiting microbial development; this later diminishes, thereby permitting formation of the rhizospheric zone by the soil microorganisms. Certain toxic substances have been isolated and identified. In the root tips of oats, Schönbeck (1958) isolated a toxic glucoside which inhibited certain fungi. Various phenolic substances (isochlorogenic acid, chlorogenic acid, gallotannic acid, and gallic acid) have also been isolated (Vanćura, 1964; Rice, 1967). These inhibiting substances play a dual role. First they protect plants against infection by pathogenic organisms such as fungi, bacteria, and possibly viruses. As shown earlier, plants that are resistant to pathogenic fungi owe this property to the nature of their exudates which inhibit the germination of spores of the parasite. This situation received further confirmation from Strzelczyk's (1961) work on flax. The varieties resistant to *Fusarium* were those which exuded substances limiting the development of the rhizospheric micropopulation. Second, competition among plants is seriously affected. Rice (1967) showed that a number of plants requiring little mineral nitrogen (*Ambrosia elatior, Euphorbia corollata,* and *Helianthus annuus*) produced inhibitors of nitrifying and nitrogen-fixing microorganisms that gave them a considerable advantage over plants whose nitrogen needs were greater. This competitive mechanism would appear to slow down the speed of succession of plant species.

8. Conclusions

From all these investigations it is clear that formation of the rhizospheric environment proceeds in ways and with mechanisms that are substantially common to the majority of plants. However, the phenomenon is not unexpectedly linked also to multiple environmental factors and to certain individual characteristics of the plant.

There is little doubt that with regard to the formation of the rhizospheric environment, a broad division could be made between mycorrhizal and nonmycorrhizal plants. The former, which are more common, certainly produce stimulatory effects in the soil microbial environment that are substantially different from the latter. The latter condition is encountered less frequently but must necessarily be taken into consideration, since several investigations have been conducted on plants grown in unnatural conditions that do not permit mycorrhizae to form.

Mycorrhizal roots could be divided further into ecto- and endo-mycorrhizae. The former include nearly all forest plants, and the rhizospheric phenomenon is typical of plants characterized by a metabolism linked to slow regular growth in equilibrium with the environment. These conditions affect a large volume of soil, and the mycorrhiza-forming fungus plays a very important role. The second group is characteristic of most herbaceous plants, and the rhizospheric phenomenon is typical of fast-growing plants that have a relatively intense metabolism in a moderate soil volume in which the mycorrhiza-forming fungus plays a lesser role.

There is little doubt that metabolism of these plants is profoundly influenced by the presence of mycorrhizae. Hence studies on the formation and composition of the rhizosphere should be made on mycorrhizal plants. Many workers, as far as can be ascertained, have transferred results obtained from laboratory studies on nonmycorrhizal plants to the natural environment, without considering that, under natural conditions, they may form mycorrhizae and as a consequence their rhizosphere environment would be completely different.

The trees forming the forests in tropical and equatorial areas must be considered separately. In these areas, the majority of the trees are characterized by having endomycorrhizae; their root systems occupy a large volume of soil and very often develop with great rapidity and thus possess a very active metabolism. The formation of the rhizospheric environment in these plants, the type of stimulation, and the interactions that develop in these ecosystems have been little studied and deserve greater attention.

B. Formation of the Rhizosphere Ambient

The rhizosphere is characterized by a particular type of microbial population (bacteria, actinomycetes, and fungi) that is unique in its quantitative and qualitative features. In order to understand the formation of this environment, a hypothetical soil will be considered that is absolutely devoid of vegetation in which the microorganisms are evenly distributed and live together in perfect ecological equilibrium. The germination of the seed of a higher plant in such a soil, and the penetration and development of a rootlet, will cause natural and obvious changes in the balance of the microbial population. These changes stem not only from the absorption of nutrient substances by the growing plant, which are taken away from the terrestrial microflora, but also from secretion of the by-products of the plant's metabolism in the soil around the root tips. These substances, or "root exudates" have, as indicated, a complex composition. They contain organic acids, vitamins, free amino acids, and large protein molecules.

The accumulation of exudates near the root tips stimulates multiplication of all the microbial forms (bacteria, actinomycetes, and fungi) present in the soil around the tips, which are able to break the protein macromolecules and to use them, at least in part, for nutritional purposes. The multiplication of these microbial forms inevitably inhibits the development of others which are unable to break down the protein macromolecules and are thus at a disadvantage in the particular nutritional environment created.

Fragments of the protein macromolecules are released after they have been partially utilized by the microbes. These substances are displaced by new exudates and by solutions circulating in the soil, and in this way they come into contact with other terrestrial microorganisms which further degrade them for nutritional purposes, thereby reducing them to still simpler forms. The whole system of rhizospheric microorganisms therefore undergoes change, starting from the root surface and gradually moving through the soil away from it, for a distance of up to a few millimeters. If these microorganisms have the capacity of using the substances normally secreted by the root tips, consideration must also be given to their distribution because of their intimately interlinked metabolic capacities.

It is easy to understand from the above how well-balanced microflora in the hypothetical soil evolved toward another form of equilibrium, regulated and conditioned by factors outside of the microbial ambient, i.e., by vegetational factors, typical of the majority of ecological relationships.

Naturally, the rhizospheric microbial groups will become more and more defined by the development of plants and, consequently, by the quantity of catabolites produced by their roots. In view of the composition of the root exudates, the ammonifying group of bacteria is normally well represented in the rhizosphere. These microorganisms are able to attack proteins, and, as the final stage in the degradation process, produce ammonia.

Fungi are also abundantly represented in the rhizosphere. Recent investigations, however, have brought to light a certain specificity between the rhizospheric fungal population and the host. In this way, each host, on the basis of the constituents of the root exudates produced, seems able to select a given fungal microflora, and among them a number of species are dominant.

C. Formation of Mycorrhizae inside the Rhizospheric Ambient[2]

The mycorrhiza-forming fungi are undoubtedly among those best able to degrade the protein molecules which comprise the root exudates. For adequate understanding of formation of mycorrhizae in the rhizosphere, the pedological ambient described previously will again be considered. Seed germination, development of vegetation, and liberation of root exudates in the zone of soil surrounding the tips gave rise, as seen, to ordered development of given microbial groups. Among the complex of terrestrial microflora there will also be mycelial fragments, and spores of mycorrhiza-forming fungi. Germination of these propagules results in production of a mycelial thread which will spread out in the surrounding soil. When the root exudates, even if considerably altered by rhizospheric microflora, come into contact with mycelium of the mycorrhiza-forming fungi, the hyphae are stimulated to develop, especially in the direction of the zones where such exudates are in the greatest concentrations. The mycorrhiza-forming hyphae thus compete with the rhizospheric microflora for these exudates and develop toward the source of these substances until the root tip is reached. Once the tip has been

[2] See Chapters 1, 4, 6, and 7.

reached, they colonize the root surface and utilize the exudates in active competition with the other microorganisms.

At this point, the fungus covers the root tips with an initially loose mycelial network, which becomes increasingly dense until it forms the "mycoclene" or fungus cover. The fungus then penetrates the root tissues of the host and starts the process of mycorrhizal symbiosis.

From this moment on the rhizosphere phenomena undergo a total change. The exudates first liberated directly in the soil around the tips are now filtered by the fungus cover and are partially utilized by the mycorrhizal fungus for nutritional purposes. The exudates are modified profoundly when they reach the soil, especially by products of metabolism of the mycorrhiza-forming fungus. In the soil around the mycorrhizal root tips very complex substances are now released which can be attacked and utilized, at least partially, only by certain microorganisms capable of metabolizing them. Parallel to these developments in the rhizospheric ambient, the first metabolites released are degraded into simpler forms that could be used by microorganisms with less demanding nutritional requirements. In this way a distribution pattern of the rhizospheric microorganisms can be imagined, beginning from the mycorrhizae, with which they are intimately connected, and extending into the surrounding soil. These microorganisms are thus in a balanced and interrelated distribution, depending on their ability to utilize root and mycorrhizal exudates.

D. Concept of the Mycorrhizosphere

Root exudates and metabolites of the mycorrhiza-forming fungus are therefore the basis of microbial stimulation which precede formation of the rhizospheric ambient of the mycorrhizae. Not many studies exist on this subject. Tribunskaya (1955), comparing the rhizosphere of mycorrhizal and nonmycorrhizal pine seedlings, identified a considerable increase and a change in composition of the rhizospheric microflora around the mycorrhizal rootlets. The author found nine to ten times as many fungi in the rhizosphere of the latter and a large number of bacterial saprophytes and bacteria capable of degrading organic substances. In both rhizospheric ambients investigated, the development of *Azotobacter* was inhibited in some way. Ivarson and Katznelson (1960) studied the rhizosphere of birch seedlings grown in forest soil under controlled conditions. They probably used mycorrhizal plants. A marked rhizospheric effect was noted, especially 28 weeks after the end of the dormancy period of the seedlings. This effect was associated with an increase in the number of ammonifying bacteria, that reduced methylene blue, and with gas-producing anaerobes. Later, Katznelson *et al.* (1962) studied the rhizosphere in mycorrhizal and nonmycorrhizal *Betula alleghaniensis* roots. In this case, also, the rhizospheric effect was greater in the mycorrhizal rootlets, and they identified a large number of bacteria and actinomycetes, as well as bacteria capable of reducing methylene blue, fermenting glucose, and promoting ammonification.

Oswald and Ferchau (1963) carried out investigations on aerobic bacteria associated with mycorrhizal and nonmycorrhizal rootlets of *Picea engelmannii, P. pungens, Pinus aristata, P. flexilis,* and *Pseudotsuga menziesii.* They identified 51 species of bacteria in 253 isolates, including 22 associated with mycorrhizae, 7 with nonmycorrhizal rootlets, and 22 common to both types of rootlets. Neal *et al.* (1964) examined the rhizospheric microflora associated with mycorrhizae of *Pseudotsuga menziesii* and found stimulation of the bacteria in various mycorrhizal forms, but compared with nonmycorrhizal and suberized rootlets, no great difference appeared, in favor of the former, regarding the morphological and physiological forms of the isolates.

Foster and Marks (1967) examined the mycorrhizospheres of *Pinus radiata* with an electron microscope, and found various morphological types of bacteria associated with different types of mycorrhizae. Rambelli (1970) identified, as noted previously, a quantitatively different population of oligonitrophilous bacteria in mycorrhizal and nonmycorrhizal rootlets of *P. radiata.* These bacteria possibly belong to a single species, and are present inside the fungus cover in mycorrhizae produced by *Boletus granulatus.* They appear capable of fixing small quantities of atmospheric nitrogen.

One of the most interesting and little understood aspects of the formation of the rhizospheric ambient of mycorrhizae, especially in cases of ectendotrophy, is the complex mechanism of antibiotic production[3] by the mycorrhiza-forming fungi. The production of antibiotics by basidiomycetes has been known for some time, and the bibliography contains long lists of species capable of liberating these complex substances in culture substrates (Wilkins and Harris, 1943a; Wilkins, 1945, 1947; Mathieson, 1946; Robbins *et al.,* 1946; Santoro and Casida, 1962; Krywolap and Casida, 1964; Sásek and Musilek, 1967, 1968; Musilek *et al.,* 1969). However, the phenomenon in nature is less known, especially when the mycorrhiza-forming fungi are associated with the root tips to form the mycorrhizae. Krywolap *et al.* (1964) discovered an antibiotic substance in extractions made with acetone or methanol from mycor-

---

[3] See Chapter 9.

rhizae formed between *Cenococcum graniforme* and the roots of white pine (*Pinus strobus*), red pine (*P. resinosa*), and Norway spruce (*Picea abies*). The authors also found these substances in the root tissues and needles of these plants, and suggested that they migrated from the mycorrhizal fungus.

Nothing, however, is known of the possible liberation of antibiotic substances by mycorrhizae in the soil around the mycorrhizal tips. This seems plausible even though, as far as is known, it has never been proven. The formation of the rhizospheric ambient around the mycorrhizal tips would, in this case, depend on two contrasting mechanisms (i.e., microbial stimulation, due to the production of root exudates liberated by the mycorrhizal fungus, and microbial inhibition, due to production of antibiotics) which would inhibit sensitive microorganisms. This dual mechanism would result in true microbial selection and formation of an undoubtedly special rhizospheric ambient.

Even if these observations are hypothetical for the moment, the mycorrhizospheric ambient seems clearly delimitable. Undoubtedly the fungal symbiont intervenes, modifying the previously formed rhizospheric ambient, changing the composition of the substances which are liberated into the soil surrounding the root tips. Even if we consider that mycorrhizae originated within the rhizosphere, through the interaction of microbial stimulation by the root exudates and by selection by microbial competition, the mycorrhizal fungus will dominate the other microorganisms and colonize the root tips. The changes wrought by this fungus on the residual rhizospheric microflora are profound and permanent. In the author's view it appears justifiable to apply the name "mycorrhizosphere" to this particular microbial ambient. Perhaps the distinction between the mycorrhizosphere and rhizosphere might also apply to the microbial ambient around the mycorrhizae. The latter will include all those microorganisms located further from the surface of the fungus cover and influenced either very slightly or not at all by the action of the substances produced by the symbiont. However, they are still stimulated to multiply by residues of substances partly utilized by the mycorrhizospheric microflora.

II. Distribution of Rhizospheric and Mycorrhizospheric Microorganisms over the Four Seasons

Reference has already been made to the work of many investigators on the qualitative distribution of rhizospheric microorganisms in various plants (Tribunskaya, 1955; Ivarson and Katznelson, 1960; Katznelson *et al.*, 1962; Neal *et al.*, 1964; Oswald and Ferchau, 1963). Skyring and Quadling (1969) developed noteworthy techniques, and they were able to characterize the rhizospheric ambient by subjecting 400 bacterial strains isolated from the rhizosphere to 98 tests. In this way they identified characteristics typical of the rhizospheric microorganisms which proved completely different from those of the bacteria in the control. De Leval and Remacle (1969) followed changes in the rhizospheric microflora of *Populus canadensis* v. *robusta* at different periods of development. Using the techniques of Pochon *et al.* (1954), they examined the total microflora, aerobic and anaerobic nitrogen fixation, proteolysis, ammonification, denitrification, amylolysis, pectinolysis, hemicellulolysis, aerobic and anaerobic cellulolysis, and other physiological properties. Rambelli *et al.* (1972), in a recent investigation of the mycorrhizosphere of *Pinus radiata*, studied the variation of the rhizospheric microbial population for the whole range of seasons. Applying the techniques of Pochon *et al.* (1954) and those of Lochhead and Chase (1943), they studied the fungal, bacterial, and actinomycetic populations by making isolations from the soil immediately around the mycorrhizal root tips (first isolation) as well as isolations from the surface and within the mycorrhizae (second isolation). A brief review of some of these results follows (Rambelli *et al.*, 1972).

A. The Fungi

1. Identification

Rambelli *et al.* (1972) noted an extremely irregular distribution of the microflora throughout the four seasons. Isolations made from the adhering rhizosphere (second isolation) produced negative results in spring and winter. The first isolation, made in spring, produced only a few fungi representing few species. In summer, however, there was a considerable group of *Penicillia*, dominated mainly by *Penicillium corylophilum*, *P. asperum*, and another unidentified species. *Penicillium corylophilum* was present in almost all the isolations made during the four seasons, it proved dominant in autumn (first and second isolation) and, together with *P. waksmanii*, *P. vinaceum*, and an undetermined species of *Penicillium*, it was dominant in the winter (first isolation) (Table V).

These results do not agree with those obtained by De Leval and Remacle (1969), Katznelson *et al.* (1962), or Tribunskaya (1955). However, it is often difficult to compare results obtained by different authors, since identification of the fungal population isolated is not always complete.

TABLE V

FUNGI ISOLATED FROM *Pinus radiata* RHIZOSPHERE IN THE FOUR SEASONS

Species	Spring		Summer		Autumn		Winter	
	I	II	I	II	I	II	I	II
Penicillium thomii	1	—	—	—	—	—	—	—
Penicillium cyclopium	—	—	2	1	3	2	1	—
Pen. cyclopium v. *echinulatum*	—	—	1	—	—	—	—	—
Penicillium frequentans	—	—	2	—	—	—	—	—
Penicillium corylophilum	1	—	18	1	8	4	23	—
Penicillium vinaceum	—	—	1	—	—	—	6	—
Penicillium raistrickii	—	—	1	—	—	—	—	—
Penicillium sclerotiorum	—	—	1	—	—	—	—	—
Penicillium janthinellum	—	—	—	—	1	—	—	—
Penicillium restrictum	—	—	—	—	1	—	1	—
Penicillium oxalicum	—	—	—	—	—	1	—	—
Penicillium lilacinum	—	—	—	—	—	—	1	—
Penicillium waksmanii	—	—	—	—	—	—	14	—
Penicillium asperum	—	—	4	—	—	—	—	—
Penicillium sp. No. 1	1	—	—	—	—	—	—	—
Penicillium sp. No. 2	1	—	—	—	—	—	—	—
Penicillium sp. No. 3	—	—	10	—	—	—	—	—
Penicillium sp. No. 4	—	—	—	—	—	1	8	—
Aspergillus fumigatus	—	—	—	—	—	—	1	—
Aspergillus versicolor	—	—	1	—	—	—	—	—
Spicaria sp. No. 1	—	—	—	—	6	—	—	—
Spicaria violacea	—	—	3	—	2	—	3	—
Beauveria bassiana	1	—	—	—	—	—	—	—
Cladosporium sp. No. 1	1	—	1	1	—	—	—	—
Cephalosporium sp. No. 1	1	—	1	—	1	—	—	—
Phoma sp. No. 1	1	—	—	—	—	—	—	—
Chaetomium spinosum	—	—	1	—	—	—	—	—
Verticillium sp. No. 1	—	—	—	—	1	—	—	—
Mortierella alpina	—	—	—	—	1	—	—	—
Rhizopus sp. No. 1	1	—	—	—	—	—	—	—
Micelia sterilia	2	—	4	—	5	—	2	—

B. THE BACTERIA

1. Nitrogen Fixation

Investigations on seasonal nitrogen fixation in the mycorrhizosphere are relatively scarce. Most of these have been conducted on the rhizosphere of legumes, probably in an attempt to identify rhizospheric stimulation by root symbiont bacteria. Rangaswami and Vasantharajan (1962a,b), who studied the rhizosphere of citrus, identified an appreci-

TABLE VI

Distribution of Rhizospheric Microorganisms with Different Physiological Characters in the Four Seasons. Data Expressed in Percentage of Total Number of Bacteria Tested

	N fixation		Ammonification		Nitrite formation		Nitrate formation		Denitrifaction		Not identified	
	I	II	I	II	I	II	I	II	I	II	I	II
Spring	18	25	69	79	0	0	0	0	9	0	4	0
Summer	5	17	38	67	0	0	0	0	4	0	53	16
Autumn	6	15	38	76	0	0	0	0	0	7	56	2
Winter	7	6	90	78	0	0	0	0	3	11	0	5

able population of nitrogen fixers, with R/S[4] indices ranging from 0.78 to 1.71 in the varieties considered. According to Tribunskaya (1955), the development of nitrogen fixers does not occur normally in the mycorrhizosphere and the rhizosphere of pine, which does not favor their multiplication. Maliszewska and Moreau (1959), who investigated the rhizosphere of *Abies alba,* did not identify any nitrogen fixers except *Clostridium* sp.

The investigations conducted by Rambelli (1970) agree well with those of Tribunskaya. Indeed, in the mycorrhizosphere of *Pinus radiata,* Rambelli found a certain percentage of microorganisms that grew well in a liquid mineral medium devoid of nitrogen. However, although he considered that these microorganisms were able to fix small amounts of atmospheric nitrogen, they were not true nitrogen fixers. In any case, the physiological characters of these bacteria are still extremely uncertain. Their quantitative distribution in the soil adhering immediately to the mycorrhizae over the four seasons is about 18% in spring, 5% in summer, 6% in autumn, and 7% in winter (Table VI). In isolations made from the actual mycorrhizal mantle, 25, 17, 15, and 6% were obtained during the four seasons. These microorganisms therefore seem to be located mainly on the surface of the fungus cover, and they probably receive positive stimulation from the fungal metabolites and root exudates.

By making isolations inside the mycorrhizae, and after sterilizing the outer surface of the fungus cover, Rambelli (1970) obtained, in a pure culture, the same microorganisms in the form of an extremely homogeneous microbial population, which showed good growth in a liquid mineral medium devoid of nitrogen as the only physiological characteristic.

2. Ammonification

This subject has received the most attention in the mycorrhizosphere (Tribunskaya, 1955; Ivarson and Katznelson, 1960; Katznelson *et al.,* 1962; Neal *et al.,* 1964; Oswald and Ferchau, 1963; Rambelli, 1966; Rambelli *et al.,* 1972). Over the four seasons, Rambelli identified the following variation in the quantitative distribution of ammonifying bacteria in the micorrhizosphere of *Pinus radiata:* spring, 69%; summer, 38%; autumn, 38%; and winter, 90% in the soil immediately adjoining the mycorrhizae, and on the surface of the fungus cover the variations were 79, 67, 76, and 78%. This group, therefore, also seems to be stimulated appreciably by fungal metabolites and root exudates. But their development, as noted earlier, should cause no surprise, in view of the enrichment of the ambient by easily metabolized sources of nitrogen through the seasons.

---

[4] R/S, rhizosphere–control soil.

3. Nitrate Formation

The interpretation of seasonal variations in nitrate formation is still open to considerable controversy. Under natural conditions the mycorrhizospheric ambient would not appear able to support any active nitrite- and nitrate-forming microflora. Remacle (1963), studying the rhizospheric effect in plants that colonize pedologically different environments, found in a number of cases a rhizospheric effect that stimulated the nitrite- and nitrate-forming microflora. In the rhizosphere of mycorrhizal plants of *Abies alba,* Maliszewska and Moreau (1959), however, did not find any nitrate-forming microorganisms. In preliminary investigations on the distribution of physiological groups in the mycorrhizosphere of *Pinus radiata,* Rambelli (1970) found a small percentage of nitrite- and nitrate-forming bacteria. In his subsequent study on the distribution of the mycorrhizospheric flora of *P. radiata* over the four seasons, the author did not find any nitrite- and nitrate-forming microorganisms (Table VI). Probably the techniques used for the study of this group of microorganisms in the rhizosphere are not the most suitable and should be appropriately modified in order to obtain more reliable results.

4. Denitrification

Physiological aspects of denitrification have not been fully studied in the rhizosphere (Ivarson and Katznelson, 1960; Katznelson *et al.,* 1962; Neal *et al.,* 1964; De Leval and Remacle, 1969; Rambelli, 1970). Rambelli *et al.* (1972) reported seasonal variation in distribution of denitrifying microorganisms in the soil surrounding the mycorrhizosphere of *P. radiata:* 9% in spring, 4% in summer, 0% in autumn, and 3% in winter. On the other hand, the following percentages of denitrifying microorganisms were isolated from the mycorrhizal mantle: 0% in spring and summer, 7% in autumn, and 1% in winter (Table VI).

5. Proteolysis

Proteolytic capacity of the rhizospheric bacterial flora has been studied by several investigators (Chalvignac, 1958; Rangaswami and Vasantharajan, 1962a,b; Remacle, 1963; De Leval and Remacle, 1969; Skyring and Quadling, 1969). However, since it is impossible to reproduce proteolytic capacity under conditions similar to nature, any practical study is nearly useless. Indeed, testing development of individual strains, isolated from the mycorrhizosphere, in the presence of fungal metabolites and root exudates, would have dubious value, because of their proteolytic action, sensitivity to growth inhibitors in fungal metabolites, and above all the difficulty of identifying the proportions of the two substances to be tested. In any case, when we consider that the protein substances of the fungal metabolites and root exudates are mainly composed of amino acids, preliminary data can be obtained by examining ammonification of the individual rhizospheric strains isolated.

C. The Actinomycetes

Proteolysis

It is known from the literature that the actinomycetes as a group are stimulated to various degrees by root exudates and fungal metabolites. Maliszewska and Moreau (1959) noted a smaller number of actinomycetes in the rhizosphere of *Abies alba* than in control soil. Neal *et al.* (1964) identified a larger actinomycete population in suberized rootlets than in the mycorrhizae of Douglas fir. The opposite situation, however, is reported by many (Tribunskaya, 1955; Ivarson and Katznelson, 1960; Katznelson *et al.*, 1962; Rangaswami and Vasantharajan, 1962b; Rambelli, 1962, 1963, 1965, 1966). Thus, the behavior of the actinomycetes inside the mycorrhizosphere is still obscure. Indeed, considerable interest is attached to their presence in the soil near the mycorrhizae in numbers often far higher than that found on the fungal cover (Rambelli, 1962, 1963).

Because of their peculiar location, Rambelli carried out investigations on the behavior of actinomycetes isolated from the mycorrhizosphere of *Pinus radiata* in the presence of the metabolites produced by *Boletus granulatus*. The results, obtained during the four seasons of the year (Rambelli *et al.*, 1972), indicate, despite the limitations noted already, an almost total incapacity to attack the symbiont's protein substances. It is thus difficult to specify the role played by these actinomycetes inside the rhizosphere and the mycorrhizosphere. In any event, the subject deserves further investigation.

III. Relationships between Mycorrhizae and Rhizosphere

Root Exudates and Symbiont Metabolites

1. Influence of Root Exudates and Symbiont Metabolites on the Rhizosphere Microflora

It has already been shown that root exudates and fungal metabolites form the basis of the stimulatory processes that lead to formation of the mycorrhizospheric ambient. They provide a selective substrate for the terrestrial microflora, favoring development of certain forms and inhibit-

336

Angelo Rambelli

TABLE VII

Total Microflora Grown with Various Percentages of Integral or Boiled Cultural Solution of *Boletus granulatus* and Added to Soil Agar

Percentages of culture solution	Raw cultural solution	Boiled cultural solution
Control	122,395,000	122,395,000
1	7,510,000	4,157,000
5	4,102,000	3,497,000
10	3,055,000	2,837,500
20	1,100,000	685,000

ing others; probably as indicated earlier, partly, through the effect of antibiotic substances produced by mycorrhizal fungi. But these substances not only inhibit given microbial groups, but also permit the indiscriminate development of others, so that, in a number of instances, there is an increase in the microbial flora in soils treated with antibiotics (Hervey, 1955).

Voznyakovskaya and Ryzhkova (1955) state, "There are certain facts

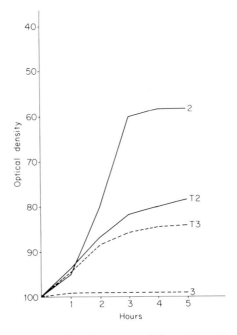

Fig. 1. Growth according to biophotometer of rhizosphere bacteria (strains 2 and 3) in a Lochhead mineral medium and root exudates (50%) of *Pinus radiata*.

indicating the dependence of mycorrhizal formation upon the presence of certain microbial associations." They consider that the prevalence of certain microbial groups in the mycorrhizosphere facilitates penetration of the symbiont into the host's root tissues. They reached this conclusion by simultaneously inoculating mycorrhiza-forming fungi and microorganisms isolated from the rhizosphere in seedlings grown in sterile soil. The attempt produced negative results. However, by adding 1 gm of forest soil to the pots containing sterile soil, mycorrhizal formation was able to proceed regularly.

Rambelli (1970) obtained interesting preliminary data on the influence of fungal metabolites on formation of the rhizospheric ambient. He tested the influence of culture solutions of *Boletus granulatus* on the microbial population of some good garden soil (Table VII). Thus, in the culture solutions of 60-day-old colonies of *B. granulatus* there are substances that inhibit microbial development. However, this is not indiscriminate but seems to take place through some type of selection process, since even 20% solutions do not inhibit many microorganisms. Consequently they could develop in the mycorrhizospheric ambient, favored also by the

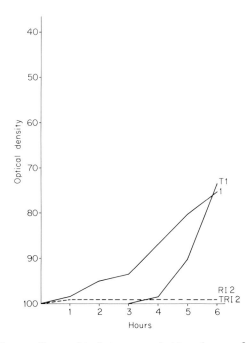

Fɪɢ. 2. Growth according to biophotometer of rhizosphere and mycorrhizosphere bacteria (strains 1 and RI2) in a Lochhead mineral medium and root exudates (50%) of *Pinus radiata*.

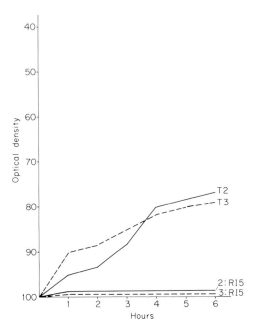

Fig. 3. Growth according to biophotometer of rhizosphere and mycorrhizosphere bacteria (strains 2, 3, and RI5) in a Lochhead mineral medium and root exudates (50%) of *Pinus radiata*.

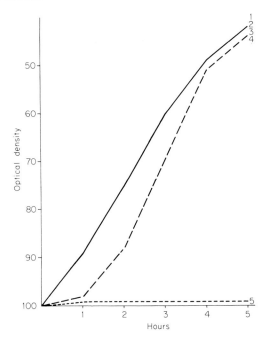

Fig. 4. Growth according to biophotometer of RI2 bacterial cells with 75% (No. 1), 50% (No. 2) and 25% (No. 3) metabolite produced by *Boletus granulatus*. Number 4 refers to the control and No. 5 to the integral metabolite.

lack of competition normally exerted by other terrestrial forms of bacteria. These results have suggested individual studies of the behavior of certain rhizospheric bacteria of *P. radiata* in the presence of root exudates and metabolites produced by *B. granulatus* (Figs. 1 and 2).

The investigations were conducted, above all, to check the influence of these substances on strain (RI2) isolated from inside the fungal mantle produced by *Boletus granulatus*, on root tips of *Pinus radiata*. Growth was measured with a biophotometer, and respiration was measured with the Warburg apparatus (Fig. 3). The root exudates generally stimulated respiration of certain rhizospheric strains and depressed others (Figs. 1–3). Regarding RI2, no stimulating action by root exudates was noted. The same strain, appeared to be stimulated considerably by large quantities of metabolites produced by *B. granulatus* (Figs. 4 and 5). These results seem to indicate that certain basic events affect the ecology of the bacterium RI2. This is a microorganism which is gram-positive, and in nitrogen-fixing agar forms round, bright, colorless,

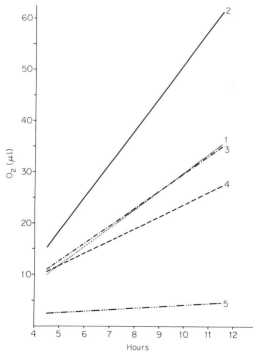

Fig. 5. Warburg respiration of RI2 bacterial cells with 75% (No. 1), 50% (No. 2), and 25% (No. 3) metabolite produced by *Boletus granulatus*. Number 4 refers to the control and No. 5 to the original metabolite.

highly mucous colonies. Under the microscope it appears in the form
of thin rodlike cells, sometimes filamentous, or coupled, with one or two
refractive polar drops. Rambelli studied its nutrition by cultivating it in
various selective substrates. It grew well in a nitrogen-fixing substratum,
and this would explain its inability to break down amino acids, cause
nitrite and nitrate formation, and induce denitrification. Clearly, it
utilized metabolites of the symbiont since it did not appear capable of
using root exudates and the atmosphere as its source of nitrogen. It is possible that it was located inside the fungal mantle because of its particular
requirements for nutrients or fungal hormones (Figs. 6 and 7).

Production of cytokinin has now been shown to occur in a number of
mycorrhiza-forming fungi (Miller, 1967), and the presence of substances
similar to kinetin in the root exudates of certain plants has been studied
(Kende, 1965). It is not yet known what action these compounds exert
on the rhizospheric microflora, nor, for that matter, is anything apparently
known about diffusion of these substances in the soil around mycorrhizal

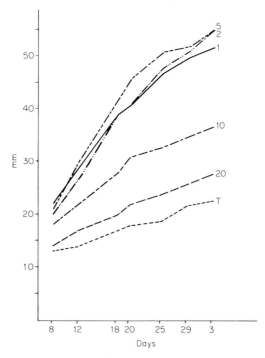

FIG. 6. Diameter increment of *Boletus granulatus* colonies in Hagem agar without
malt, with varying percentages of metabolite obtained from strain RI2.

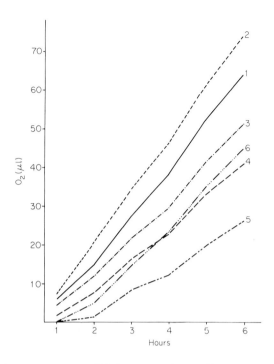

FIG. 7. Warburg respiration of *Boletus granulatus* cells with 20% RI2 metabolite (No. 1), 10% (No. 2), 5% (No. 3), 2% (No. 4), 1% (No. 5), and 0% (No. 6).

and nonmycorrhizal root tips. There is little doubt, that if diffusion occurs, these substances certainly exert some and perhaps considerable action on the rhizospheric microflora (Fig. 8).

2. Influence of Rhizospheric Microflora on Symbiont Growth

Despite the fact that all the literature on this subject was not examined, it is apparent that it has not received the attention of many researchers. There is no lack of research on the metabolism of rhizospheric microorganisms, especially on transformation of the soil's organic substance and their influence on plant development (Clark, 1949; Gerretsen, 1948; Katznelson and Bose, 1959; Katznelson *et al.*, 1948; Krasilnikov and Kotelev, 1956; Louw and Webley, 1959; Levishon, 1952; Pantos, 1956; Rovira and Bowen, 1960; Sperber, 1958; Starkey, 1958; Timonin, 1946; McCalla and Haskins, 1964; Norman, 1959; Parkinson, 1967; Rovira, 1965b; Rovira and Bowen, 1966), but studies on the direct influence of the rhizospheric microflora, or terrestrial microflora, in general, on the

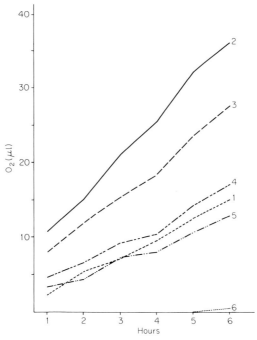

FIG. 8. Warburg respiration of *Boletus granulatus* cells in medium without nitro-gen and with the following percentages of metabolite produced by RI2: 20% (No. 1), 10% (No. 2), 5% (No. 3), 2% (No. 4), 1% (No. 5), and 0% (No. 6).

plant symbiont or symbionts are very scarce. This influence, nevertheless, cannot be denied. Wilde as far back as in 1954 stated: "It appears highly probable that in grassland soils the survival of mycorrhizal fungi without their woody symbionts is precluded by toxic and antibiotic substances emitted by roots of prairie plants." But certainly, it is not only the toxic substances released by herbaceous plants that influence the viability of the symbionts. Without doubt great importance should also be attached to all the interactions of antagonism and competition that normally occur in the soil. The mechanisms described above create a special microbial ambient of special significance around the mycorrhizal root tips. These microorganisms associated with the mycorrhizae have an important task in maintaining symbiosis and survival of the fungus. This is supported by observations that in many environments the symbiont dies when the young abundantly mycorrhizal seedlings of *Pinus radiata* are trans-planted. Undoubtedly many factors are involved in this phenomenon, but the contribution made by the rhizospheric microflora to the stability of the mycorrhizal association should not be underestimated.

References

Adati, M. (1939). Untersuchungen über die Rhizosphäre der Pflanzen. *J. Soc. Trop. Agr.* (*Taiwan*) 11, 57.

Barker, H. A., and Broyer, T. C. (1942). Notes on the influence of microorganisms on growth of squash plants in water culture with particular reference to manganese nutrition. *Soil. Sci.* 53, 467.

Beijerinck, M. W. (1888). Die Bacterien der Papilionaceen-Knöllchen. *Bot. Ztg.* 46, 725.

Berezova, E. F. (1941). Microflora of the rhizosphere of flax. *Microbiology* (*USSR*) 10, 918.

Bochow, H. (1965). *In* "Plant Microbes Relationships" (J. Macura and V. Vančura, eds.), pp. 296–299. Czech. Acad. Sci., Prague.

Bowen, G. D. (1969). The uptake of orthophosphate and its incorporation into organic phosphates along roots of *Pinus radiata. Aust. J. Biol. Sci.* 22, 1125.

Buxton, E. W. (1957). Some effects of pea root exudates on physiologic races of *Fusarium oxysporum* Fr. f. *pisi. Trans. Brit. Mycol. Soc.* 40, 145.

Cátská, V. (1965). *In* "Plant Microbes Relationships" (J. Macura and V. Vančura, eds.), pp. 60–68. Czech. Acad. Sci., Prague.

Chalvignac, M. A. (1958). Effet rhizosphère comparé du lin en culture hydroponique et en terre. *Ann. Inst. Pasteur, Paris* 95, 474.

Chalvignac, M. A. (1966). Contribution à l'étude de la rhizosphère du lin. *Ann. Inst. Pasteur, Paris* 110, 603.

Chan, E. C. S., and Katznelson, H. (1961). Growth interactions of *Arthrobacter globiformis* and *Pseudomonas* sp. in relation to rhizosphere effect. *Can. J. Microbiol.* 7, 759.

Chan, E. C. S., Katznelson, H., and Rouatt, J. W. (1963). The influence of soil and root extracts on the associative growth of selected soil bacteria. *Can. J. Microbiol.* 9, 187.

Clark, F. E. (1940a). Effects of soil amendments upon the bacterial populations associated with roots of wheat. *Trans. Kans. Acad. Sci.* 42, 91.

Clark, F. E. (1940b). Notes on types of bacteria associated with plant roots. *Trans. Kans. Acad. Sci.* 43, 75.

Clark, F. E. (1949). Soil microorganisms and plant growth. *Advan. Agron.* 1, 241.

Dart, P. J., and Mercer, F. V. (1964). The legume rhizosphere. *Arch. Mikrobiol.* 47, 344.

Daste, P. (1956). Recherches sur l'écologie bactérienne dans la rhizopshère de quelques plantes supérieures. *Rev. Cytol. Biol. Veg.* 19, 251.

De Leval, J., and Remacle, J. (1969). A microbiological study of the rhizosphere of poplar. *Plant Soil* 31, 31.

Foster, R. C., and Marks, G. C. (1967). Observations on the mycorrhizas of forest trees. II. The rhizosphere of *Pinus radiata* D. Don. *Aust. J. Biol. Sci.* 20, 915.

Gerretsen, F. C. (1948). The influence of microorganisms on the phosphate intake by the plant. *Plant Soil* 1, 51.

Gräf, G. (1930). Über den Einfluss des Pflanzenwachstums auf die Bakterien im Wurzelbereich. *Centralbl. Bakteriol.* 82, 44.

Harley, J. L., and Waid, J. S. (1955). A method of studying active mycelia on living roots and other surfaces in the soil. *Trans. Brit. Mycol. Soc.* 38, 104.

Harmsen, G. W., and Jager, G. (1963). Determination of the quantity of carbon and nitrogen in the rhizosphere of young plants. *In* "Soil Organisms" (J. Doeksen and J. Van der Drift, eds.), pp. 245–251. North-Holland Publ., Amsterdam.

Hervey, R. J. (1955). Stimulation of soil microorganisms by antibiotics. *Antibiot. Chemother.* (*Washington, D. C.*) **5**, 96.

Hiltner, L. (1904). Uber neuere Erfahrungen und Probleme auf dem Gebiet der Bodenbakteriologie und unter besonderer Berücksichtigung der Gründüngung und Brache. *Arb. Deut. Landwirtschaftsges.* **98**, 59.

Hulpoi, N. (1936). Demonstration von Mikroorganismen der Rhizosphäre vermittels der Aufwuchsplattenmethode nach Cholodny. *Arch. Mikrobiol.* **7**, 759.

Isakova, A. A. (1938). Experimental application of the Rossi-Cholodny method for the study of bacteriorrhiza of various plants. *Bull. Acad. Sci. USSR* **2**, 517.

Ivarson, K. C., and Katznelson, H. (1960). Studies on the rhizosphere microflora of yellow birch seedlings. *Plant Soil* **12**, 30.

Jenny, H., and Grossenbacher, K. (1963). Root-soil boundary zones as seen in the electron microscope. *Soil Sci. Soc. Amer., Proc.* **27**, 273.

Katznelson, H. (1946). The rhizosphere effect of mangels on certain groups of soil microorganisms. *Soil Sci.* **62**, 343.

Katznelson, H. (1965). Nature and importance of the rhizosphere. *In* "Ecology of Soil-borne Plant Pathogens" (K. F. Baker and W. C. Snyder, eds.), pp. 187–209. Univ. of California Press, Berkeley.

Katznelson, H., and Bose, B. (1959). Metabolic activity and phosphate dissolving capability of bacteria isolated from wheat roots, rhizosphere and non-rhizosphere soil. *Can. J. Microbiol.* **5**, 79.

Katznelson, H., and Richardson, L. T. (1943). The microflora of the rhizosphere of tomato plants in relation to soil sterilization. *Can. J. Res.* **21**, 249.

Katznelson, H., Lochhead, A. G., and Timonin, M. I. (1948). Soil microorganisms and the rhizosphere. *Bot. Rev.* **14**, 543.

Katznelson, H., and Strzelczyk, E. (1961). Studies on the interaction of plants and free-living nitrogen-fixing microorganisms. I. Occurrence of *Azotobacter* in the rhizosphere of crop plants. *Can. J. Microbiol.* **7**, 437.

Katznelson, H., Rouatt, J. W., and Payne, T. M. B. (1955). The liberation of amino acids and reducing compounds by plant roots. *Plant Soil* **7**, 35.

Katznelson, H., Rouatt, J. W., and Peterson, E. A. (1962). The rhizosphere effect of mycorrhizal and non-mycorrhizal roots of yellow birch seedlings. *Can. J. Bot.* **40**, 377.

Kende, H. (1965). Kinetinlike factors in the root exudate of sunflowers. *Proc. Nat. Acad. Sci. U. S.* **53**, 1302.

Knudson, L. (1920). The secretion of invertase by plant roots. *Amer. J. Bot.* **7**, 371.

Kostychev, S. P., Sheloumova, A., and Shulgina, O. (1926). Microbiological characteristics of southern soils. I. Nitrogen in soils of southern shores of Crimea. *Sov. Agron.* **1**, 5.

Krasilnikov, N. A. (1934). The effect of root excretions on the growth of the Azotobacter and other soil microbes. *Mikrobiologiya* **3**, 3.

Krasilnikov, N. A. (1936). Focal distribution of microorganisms in soil. *Izv. Akad. Nauk SSSR, Ser. Biol.* **1**, 192.

Krasilnikov, N. A. (1952). Enzyme release by plant roots. *Rep. Acad. Sci., USSR* **87**, 2.

Krasilnikov, N. A., and Koreniako, A. I. (1946). The effect of nonroot nodule bacteria on growth and nitrogen fixation by legumines. *Microbiology* (*USSR*) **15**, 421.

Krasilnikov, N. A., and Kotelev, V. V. (1956). The effect of soil bacteria on the assimilation of phosphorous compounds by plants. *Rep. Acad. Sci., USSR* **60,** 5.

Krasilnikov, N. A., and Kotelev, V. V. (1959). Adsorption of phosphatases of soil microorganisms by corn roots. *Mikrobiologiya* **28,** 515.

Krasilnikov, N. A., Kriss, A. E., and Litvinov, M. A. (1936). The effect of the root system on soil microorganisms. *Microbiology (USSR)* **5,** 270.

Krywolap, G. N., and Casida, L. E. (1964). An antibiotic produced by the mycorrhizal fungus *Cenococcum graniforme. Can. J. Microbiol.* **10,** 365.

Krywolap, G. N., Grand, L. F., and Casida, L. E. (1964). The natural occurrence of an antibiotic in the mycorrhizal fungus *Cenococcum graniforme. Can. J. Microbiol.* **10,** 323.

Levishon, I. (1952). Forking in pine roots. *Nature (London)* **169,** 715.

Linford, M. B. (1939). Attractiveness of roots and excised shoot tissues to certain nematodes. *Helminthol. Soc. Wash.* **6,** 11.

Lochhead, A. G. (1940). Qualitatives studies of soil microorganisms. III. Influence of plant growth on the character of the bacterial flora. *Can. J. Res.* **18,** 42.

Lochhead, A. G., and Chase, F. E. (1943). Qualitative studies of soil microorganisms. V. Nutritional requirements of the predominant bacterial flora. *Soil Sci.* **55,** 185.

Lochhead, A. G., and Thexton, R. H. (1947). Qualitative studies of soil microorganisms. VII. The "rhizosphere effect" in relation to the amino acids nutrition of bacteria. *Can. J. Res.* **25,** 20.

Louw, H. A., and Webley, D. M. (1959). A study of soil bacteria dissolving certain mineral phosphate fertilizers and related compounds. *J. Appl. Bacteriol.* **22,** 227.

McCalla, T. M., and Haskins, F. A. (1964). Phytotoxic substances from soil microorganisms and crop residues. *Bacteriol. Rev.* **28,** 181.

Macura, J. (1966). Interactions nutritionelles plantes bactéries et base expérimentales de la bactérisation des graines. *Ann. Inst. Pasteur, Paris* **111,** 9.

Maliszewska, W., and Moreau, R. (1959). Sur la rhizosphère du sapin blanc (Abies alba Mill.). *C. R. Acad. Sci.* **249,** 303.

Mathieson, J. (1946). Antibiotics from Victorian Basidiomycetes. *Aust. J. Biol. Med. Sci.* **24,** 57.

Melin, E. (1962). Physiological aspects of mycorrhizae of forest trees. *In* "Tree Growth" (T. T. Kozlowski, ed.), pp. 247–263. Ronald Press, New York.

Miller, C. O. (1967). Zeatin and zeatin riboside from a mycorrhizal fungus. *Science* **157,** 1055.

Molina, J. A. E., and Rovira, A. D. (1964). The influence of plant roots on autotrophic nitrifying bacteria. *Can. J. Microbiol.* **10,** 249.

Musilek, V., Cerna', J., Sásek, V., Semerdzieva, M., and Vondracek, M. (1969). Antifungal antibiotic of the Basidiomycete *Oudemansiella mucida. Foliax. Microbiol. (Prague)* **14,** 377.

Neal, J. L., Bollen, W. B., and Zak, B. (1964). Rhizosphere microflora associated with mycorrhizae of Douglas Fir. *Can. J. Microbiol.* **10,** 259.

Norman, A. G. (1959). Inhibition of root growth and cation uptake by antibiotics. *Soil Sci. Soc. Amer., Proc.* **23,** 368.

Nutman, P. S. (1957). Studies on the physiology of nodule formation. V. Further experiments on the stimulating and inhibitory effects of roots secretions. *Ann. Bot. (London)* [N.S.] **21,** 321.

Obraztsova, A. A. (1935). The rhizosphere microorganisms of the Batum red soils. *Dokl. Akad. Nauk SSSR* **1,** 70.

Oswald, E. T., and Ferchau, H. A. (1963). Bacterial associations of coniferous mycorrhizae. *Plant Soil* **28**, 187.

Pantos, G. (1956). Principal aspects of corn (wheat) rhizosphere bacteria and action of this plant on microflora. *Trans. Int. Congr. Soil Sci., 6th, 1956* Part C, p. 225.

Parkinson, D. (1967). Soil microorganisms and plant roots. *In* "Soil Biology" (N. A. Burges and F. Raw, eds.), pp. 449–478. Academic Press, New York.

Parkinson, D., and Thomas, A. (1969). Quantitative study of fungi in the rhizosphere. *Can. J. Microbiol.* **15**, 875.

Parkinson, D., Taylor, G. S., and Pearson, R. (1963). Studies on fungi in the root region. I. The development of fungi on young roots. *Plant Soil* **19**, 332.

Paul, E. A., and Schmidt, E. L. (1961). Formation of free amino acids in rhizosphere and non rhizosphere soil. *Soil Sci. Soc. Amer., Proc.* **25**, 359.

Pochon, J., Augier, J., de Barjac, H., Martre-Coppier, O., Chalvignac, M. A., and Lajudie, J. (1954). "Manual technique d'analyse microbiologique du sol." Masson, Paris.

Poschenrieder, H. (1930). Uber die Verbreitung des *Azotobacter* in Wurzelbereicte der Pflanzen. *Zentralbl. Bakteriol., Parasitenk. Infektionskr., Abt. 2* **80**, 369.

Rakhtcenko, I. N. (1958). Seasonal cycle of absorption and excretion of mineral nutrients by roots of woody species. *Fiziol. Rast.* **5**, 447.

Rambelli, A. (1962). Ricerche sulla rizosfera dell'eucalitto. *Pubbl. Cent. Sper. Agr. Forest., Rome* **6**, 83.

Rambelli, A. (1963). Ricerche sulla rizosfera del *Pinus pinea*. *Pubbl. Cent. Sper. Agr. Forest., Rome* **7**, 117.

Rambelli, A. (1965). Ricerche sulla rizosfera del pioppo. *Pubbl. Cent. Sper. Agr. Forest., Rome* **8**, 27.

Rambelli, A. (1966). Il *Pinus radiata* D. Don nei suoi rapporti con l'ambiente microbico del terreno. *Pubbl. Cent. Sper. Agr. Forest., Rome* **9**, 31.

Rambelli, A. (1970). Rapporti tra micorrizia e micorrizosfera. *Atti Accad. Sci. Forest.* **19**, 393.

Rambelli, A., Freccero, V., and Fanelli, C. (1972). Indagini sulla micorrizosfera di *Pinus radiata* D. Don. *Pubbl. Cent. Sper. Agr. Forest., Rome* **11**, 271.

Rangaswami, G., and Vasantharajan, V. N. (1962a). Studies on the rhizosphere microflora of Citrus tree. II. Qualitative distribution of the bacterial flora. *Can. J. Microbiol.* **8**, 479.

Rangaswami, G., and Vasantharajan, V. N. (1962b). Studies on the rhizosphere microflora of Citrus tree. III. Fungal and actinomycete flora of the rhizosphere. *Can. J. Microbiol.* **8**, 485.

Remacle, J. (1963). Contributions à la microbiologie du sol. Introduction general à l'écologie des micromycetes du sol. Techniques d'étude des champignons du sol. *Lejeunia* **19**, 1.

Rice, E. L. (1967). Chemical warfare between plants. *Bios* **38**, 67.

Rivière, J. (1959). Contribution à l'étude de la rhizosphère du blé. D.Sc. Thesis, University of Paris.

Rivière, J. (1960). Etude de la rhizosphère du blé. *Ann. Agron.* **11**, 397.

Robbins, W. J., Kavanagh, F., and Hervey, A. (1946). Production of antibiotic substances by Basidiomycetes. *Ann. N. Y. Acad. Sci.* **1**, 31.

Rokitzkaya, A. I. (1932). Soil microflora in the root system zone of plant in the region of forest-steppe and chernozem in Ukraine. *Proc. Int. Congr. Soil Sci., 2nd, 1930*, p. 260.

Rovira, A. D. (1956). Plant root excretions in relation to the rhizosphere effect. I. The nature of root exudates from oats and peas. *Plant Soil* **7**, 178.

Rovira, A. D. (1959). Root excretions in relation to the rhizosphere effect. IV. Influence of plant species, age of plant, light, temperature and calcium nutrition on exudation. *Plant Soil* **11**, 53.

Rovira, A. D. (1965a). Plant root exudates and their influence upon soil microorganisms. *In* "Ecology of Soil-borne Plant Pathogens" (K. F. Baker and W. C. Snyder, eds.), pp. 170–186. Univ. of California Press, Berkeley.

Rovira, A. D. (1965b). Interactions between plant roots and soil microorganisms. *Annu. Rev. Microbiol.* **19**, 241.

Rovira, A. D. (1969). Plant root exudates. *Bot. Rev.* **35**, 35.

Rovira, A. D., and Bowen, G. D. (1960). Effect of microorganisms in the development of roots and root hairs of Subterranean clover (*T. subterraneum*). *Nature (London)* **185**, 260.

Rovira, A. D., and Bowen, G. D. (1966). The effects of microorganisms upon plant growth. II. Detoxication of heat-sterilized soils by fungi and bacteria. *Plant Soil* **25**, 129.

Rubenchik, L. I. (1960). "*Azotobacter* and its Use in Agriculture." (Israel Program for Scientific Translations, Jerusalem, 1963.)

Sabinin, D. A., and Minina, E. G. (1932). Das mikrobiologische Bodenprofil als zonales Kennzeichen. *Proc. Int. Congr. Soil Sci., 2nd, 1930*, p. 224.

Samtsevich, S. A. (1965). *In* "Plant Microbes Relationships" (J. Macura and V. Vančura, eds.), pp. 48–53. Czech. Acad. Sci., Prague.

Santoro, T., and Casida, L. E. (1962). Elaboration of antibiotics by *Boletus luteus* and certain other mycorrhizal fungi. *Can. J. Microbiol.* **8**, 43.

Sásek, V., and Musilek, V. (1967). Cultivation and antibiotic activity of mycorrhizal Basidiomycetes. *Folia Microbiol. (Prague)* **12**, 515.

Sásek, V., and Musilek, V. (1968). Two antibiotic compounds from mycorrhizal Basidiomycetes. *Folia Microbiol. (Prague)* **13**, 43.

Scheffer, F., Kickuth, R., and Stricker, G. (1964). Organische Verbindungen aus dem Wurzelraum von Triticum-Arten und-Sorten. *Z. Pflanzenernachr., Dueng., Bodenk.* **105**, 13.

Schippers, B., Schroth, M. N., and Hildebrand, D. C. (1967). Emanation of water from underground plant parts. *Plant Soil* **27**, 81.

Schönbeck, F. (1958). Untersuchungen über den Einfluss von Wurzelausscheidungen auf die Entwicklung von Bodenpilzen. *Naturwissenschaften* **45**, 63.

Schroth, M. N., and Hendrix, F. F. (1962). Influence of non-susceptible plants on the survival of *Fusarium solani* f. *phaseoli* in soil. *Phytopathology* **52**, 906.

Skyring, G. W., and Quadling, C. (1969). Soil bacteria: Comparisons of rhizosphere and non-rhizosphere populations. *Can. J. Microbiol.* **15**, 473.

Slankis, V., Runeckles, V. C., and Krotkov, G. (1964). Metabolites liberated by roots of white pine (*Pinus strobus* L.) seedlings. *Physiol. Plant.* **17**, 301.

Sperber, J. I. (1958). The incidence of apatite-solubilizing organisms in the rhizosphere and soil. *Austr. J. Agr. Res.* **6**, 778.

Starkey, R. L. (1929a). Some influences of the development of higher plants upon the microorganisms in the soil. I. Historical and introductory. *Soil Sci.* **27**, 319.

Starkey, R. L. (1929b). Some influences of the development of higher plants upon the microorganisms in the soil. II. Influence of the stage of plant growth upon abundance of organisms. *Soil Sci.* **27**, 355.

Starkey, R. L. (1929c). Some influences of the development of higher plants upon the microorganisms in the soil. III. Influence of the stage of plant growth upon some activities of the organisms. *Soil Sci.* **27**, 433.

Starkey, R. L. (1931). Some influences of the development of higher plants upon the microorganisms in the soil. IV. Influence of proximity to roots on abundance and activity of microorganisms. *Soil Sci.* **32**, 367.

Starkey, R. L. (1938). Some influences of the development of higher plants upon the microorganisms in the soil. VI. Microscopic examination of the rhizosphere. *Soil Sci.* **45**, 207.

Starkey, R. L. (1958). Interrelations between microorganisms and plant roots in the rhizosphere. *Bacteriol. Rev.* **22**, 154.

Stille, B. (1938). Untersuchungen über die Bedeutung der Rhizosphäre. *Arch. Mikrobiol.* **9**, 477.

Strzelczyk, E. (1958). Development of *Azotobacter* and *Clostridium* in the rhizosphere of various crop plants. *Acta Microbiol. Pol.* **7**, 115.

Strzelczyk, E. (1961). Studies on the interaction of plants and free-living nitrogen-fixing microorganisms. II. Development of antagonists of *Azotobacter* in the rhizosphere of plants at different stages of growth in two soils. *Can. J. Microbiol.* **7**, 507.

Strzelczyk, E. (1966). Effect of associated growth of bacteria from rhizosphere and non-rhizosphere soil on growth of fungi. *Ann. Inst. Pasteur, Paris* **111**, 314.

Sulochana, C. B. (1962). Amino acids in root exudates of cotton. *Plant Soil* **16**, 312.

Tardieux, P., Chalvignac, A., and Charpentier, M. (1961). Interactions microorganisms-maïs en culture hydroponique aux premiers stades de croissance. *Ann. Inst. Pasteur, Paris* **100**, 243.

Taylor, C. B., and Lochhead, A. G. (1938). Qualitative studies of soil microorganisms. II. A survey of the bacterial flora of soils differing in fertility. *Can. J. Res.* **16**, 162.

Thom, C. (1935). Micropopulations correlated to decomposition processes. *Trans. Int. Congr. Soil Sci., 3, 1935,* Vol. **1**, p. 160.

Thom, C., and Humfeld, H. (1932). Notes on the association of microorganisms and root. *Soil Sci.* **34**, 29.

Timonin, M. I. (1940). The interaction of higher plants and soil microorganisms. I. Microbial population of rhizosphere of seedlings of certain cultivated plants. *Can. J. Res.* **18**, 307.

Timonin, M. I. (1941). The interaction of higher plants and soil microorganisms. III. Effect of by-products of plant growth on activity of fungi and actinomycetes. *Soil. Sci.* **52**, 395.

Timonin, M. I. (1946). Microflora of the rhizosphere in relation to the manganese-deficiency disease of oats. *Soil Sci. Soc. Amer., Proc.* **11**, 284.

Timonin, M. I. (1964). Interaction of seed-coat microflora and soil microorganisms and its effects on pre- and post-emergence of some conifer seedlings. *Can. J. Microbiol.* **10**, 17.

Tribunskaya, A. J. (1955). Investigation on the microflora of the rhizosphere of pine seedlings. *Mikrobiologiya* **24**, 188.

Truffaut, G., and Vladykov, V. (1930). La microflore de la rhizosphère du blé. *C. R. Acad. Sci.* **190**, 977.

Vančura, V. (1961). Detection of gibberellic acid in *Azotobacter* cultures. *Nature (London)* **192**, 88.

Vančura, V. (1964). Root exudates of plants. I. Analysis of root exudates of barley and wheat in their initial phases of growth. *Plant Soil* **21**, 231.

Vančura, V. (1967). Root exudates of plants. III. Effect of temperature and "cold shock" on the exudation of various compounds from seeds and seedlings of maize and cucumber. *Plant Soil* **27**, 319.

Vančura, V., and Hovadik, A. (1965). Root exudates of plants. II. Composition of root exudates of some vegetables. *Plant Soil* **22**, 21.

Vančura, V., Abd-el-Malek, Y., and Zayed, M. N. (1965). *Azotobacter* and *Beijerinckia* in the soils and rhizosphere of plants in Egypt. *Folia Mikrobiol.* (*Prague*) **10**, 224.

Velich, A. (1903). Bakteriologische Untersuchung der Zuckerrübenwurzelfasern. *Z. Zuckerind. Böehmen* **27**, 975.

Voznyakovskaya, M., and Ryzhkova, A. S. (1955). Microflora-accompanying mycorrhizas. *In* "Mycotrophy in Plants" (A. A. Imshenetskii, ed.), pp. 320–323. Israel Program for Scientific Translations. 1967.

Waksman, S. A. (1931). Decomposition of the various chemical constituents etc. of complex plant materials by pure cultures of fungi and bacteria. *Arch. Mikrobiol.* **2**, 136.

Wallace, R. H. (1947). Unpublished data.

Wallace, R. H., and Lochhead, A. G. (1950). Qualitative studies of soil microorganisms. IX. Amino acid requirements of rhizosphere bacteria. *Can. J. Res.* **28**, 1.

West, P. M., and Lochhead, A. G. (1940). Qualitative studies of soil microorganisms. IV. The rhizosphere in relation to the nutritive requirements of soil bacteria. *Can. J. Res.* **18**, 129.

Wilde, S. A. (1954). Mycorrhizal fungi: Their distribution and effect on tree growth. *Soil Sci.* **78**, 23.

Wilkins, W. H. (1945). Investigation into the production of bacteriostatic substances by fungi. Cultural work on Basidiomycetes. *Trans. Brit. Mycol. Soc.* **28**, 110.

Wilkins, W. H. (1947). Investigation into the production of bacteriostatic substances by fungi. Preliminary investigation of the sixth 100 species, more Basidiomycetes of the wood-destroying type. *Brit. J. Exp. Pathol.* **28**, 53.

Wilkins, W. H., and Harris, G. C. M. (1943). Investigation into the production of bacteriostatic substances by fungi. II. A method for estimating the potency and specificity of the substances produced. *Ann. Appl. Biol.* **30**, 226.

Zukovskaya, P. W. (1941). Changes in bacteriorrhiza of cultivated plants. *Microbiology* (*USSR*) **10**, 919.

CHAPTER 9

Mycorrhizae and Feeder Root Diseases

DONALD H. MARX

I. Introduction

Mycorrhizal associations and diseases of feeder roots of plants have one important similarity in that both are types of parasitism that are intimately involved with the succulent fine feeder roots of their hosts. As discussed in previous chapters, mycorrhizal fungi are stimulated by host roots, symbiotically infect and eventually transform the feeder roots into dual organs in which the cortex cells are enclosed in the Hartig net, and are isolated from direct contact with the soil by the fungal mantle. During synthesis of mycorrhizae, the host responds physiologically to the infection and the fungal symbionts undergo certain transformations. Similarly, feeder root pathogens—species of *Phytophthora, Pythium, Rhizoctonia,* and *Fusarium*—are stimulated by feeder roots, pathogenically infect the meristematic and immature primary cortex tissues, ramify further through tissues, and eventually cause either limited or extensive necrosis. Only in advanced stages of necrosis of feeder roots do these pathogens spread into vascular tissues. It appears that the highly suberized endodermis of roots by either chemical or physical means limits the spread of infection of both symbionts and pathogens into vascular tissues.

If a pathogen infects and destroys a feeder root prior to infection of this root by a mycorrhizal fungus, it is obvious that mycorrhizal development on this root cannot take place. But, if the sequence of parasitic attack is reversed (i.e., if a symbiont infects the feeder root and synthesizes a mycorrhiza prior to infection of this root by the pathogen) are the succulent root tissues in this transformed root still susceptible to infection by a feeder root pathogen? One must consider that a pathogen infecting a nonmycorrhizal feeder root is initially confronted externally only with succulent, thin-walled epidermal cells with or without root hairs and internally with cortex cells which, in most instances, have not undergone secondary cell-wall thickening. On the other hand, a pathogen attacking ectomycorrhizae is initially confronted externally with the tightly interwoven network of hyphae that makes up the fungal mantle, and then internally with cortex cells whose cell walls are surrounded by the Hartig net hyphal tissues. It would appear, therefore, that a pathogen of mycorrhizae must have the physical and chemical ability to penetrate the fungal mantle and Hartig net in order initially to establish a successful pathogenic relationship in the root tissues of the host plant. However, few feeder root pathogens have the chemical or enzymatic ability to be hyperparasitic on other fungi or, in this instance, on fungal symbionts (Boosalis, 1964).

If feeder roots in the ectomycorrhizal condition are altered to such a degree that they are resistant to pathogenic infections then, in addition to the well-documented, physiological benefit of mycorrhizae to plants, they are also beneficial to plant health as biological deterrents to feeder root infection by pathogens. Garrett (1960) is in agreement with this view, since he concluded that a mycorrhizal association is not simply a symbiosis for nutrition of the host but also one for defense against pathogens. He based this conclusion on observations of various workers that mycorrhizae seem to remain functional longer and are less subject to certain types of pathogenic infection than are nonmycorrhizal roots.

The purpose of this chapter is to review the literature pertinent to the role of mycorrhizae in feeder root diseases and the purported mechanisms for their resistance.

II. Field Observations Relating Mycorrhizae to Decreases in Diseases of Feeder Roots

Several workers have observed that tree seedlings with mycorrhizae were more resistant to feeder root infections caused by fungal pathogens than were seedlings with few or no mycorrhizae. Davis *et al.* (1942), after a thorough examination of root diseases of forest tree nursery stock, suggested that mycorrhizae are beneficial to tree seedlings by preventing infection of feeder roots by pseudomycorrhizal fungi. They also observed that mycorrhizal roots appear to be less susceptible to attack by root pathogens than are nonmycorrhizal roots. Levisohn (1954) reported that feeder roots of various *Pinus* spp. and Sitka spruce seedlings were resistant to infection by a *Rhizoctonia* sp. and that the pathogen readily infected nonmycorrhizal feeder roots. She concluded that mycorrhizae on the tree seedlings functioned as biological controls against the root pathogen and that soil conditions inhibitory to mycorrhizal development stimulated root infections by the pathogen. More recently, Powell *et al.* (1968), after application of various nematicides and fungicides to soil around pecan trees (*Carya illinoensis*) with symptoms of feeder root necrosis, observed an enormous and prompt increase in mycorrhizal development by *Scleroderma bovista*. Most chemicals did not significantly reduce populations of *Pythium* spp. or nematodes which were the cause of feeder root necrosis, but foliar and root symptoms gradually disappeared. These authors concluded that populations of competing soil microorganisms were reduced by the chemicals, which in turn caused a stimulation in mycorrhizal development by *S. bovista*. Furthermore, they concluded that the increase in mycorrhizal development increased nutrient absorption by the trees and, more importantly, that the mycorrhizae functioned also as deterrents to infection of feeder roots by pathogens which were still present in significant numbers in the soil. Both functions of mycorrhizae, according to these authors, accounted for the disappearance of feeder root disease symptoms on the trees. Corte (1969) observed that mycorrhizae formed by *Suillus granulatus* appeared to protect seedlings of *Pinus excelsa* from root rot caused by a *Rhizoctonia* sp. since the incidence of root rot was less on seedlings with mycorrhizae. Napier (1969), after an assessment of mycorrhizae on loblolly pine (*Pinus taeda*) growing vigorously on pimple mounds and poorly on flats in lowland areas, reported that trees in decline on the latter sites had significantly fewer mycorrhizae than healthy trees on the mounds. She concluded that the low numbers of mycorrhizae on trees in decline contributed little in the way of defense against the attack by *Phytophthora cinnamomi* and

Pythium spp. These pathogens had been implicated previously with the decline of loblolly pines in these lowland areas (Lorio, 1966).

The major difficulty in assessing the significance of the above reports is that it is nearly impossible to separate cause from effect. One cannot be sure that the presence of mycorrhizae on seedlings or trees has not simply brought about a favorable physiological state in these plants which may have caused a masking of symptoms of feeder root disease. Additionally, since the above observations were made under field conditions, comparisons of mycorrhizal with nonmycorrhizal plants are, at best, questionable. In many instances conditions such as poor soil aeration, low organic matter, and others that contribute to feeder root disease development are also conditions inhibitory to mycorrhizal development and vice versa.

III. Systematic Research Relating Mycorrhizae to Control of Diseases of Feeder Roots

In recent years, a few researchers have systematically investigated the role of mycorrhizal roots as opposed to nonmycorrhizal roots in the resistance of plants to feeder root diseases. Even though only a few reports are available, the results show that mycorrhizae on plants decrease the incidence of feeder root disease.

Wingfield (1968) worked with aseptic seedlings of loblolly pine and observed that mycorrhizae formed by *Pisolithus tinctorius* enhanced survival of seedlings growing with the root pathogen *Rhizoctonia solani*. Pine seedlings without mycorrhizae and inoculated with the pathogen exhibited significantly lower survival and vigor. Richard *et al.* (1971) studied the interaction between the mycorrhizal fungus *Suillus granulatus* and the root pathogen *Mycelium radicis atrovirens* on aseptic seedlings of *Picea mariana*. Seedlings inoculated with only *S. granulatus* grew well, whereas those inoculated with only *M. radicis atrovirens* were chlorotic and severely stunted. Observations on the seedlings in the latter group revealed that the pathogen initially infected the root collar and eventually was detected in the lateral roots, short roots, and root hairs. The fungus was detected in the feeder root cortex of these seedlings, often penetrating to the endodermis of the roots. However, when *S. granulatus* was inoculated simultaneously with the root pathogen, the chlorosis and stunting of seedlings caused by the pathogen were completely eliminated. These seedlings grew just as well as those with only the mycorrhizal fungus.

Ross and Marx (1972) recently found that seedlings of the Ocala race of sand pine (*Pinus clausa*) were protected against *Phytophthora cinnamomi* by the presence of mycorrhizae formed by *Pisolithus tinctorius*. Nonmycorrhizal pine seedlings infected by *P. cinnamomi* exhibited massive feeder root necrosis and only 40% survival after 2 months. Nonmycorrhizal roots on pine seedlings with mycorrhizae formed by *P. tinctorius* were also infected by the pathogen. However, 25% of the feeder roots were mycorrhizal, thus reducing the amount of susceptible root tissue exposed to the pathogen which contributed to nearly 70% survival of test seedlings. Cortical tissues in the mycorrhizal roots were free of *P. cinnamomi*, verifying their resistance to attack by this pathogen. These authors found that the Choctawhatchee race of sand pine was killed by *P. cinnamomi* more rapidly than the Ocala race. They reasoned that the Choctawhatchee race responded to fertility more vigorously than did the Ocala race as expressed by rapid production and elongation of lateral roots which outgrew the mycorrhizal fungus, and thus outgrew its potential protection against *P. cinnamomi*. Wilcox (1968) also observed that rapidly growing roots, especially those breaking from dormancy, would outgrow ectomycorrhizal fungi and remain nonmycorrhizal for certain periods of time.

In a study similar to that on sand pine, Marx (1973) found that shortleaf pine (*Pinus echinata*) seedlings with mycorrhizae were not significantly reduced in growth by *P. cinnamomi* (Table I). Nonmycorrhizal shortleaf pine seedlings exposed to the pathogen were significantly lighter in foliar stem and root dry weights, as well as having significantly fewer new lateral roots than nonmycorrhizal seedlings grown in the absence of the pathogen. The inoculum densities of *P. cinnamomi* in soil with the nonmycorrhizal seedlings at the start and at the end of the experiment did not change significantly. Shortleaf pine seedlings with mycorrhizae formed by either *Pisolithus tinctorius* or *Cenococcum graniforme* did not exhibit reduction in foliar stem or root weights or in development of new lateral roots in the presence of *P. cinnamomi*. The significant reduction in inoculum densities of *P. cinnamomi* in soil with all mycorrhizal seedlings at the end of the study warrants special consideration. Apparently the high degree of mycorrhizal development (70–89%) reduced the amount of susceptible tissue available for attack by *P. cinnamomi*, which in turn caused a decrease in the inoculum density of the pathogen and a decrease in the incidence of disease development. Neither of the two mycorrhizal fungi used in this test produce antibiotics effective against *P. cinnamomi*.

The above reports show rather convincingly that plants with mycorrhizae do not exhibit reduced top growth, chlorosis, restricted root development, and eventually death, and are therefore more tolerant of

TABLE I

Growth of *Pinus echinata* Seedlings with and without Ectomycorrhizae in the Presence and Absence of *Phytophthora cinnamomi*[a]

Measurements	Nonmycorrhizal		Mycorrhizal with *Pisolithus tinctorius*		Mycorrhizal with *Cenococcum graniforme*	
	Without *P. cinnamomi*	With *P. cinnamomi*	Without *P. cinnamomi*	With *P. cinnamomi*	Without *P. cinnamomi*	With *P. cinnamomi*
Height (cm)	5.5 (A,1)	5.0 (A,1)	7.4 (A,2)	7.8 (A,2)	6.4 (A,3)	6.5 (A,3)
Foliar stem dry wt (mg)	99 (A,1)	81 (B,1)	185 (A,2)	203 (A,2)	115 (A,3)	126 (A,3)
Root dry wt (mg)	124 (A,1)	86 (B,1)	131 (A,1)	134 (A,2)	137 (A,1)	141 (A,2)
Number of lateral roots	22 (A,1)	10 (B,1)	23 (A,1)	21 (A,2)	19 (A,1)	17 (A,2)
% Ectomycorrhizae	—	—	86 (A,1)	89 (A,1)	70 (A,2)	76 (A,2)

	Original	Final	Original	Final	Original	Final
Prop./gm of *P. cinnamomi*	21 (A,1)	19 (A,1)	19 (A,1)	11 (B,2)	20 (A,1)	13 (B,2)

[a] Means of measurements within a mycorrhizal condition but between *P. cinnamomi* treatments with the common letter A or B and means between mycorrhizal conditions but within a *P. cinnamomi* treatment with the common number 1, 2, or 3 are not significantly different ($P \leq 0.01$) (after Marx, 1973).

feeder root diseases than are nonmycorrhizal plants. The greater tolerance of mycorrhizal plants to feeder root disease can only be comprehended after a thorough understanding of the possible mechanisms of resistance of mycorrhizae to attack by pathogens.

IV. Mechanisms of Resistance of Mycorrhizae to Pathogenic Infections

Zak (1964) discussed the role of mycorrhizae in feeder root disease and postulated several mechanisms by which mycorrhizal fungi may afford disease protection to feeder roots of plants. He suggested that mycorrhizal fungi may (a) utilize surplus carbohydrates in the root, thereby reducing the amount of chemicals stimulatory to pathogens, (b) provide a physical barrier, i.e., the fungus mantle, to penetration by the pathogen, (c) secrete antibiotics inhibitory to pathogens, and (d) support, along with the root, a protective microbial rhizosphere population. In addition to Zak's mechanisms, Marx (1969a) suggested that (e) inhibitors produced by host cortex cells in response to symbiotic infection (MacDougal and Dufrenoy, 1944) may also function as inhibitors to infection and spread of pathogens in mycorrhizal roots. Since Zak's review, numerous researchers have contributed significantly to our understanding of this area of biological control.

A. PRODUCTION OF ANTIBIOTICS BY FUNGAL SYMBIONTS

The production of antibiotics by soil fungi, actinomycetes, and bacteria has been known for several decades. It was demonstrated by Wright (1956a,b) and others that various saprophytic fungi can produce antibiotics in such restricted sites as pieces of straw and seed coats. It is generally accepted (Brian, 1957; Garrett, 1960; Jackson, 1965) that the resulting antibiotic concentrations are sufficient to influence significantly the pattern of saprophytic microbial colonization of these sites. However, the significance of antibiotic production by saprophytes in reducing the inoculum potential of root pathogens and subsequent root disease development is poorly understood. Most attempts at controlling the activities of root pathogens in soil by infestations of soil with antibiotic-producing saprophytes have failed. The most acceptable explanation for these failures is that antibiotic production is thought to be limited to the immediate substrate or "ecological niche" of the saprophyte (Garrett, 1960). This restricted site of antibiotic production apparently is not of major significance in reducing pathogen inoculum potential in other than this immediate location.

Theoretically, this need not be the fate of antibiotics produced by mycorrhizal fungi. The "ecological niche" of these specialized root parasites is the host root. These fungi, while in mycorrhizal association with roots, are ensured of essential metabolites (e.g., carbohydrates, vitamins, etc.) for which they need exert only minimal competitive efforts. Any antibiotic thus produced in this niche should be ideally located to produce inhibitory effects on pathogens attempting infection of these mycorrhizal and perhaps adjacent nonmycorrhizal roots.

1. Production of Antibiotics in Pure Culture

Several workers from different parts of the world have investigated antibiotic production by the higher basidiomycetes in pure culture. Many of these fungi have been associated with mycorrhizal relationships by Trappe (1962), although most authors investigating antibiotic production made no inference to the possible symbiotic nature of the fungi. Table II is a list of mycorrhizal fungi that reportedly produce antibiotics either in pure culture or in basidiocarps. This table contains only those fungi which either have been experimentally proven to be symbionts or have been associated with mycorrhizae according to Trappe (1962).

It is obvious from Table II that production of antibiotics effective against bacteria, in most instances *Staphylococcus aureus* and *Escherichia coli,* is the most common. This is misleading, however, since in a large proportion of the reports an examination for antifungal activity was not attempted. Since most feeder root diseases of ectomycorrhizal plants are caused by fungi, it is difficult to implicate antibacterial antibiotics in limiting feeder root diseases. However, assuming that these antibiotics are produced by the symbionts while in mycorrhizal associations, they could have a selective influence on bacterial populations in the rhizosphere of mycorrhizae. This effect, in turn, could have a direct or indirect influence on fungal pathogens in the soil (see Chapter 8).

Wilkins and Harris (1944) made extracts from basidiocarps of over 700 species of higher basidiomycetes and found that over 24% contained antibacterial activity. Many of these fungi have been associated with ectomycorrhizae of trees. It is interesting that of the seven *Lactarius* spp. reported active by these workers, all are possible mycorrhizal associates. Also, approximately 80% of the *Tricholoma* spp., 60% of the *Cortinarius,* and 55% of the *Hygrophorus* spp. which produced antibiotics are also mycorrhizal associates. The opposite of this relationship is also interesting; i.e., of 43 *Russula* spp. tested by these workers, none were found to be active and nearly 80% are probably mycorrhizal fungi. In examining the literature, the genus *Russula,* which contains many symbiont species (Trappe, 1962) apparently does not include species which produce antibiotics. One can therefore anticipate that certain genera will contain species which are both capable of forming mycorrhizae and producing antibiotics, whereas other genera may not have species which produce them at all.

Several workers have looked for antifungal activity in mycorrhizal fungi and many have used fungi pathogenic on feeder roots as bioassay organisms. Sasek and Musilek (1967, 1968a,b) found that certain mycorrhizal fungi (see Table II) inhibited *Rhizoctonia solani, Pythium debaryanum,* and *Fusarium oxysporum.* They reported that certain strains of symbiotic fungi, mainly species of *Suillus,* produced antifungal compounds while other strains did not. Sasek (1967) grew pine seedlings on polyurethane disks floating in liquid medium with several mycorrhizal fungi and pathogens and found that antifungal compounds produced by the symbionts decreased damping off caused by the pathogens. Mycorrhizae were not formed under this culture condition. *Tricholoma saponaceum* decreased damping off caused by *R. solani, P. debaryanum,* and *F. oxysporum; Scleroderma aurantium* decreased damping off caused by the latter two pathogens. *Suillus bovinus* was only effective in limiting damping off caused by *R. solani,* while *Amanita citrina, Lactarius helvus,* and *Russula fragilis* were only effective against *P. debaryanum.* Although *Fomes annosus* is not considered to be a feeder root pathogen (Hodges *et al.,* 1970), several workers have found it to be inhibited by antibiotics produced by certain mycorrhizal fungi. Sasek and Musilek (1968a) found that *F. annosus* was only weakly inhibited by a few mycorrhizal fungi in pure culture. However, Hyppel (1968a) tested 85 isolates of some 42 different species of mycorrhizal fungi and found that over 40% of them inhibited *F. annosus* in dual cultures. He also detected variation in antibiotic production by different strains of the same species of mycorrhizal fungus. Hyppel (1968b) demonstrated also that *Boletus bovinus,* a symbiont of Norway spruce and other tree species, protected spruce seedlings from attack by *F. annosus* in greenhouse studies. Although *B. bovinus* did not enter into mycorrhizal association with the seedlings owing to the short duration of the study, a water-soluble antifungal metabolite produced by the symbiont had an inhibiting effect on *F. annosus* and thereby considerably reduced seedling mortality caused by the pathogen. Marx (1969a) found that *F. annosus* was weakly inhibited by *Leucopaxillus cerealis* var. *piceina.* However, he reported that *L. cerealis* var. *piceina* strongly inhibited *Cylindrocladium scoparium,* 9 species of *Phytophthora, Polyporus tomentosus* var. *circinatus, Poria weirii,* 24 species of *Pythium,* 5 species of *Rhizoctonia* or *Thanatephorus,* and *Sclerotium bataticola.* The only root pathogens tested which were not inhibited by this symbiont were *Armillaria mellea,*

TABLE II

Mycorrhizal Fungi Reported to Produce Antibiotics either in Basidiocarps or in Pure Culture

Mycorrhizal fungus	Antibiotic activity	References
Amanita caesaria	Antibacterial	Santoro and Casida (1959, 1962)
A. citrina	Antifungal, antibacterial	Sasek and Musilek (1967, 1968a)
A. muscaria	Antifungal, antibacterial, antiviral	Hyppel (1968a), Santoro and Casida (1959, 1962), Utech and Johnson (1950)
A. pantherina	Antifungal, antibacterial	Wilkins and Harris (1944), Hyppel (1968a)
A. phalloides	Antiviral	Utech and Johnson (1950)
A. rubescens	Antifungal, antibacterial	Santoro and Casida (1959, 1962)
A. solitaria	Antibacterial	Sevilla-Santos and Encinas (1964)
A. strobiliformis	Antibacterial	Wilkins and Harris (1944)
A. vaginata	Antifungal, antibacterial	Hyppel (1968a), Sevilla-Santos and Encinas (1964)
A. virosa	Antibacterial	Wilkins and Harris (1944)
Boletinus pictus	Antiviral	Utech and Johnson (1950)
Boletus bicolor	Antifungal, antibacterial	Santoro and Casida (1959, 1962)
B. bovinus (Suillus)	Antifungal	Hyppel (1968a,b), Sasek and Musilek (1968a)
B. calopus	Antibacterial	Wilkins and Harris (1944)
B. edulis	Antifungal, antibacterial	Wilkins and Harris (1944), Hyppel (1968a)
B. elegans	Antifungal	Hyppel (1968a)
B. granulatus	Antifungal	Hyppel (1968a)
B. luteus (Suillus)	Antifungal, antibacterial	Sasek and Musilek (1968a), Marx (1969a), Santoro and Casida (1959, 1962), Pratt (1971)
B. rubellus	Antifungal, antibacterial	Santoro and Casida (1959, 1962)
B. santanus	Antibacterial	Wilkins and Harris (1944)
B. scaber	Antifungal	Hyppel (1968a)
B. subtomentosus	Antifungal	Hyppel (1968a)
B. variegatus (Suillus)	Antifungal	Hyppel (1968a,b), Sasek and Musilek (1968a), Vaartaja and Salisbury (1965), Rypáček (1960)

Species	Activity	Reference
Cantharellus cibarius	Antibacterial	Wilkins and Harris (1944)
C. tubaeformis	Antibacterial	Wilkins and Harris (1944)
Cenococcum graniforme	Antifungal, antibacterial	Marx and Davey (1969a) and Krywolap et al. (1964)
Clitocybe aurantiaca	Antifungal	Hyppel (1968a)
C. candicans	Antibacterial	Wilkins and Harris (1944) and Sevilla-Santos and Encinas (1964)
C. diatreta	Antifungal, antibacterial	Anchel et al. (1962)
C. laccata (*Laccaria*)	Antifungal, antibacterial	Sevilla-Santos and Encinas (1964) and Marx (1969a)
C. nebuleris	Antifungal	Hyppel (1968a)
C. odora	Antifungal, antibacterial	Wilkins and Harris (1944) and Anchel et al. (1962)
C. rivulosa	Antifungal, antibacterial	Hyppel (1968a), Wilkins (1946), and Anchel et al. (1962)
Clitopilus prunulus	Antifungal, antibacterial	Wilkins and Harris (1944), Bohus et al. (1961), and Sasek and Musilek (1967)
Collybia abutyracea	Antifungal	Pratt (1971)
C. asema	Antifungal	Hyppel (1968a)
Cortinarius anomalus	Antifungal, antibacterial	Wilkins and Harris (1944) and Hyppel (1968a)
C. armeniacus	Antibacterial	Wilkins and Harris (1944)
C. armillatus	Antibacterial	Wilkins and Harris (1944)
C. bolaris	Antibacterial	Wilkins and Harris (1944)
C. caesiocanescens	Antibacterial	Wilkins and Harris (1944)
C. callisteus	Antibacterial	Sevilla-Santos and Encinas (1964)
C. calochrous	Antibacterial	Wilkins and Harris (1944)
C. cinnabarinus	Antibacterial	Mathieson (1947)
C. collinithus	Antibacterial	Wilkins and Harris (1944)
C. orichalceus	Antibacterial	Wilkins and Harris (1944)
C. rotundisporus	Antibacterial	Atkinson (1946)
C. violaceus	Antibacterial	Sevilla-Santos and Encinas (1964)
Hebeloma crustuliniforme	Antibacterial	Wilkins and Harris (1944)
H. mesophaeum	Antibacterial	Wilkins and Harris (1944)
H. sacchariolens	Antibacterial	Wilkins and Harris (1944)
H. strophosum	Antibacterial	Wilkins and Harris (1944)

(Continued)

TABLE II (*Continued*)

Mycorrhizal fungus	Antibiotic activity	References
H. imbricatum	Antibacterial	Wilkins and Harris (1944)
H. repandum	Antibacterial	Wilkins and Harris (1944) and Wilkins (1946)
Hygrophorus chrysodon	Antibacterial	Wilkins and Harris (1944)
H. eburneus	Antibacterial	Sevilla-Santos and Encinas (1964) and Wilkins and Harris (1944)
H. nemoreus	Antibacterial	Wilkins and Harris (1944)
H. penarius	Antibacterial	Wilkins and Harris (1944)
H. virgineus	Antibacterial	Wilkins and Harris (1944)
Inocybe obscura	Antibacterial	Wilkins and Harris (1944)
Lactarius aspideus	Antibacterial	Wilkins and Harris (1944)
L. chrysorheus	Antibacterial	Wilkins and Harris (1944)
L. controversus	Antibacterial	Wilkins and Harris (1944)
L. deliciosus	Antifungal, antibacterial	Wilkins and Harris (1944), Marx (1969a), and Pratt (1971)
L. helvus	Antifungal	Sasek and Musilek (1967, 1968a)
L. necator	Antibacterial	Wilkins and Harris (1944)
L. pallidus	Antibacterial	Wilkins and Harris (1944)
L. quietus	Antibacterial	Wilkins and Harris (1944)
L. vellereus	Antibacterial	Wilkins and Harris (1944)
Lactarius spp.	Antifungal, antibacterial	Morimoto et al. (1954) and Park (1970)
Lepista nuda	Antifungal, antibacterial	Wilkins and Harris (1944), Hervey (1947), Anchel et al. (1962), and Hyppel (1968a)
L. personata	Antibacterial	Wilkins and Harris (1944)
Leucopaxillus cerealis var. piceina	Antifungal, antibacterial	Marx (1969a,b)
M. scorodonius	Antibacterial	Melin et al. (1947)
Paxillus involutus	Antibacterial	Robbins et al. (1945)
Rhizopogon roseolus	Antifungal, antibacterial	Sasek and Musilek (1967, 1968a,b), Hyppel (1968a)
Rhodophyllis clypeatus	Antibacterial	Wilkins and Harris (1944)
Russula atropurpurea	Antibacterial	Sevilla-Santos and Encinas (1964)

R. fragilis	Antifungal, antibacterial	Sasek and Musilek (1967, 1968a)
R. sanguinea	Antibacterial	Sevilla-Santos and Encinas (1964)
Scleroderma aurantium	Antifungal	Sasek and Musilek (1968a)
S. bovista	Antifungal	Marx and Bryan (1969)
Thelephora terrestris	Antifungal	Hyppel (1968a) and Marx *et al.* (1970)
Tricholoma albobrunneum	Antifungal, antibacterial	Wilkins and Harris (1944) and Sasek and Musilek (1967, 1968a)
T. equestre	Antibacterial	Sasek and Musilek (1967)
T. flavobrunneum	Antiviral	Utech and Johnson (1950)
T. imbricatum	Antibacterial	Wilkins and Harris (1944) and Sasek and Musilek (1967)
T. irinum	Antibacterial	Wilkins and Harris (1944)
T. pessundatum	Antifungal, antibacterial	Sasek and Musilek (1967) and Hyppel (1968a)
T. psammopodum	Antibacterial	Wilkins and Harris (1944)
T. saponaceum	Antifungal, antibacterial	Wilkins and Harris (1944) and Sasek and Musilek (1967, 1968a,b)
Tricholoma sp.	Antifungal	Pratt (1971)
T. ustale	Antibacterial	Wilkins and Harris (1944)
T. vaccinum	Antibacterial	Hutinel and Oddoux (1954) and Sasek and Musilek (1967)
Unidentified	Antifungal	Zak (1964), Hyppel (1968a), Marx *et al.* (1970), and Pratt (1971)

Fusarium oxysporum f. *pini, Rhizoctonia crocorum,* and *Thanatephorus cucumeris.*

The antibiotic produced by *L. cerealis* var. *piceina* was identified as diatretyne nitrile, a polyacetylene (Marx, 1969b). Diatretyne nitrile inhibited germination of zoospores of *Phytophthora cinnamomi* at a concentration of 50–70 parts per billion and killed zoospores at 2 parts per million (ppm). The antibiotic inhibited bacteria from forest soil at 0.5 ppm. *L. cerealis* var. *piceina* also produced diatretyne amide and diatretyne 3 which are reduction products of the nitrile. Both of these diatretynes are antibacterial only. Anchel (1952, 1953, 1958, 1959) and associates (Anchel *et al.,* 1962) originally identified the diatretyne antibiotics from culture filtrates of *Clitocybe diatreta, C. odora,* and *Lepista nuda* which are known mycorrhizal associates of trees. At least 10 other hymenomycetous fungi produce polyacetylene antibiotics but they have not been associated with mycorrhizae.

In addition to reporting antibiotic production by *L. cerealis* var. *piceina,* Marx (1969a) also found other symbionts of pines which could inhibit root pathogens. An interesting aspect of this report was the difference in biological activity of the inhibitors produced by various symbionts. *Laccaria laccata* inhibited 16 of the 21 species of *Pythium* tested, but only 1 of the 9 species of *Phytophthora.* A near reversal of activity was detected with *Lactarius deliciosus.* This symbiont did not inhibit any species of *Pythium* but it did inhibit 6 of the 8 species of *Phytophthora.* Extremes in activity were also found. *Pisolithus tinctorius* did not inhibit any of the 48 root pathogens tested and, as mentioned earlier, *L. cerealis* var. *piceina* inhibited nearly all (85%) of the pathogens. Somewhat intermediate in the biological spectrum of antibiotic activity was *Suillus luteus* which inhibited over 70% of the pathogens including all the *Phytophthora* spp. and most of the *Pythium* spp. Marx and Bryan (1969) found that *Scleroderma bovista,* a symbiont of pine and pecan (*Carya illinoensis*), inhibited 5 *Phytophthora* spp. and 4 *Pythium* spp. which are associated with feeder root necrosis of pecan. The presence of this antibiotic in mycorrhizal roots of pecans was implicated in feeder root disease control caused by the *Pythium* spp. Park (1970) reported that an unidentified *Lactarius* sp., symbiotic on basswood (*Tilia*) seedlings, inhibited several pathogens including the feeder root pathogens *Pythium irregulare, Phytophthora infestans, Fusarium solani, F. oxysporum, Cylindrocladium scoparium, Rhizoctonia solani, R. praticola,* and *Sclerotium rolfsii.* He further demonstrated that the culture filtrate of the *Lactarius* sp. was active against damping off of *Pinus resinosa* seed caused by *Pythium irregulare* and *Rhizoctonia praticola.* Seed presoaked for 12 hours in the antifungal filtrate had 93% germination and emergence in contrast to nontreated seed of which only 7% germinated and emerged in the presence of these pathogens. Marx *et al.* (1970) studied isolates of *Thelephora terrestris*, a widespread symbiont on many tree species, and found that certain isolates inhibited *Pythium aphanidermatum*, *P. irregulare*, and *P. spinosum* but not *P. vexans* or 4 species of *Phytophthora*. Certain unidentified symbionts of *Pinus echinata* isolated by these authors also inhibited the same species of *Pythium*. Pratt (1971) examined nearly 50 basidiomycetous fungi collected from eucalypt forests in Australia for antagonism against *P. cinnamomi*. Several of these fungi are known mycorrhizal symbionts, such as *Boletus* (*Suillus*) *luteus* and *Lactarius deliciosus*, and they were found to strongly inhibit vegetative mycelial growth of *P. cinnamomi* in agar plate studies. He discussed the possible role of these antibiotics in reducing feeder root disease of *Eucalyptus marginata* caused by *P. cinnamomi* in Australia.

It is obvious that numerous mycorrhizal fungi have the capacity to produce antibiotics in pure culture or in their basidiocarps. More specifically, many mycorrhizal fungi produce antifungal compounds in pure culture which inhibit a broad spectrum of pathogens that can cause feeder root diseases. However, a prerequisite to implicating these antibiotics in feeder root disease control is to show that these antibiotics are produced by the symbionts while in mycorrhizal association with their hosts.

2. Production of Antibiotics in Mycorrhizal Association

Only a few mycorrhizal fungi have been examined for production of antibiotics while in mycorrhizal association. Krywolap *et al.* (1964) extracted an antibiotic from mycorrhizae formed by *Cenococcum graniforme* on *Pinus strobus*, *P. resinosa*, and *Picea abies*. This antibiotic, which was active against bacteria but not fungi, was similar in chromatographic and ultraviolet fluorescence to the antibiotic produced in the mycelium of pure cultures of *C. graniforme*. Foliar extracts of the pines with mycorrhizae formed by *C. graniforme* under forest soil conditions contained the antibiotic, indicating that it was readily translocated. However, the antibiotic was detected only in roots but not in foliage of trembling aspen (*Populus tremuloides*), which suggested a lack of translocation in this tree species. They also found that nursery-grown pine seedlings without mycorrhizae formed by *C. graniforme* contained similar antibiotic activity. They concluded that the seedlings absorbed the antibiotic from sclerotia or hyphae of *C. graniforme* present in great abundance in the nursery soil. These authors suggested that this antibacterial compound may confer some degree of protection to trees against bacterial pathogens. Grand and Ward (1969) also reported this antibiotic in foliage of these same tree species, but they could not find a relationship between the number of mycorrhizae formed by *C. graniforme* in two different soil types and the amount of antibiotic activity from the foliage of a tree species. They attributed the lack of correlation to physiological differences between species of hosts which could have inactivated the antibiotic.

Marx and Davey (1969a) extracted diatretyne nitrile and diatretyne 3 from mycorrhizae formed by *L. cerealis* var. *piceina* and from the rhizosphere substrate of the mycorrhizae on aseptic *Pinus echinata* seedlings. Neither short roots nor the substrate adjacent to short roots on nonmycorrhizal pine seedlings contained the diatretynes. No attempt was made to detect the diatretyne antibiotics in foliage of seedlings with mycorrhizae formed by this symbiont. However, in earlier work (Marx, 1969b), diatretyne nitrile was not detected in foliage of young pine seedlings with roots exposed to the antibiotic for 40 days. In experiments designed to determine the susceptibility or resistance of mycorrhizae to infection by a pathogen, Marx and Davey (1969a) were able to demonstrate that the diatretynes present in mycorrhizae formed by *L. cerealis* var. *piceina* were functional in the resistance of feeder roots to infection by *Phytophthora cinnamomi*. Nonmycorrhizal short roots adjacent to mycorrhizae formed by *L. cerealis* var. *piceina* which contained the diatretyne antibiotics were only 25% susceptible to infection by zoospores of *P. cinnamomi*. Short roots on control seedlings and on seedlings with mycorrhizae formed by either *Laccaria laccata* or *Pisolithus tinctorius* were 100% susceptible to infection. It was not determined whether the diatretynes were translocated to short roots from adjacent mycorrhizae or simply absorbed from the rhizosphere. These authors also observed that instead of 100% infection only 77% of short roots on *Pinus echinata* and 85% of short roots on *Pinus taeda* seedlings, adjacent to mycorrhizae formed by *Suillus luteus* were infected by *P. cinnamomi*. They attributed some antibiotic protection here also, since *S. luteus* was found to be antagonistic to *P. cinnamomi* and other related fungi in earlier pure-culture tests (Marx, 1969a).

B. MECHANICAL BARRIER CREATED BY FUNGAL MANTLE

A mechanical barrier creates a physical rather than a chemical impediment to the entrance or spread of a pathogen. The significance of mechanical barriers to host defense against pathogenic attack has received attention in many plant disease investigations. Several external and internal morphological barriers of a mechanical nature are found in plants which may influence either pathogen entrance or its spread in host tissue. Tough outer walls of epidermal cells, suberized root periderm and endodermis, and thick cuticles of leaves have been reported to impede direct penetration by pathogens and may function as mechanical barriers. Other mechanical barriers are often formed after the establishment of infection. In some plants, suberized wound tissue or cicatricial layers develop, localizing the infection. Abscission layers, tyloses, gum deposits, callosites, cellulosic coverings over infective hyphae, and other barriers have been reported to physically impede spread of certain pathogens. In assessing the role of these various mechanical barriers, especially those of external origin, in the protection of plants in general, it has been concluded by several investigators (Akai, 1959; Bürstrom, 1965; Dickenson, 1960; Martin, 1964) that they are relatively ineffectual.

However, the fungal mantle of mycorrhizae creates a unique and totally different type of obstruction to pathogens attempting penetration. In mature mycorrhizae, the mantles are composed of tightly interwoven hyphae, often in well-defined layers, which usually completely cover the root meristem and cortical tissues. This hyphal network usually is complete, i.e., relatively free of voids, which preclude exposure of root tissue to direct contact with the rhizosphere. The thickness of the mantle is variable, depending on fungus associate, nutrition, temperature, and other factors (Harley, 1969). Normally, however, mantle thicknesses of 20–80 μm are encountered (Marx and Bryan, 1970).

Marx and Davey (1969a,b) and Marx (1970) have concluded that the fungus mantles of mycorrhizae are formidable physical barriers to penetration by *Phytophthora cinnamomi* (Fig. 1). They found that fungus mantles of mycorrhizae formed by non-antibiotic-producing fungal symbionts which passively covered adjacent nonmycorrhizal root initials on pine seedlings protected the root initials from penetration by *P. cinnamomi*. Close, histological examination of the root initials revealed the complete absence of symbiont infection and the apparent passive protection afforded by the mantle. Additional observation on covering of root meristems provided further evidence of the barrier effect against pathogen penetration. These workers found that meristems of mycorrhizae of *Pinus echinata* and *P. taeda* were readily infected by *P. cinnamomi* when the mantle covering was either incompletely formed over the root tip or artificially removed. Infection, however, did not take place in the meristem tissues when the root tips were covered by a fungal mantle. These authors further suggested that the Hartig net surrounding the cortical cells may function as an additional physical barrier. They observed that spread of *P. cinnamomi*, which either originated from infections of nonprotected meristem tissues without mantle coverings or from infections through artificially excised root tips, was

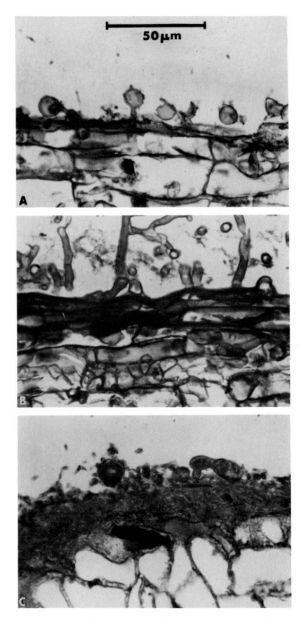

Fig. 1. Susceptibility and resistance of nonmycorrhizal and mycorrhizal feeder roots of pine to *Phytophthora cinnamomi*. Direct penetration of zoospores (A) and vegetative hyphae (B) of *P. cinnamomi* into nonmycorrhizal feeder roots with intracellular cortex infection. Note vesicles and hyphae of *P. cinnamomi* on mantle barrier of mycorrhizae in (C) and absence of cortex infection (after Marx, 1970; and Marx and Bryan, 1969).

blocked at the Hartig net region located several cortex cells behind the root tips at the region of cell maturation (Fig. 2). These authors, however, were unable to separate the possible indirect chemical effect of the Hartig net, i.e., inducing the production of chemical inhibitors in cortex cells, from the suggested mechanical effect of this hyphal network (see Chapter 1, on formation of Hartig net).

C. CHEMICAL INHIBITORS PRODUCED BY HOST

Most plant cells are capable of elaborating inhibitory substances during their metabolic response to pathogenic attack. Phenols, quinones,

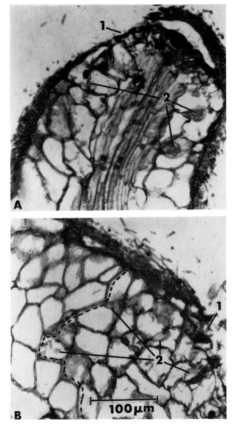

FIG. 2. *Phytophthora cinnamomi* infection of incomplete mycorrhiza of pine. (A) and (B) are sections of the same mycorrhiza taken 60 μm apart. Note void in fungal mantle (1), intracellular hyphae of *P. cinnamomi* (2), and boundary of Hartig net (dashed line) which blocked further spread of *P. cinnamomi* (after Marx and Davey, 1969a).

various phytoalexins, and numerous other compounds have been found in tissues of a variety of plants during pathogenesis. Many of these were found to be inhibitory to the pathogen and are considered by many authors to be important in disease resistance (Cruickshank, 1963; Tomiyama, 1963).

Plant cells exposed to symbiotic parasitic invasion have also been reported to respond by producing substances inhibitory to the fungal symbiont. Bernard (1911) concluded, from the results of studies on the symbiosis between *Rhizoctonia repens* and orchid tubers, that an antifungal compound was formed in response to infection by *R. repens*. Subsequent research by Gaümann (1960) and associates (Gaümann and Kern, 1959; Gaümann *et al.*, 1960) revealed that tubers of several species of orchid produce orchinol, coumarin, and an unidentified phenolic compound in response to parasitism by *R. repens*. Orchinol is also produced by orchid tubers in response to infection by several other species of *Rhizoctonia* and other endomycorrhizal and pathogenic fungi. Orchinol could not be found in noninfected tubers.

The production of these inhibitory compounds is considered to be the defense mechanism of the orchid which maintains the fungus in a symbiotic state. Without this defense mechanism, the fungus could be pathogenic on the orchids. Another consequence of tuber synthesis of these inhibitory compounds, which extends throughout the tuber, is that their presence protects the tissue not only against reinfection by the fungal symbiont but also against infection by pathogenic organisms.

The mycorrhizae of orchids are endomycorrhizal, i.e., the hyphae of the symbiont penetrate the cells and are in close contact with the cytoplasm. Obviously, endomycorrhizal fungi have the enzymatic capacity to degrade cellulosic cell walls. Few ectomycorrhizal fungi have this capacity (Norkrans, 1950) and most are limited to utilizing a few relatively simple carbohydrates (Hacskaylo, 1971). The inability of ectomycorrhizal fungi to digest cellulosic cell walls may be sufficient to explain their inability to penetrate cell walls of their hosts. However, Foster and Marks (1966) have presented electron micrographic evidence which suggests that mycorrhizal fungi exert not only enzymatic action in forming the Hartig net between cortex cells but also detectable mechanical pressure. Why the chemical and mechanical penetration of cortical cell walls by mycorrhizal fungi does not occur is conjecture. Inhibitors of host origin may be induced in response to incipient Hartig net development that restricts the ectomycorrhizal symbiont to the balanced symbiotic state in much the same manner as host origin inhibitors apparently do, in endomycorrhizal orchid tubers. The cellulases produced by the mycorrhizal fungus may also be inhibited by a tannin fraction containing polyphenols present in epidermal cell remnants secreted by the host (Foster and Marks, 1967). This tannin fraction must be specific for cellulases and not inhibit pectin-degrading enzymes, since the latter enzymes are functional in forming the Hartig net hyphal development during synthesis of mycorrhizae. Also, it is known that cellulase is an inducible enzyme system (Mandels and Reese, 1965) and it is possible that as long as available carbohydrates are present its synthesis is suppressed. Hyphae of mycorrhizal fungi in the Hartig net contain high quantities of glycogen (Foster and Marks, 1966) suggesting that available carbohydrates are not limiting in this region. Although Foster and Marks (1966) could not detect apposition of materials by the host cytoplasm that restricted hyphae of the fungal symbiont as found in pathogenic associations, they did find disorganized cytoplasmic organelles and abundant tannin-like materials in cortical cells of mycorrhizae. They also observed that there was a very large vacuole in each cortex cell, and often it was lined with polyphenol materials aggregated into large, dense masses. Frequently the nuclei of these cells were very large and rounded, without their usual invaginations, and the chromatin bodies were unusually heavily stained. The cortex cells were also usually devoid of starch and contained amyloplasts. Mitochondria in these cells were elongated and covered with abundant cristae. These observations on cortex cells in the Hartig net association of mycorrhizae were quite unlike those of nonmycorrhizal roots.

From the observations of Foster and Marks (1966, 1967) and the report of MacDougal and Dufrenoy (1944), it is obvious that there is a biochemical and cytological reaction of the cortex cells to infection by mycorrhizal fungi. Krupa and Fries (1971) found chemical evidence in support of these conclusions. These workers found that the fungus symbiont *Boletus variegatus* in pure culture produced isobutanol and isobutyric acid which are fungistatic, volatile compounds. Other volatile compounds were also identified. Volatile organic compounds were also extracted from the roots of intact seedlings of *Pinus sylvestris* grown in pure culture with and without *B. variegatus*. However, infection of the roots by the fungal symbiont resulted in the production and accumulation of volatile terpenes and sesquiterpenes in concentrations up to eight times greater than that found in nonmycorrhizal roots. Many of these terpenes and sesquiterpenes are fungistatic and are considered by these investigators to be produced as a nonspecific response of the host cells to symbiotic infection. These authors concluded that the nonspecific response of the host to infection by symbiotic fungi results in the increased production and accumulation of native volatile (terpenes, etc.) and nonvolatile (Hillis and Ishikura, 1969) substances. These substances when present in sufficient concentrations, may restrict the growth of the mycorrhizal fungi within the host tissue, resulting finally in the symbiotic state. Furthermore, volatile and nonvolatile substances could inhibit pathogens in the root as volatile substances inhibit the pathogens in the rhizosphere. Krupa (1970) found that several monoterpenes extracted from mycorrhizae inhibited vegetative growth of *P. cinnamomi* and *F. annosus* by 50% when vapors from 10 μl of the substances were used. Catalfomo and Trappe (1970) also found that certain mycorrhizal fungi from the genera *Amanita* and *Rhizopogon* and the family Boletaceae produced terpenes which, in the opinion of these investigators, are involved in the protective role of mycorrhizae.

There appears to be sufficient cytological and biochemical evidence to support the viewpoint that host roots respond to symbiotic infection. However, only indirect evidence is available relating this host reaction to its role in the resistance of mycorrhizae to pathogenic infection.

D. DIFFERENCES IN CHEMICAL EXUDATION OF MYCORRHIZAE

The significance of root exudates in affecting root diseases has been investigated for a variety of crop plants. The chemical composition of root exudates of herbaceous plants includes carbohydrates, amino acids, vitamins, organic acids, nucleotides, flavonones, enzymes, and a wide variety of miscellaneous compounds, such as hydrocyanic acid, glycosides, and saponins. These exudates have either stimulating or inhibiting effects on certain root pathogens. Exudates from certain plants stimulate mycelial growth, microsclerotia germination, tactic zoospore activity, and pathogenicity. Root exudates from other plants contain chemicals which inhibit the same processes (Rovira, 1965).

The chemical nature of tree root exudates has received little attention. Slankis and co-workers (Slankis, 1958; Slankis *et al.*, 1964) reported that roots of aseptic *Pinus strobus* seedlings with needles exposed to $^{14}CO_2$ liberated a complex mixture of more than 35 radioactive sugars, amides, and organic acids. Agnihotri and Vaartaja (1967) isolated and identified 3 sugars and 13 amino acids from exudates of young radicles of *Pinus resinosa* seedlings. These workers, as did Slankis and his co-workers, found glucose, fructose, arabinose, asparagine, glutamine, and several amino acids to be the most prevalent chemicals in pine exudates. Smith (1969) also found similar carbohydrates and amino acids as well as vitamins and certain organic acids in exudates of very young roots of pines and *Robinia pseudoacacia*. He also demonstrated qualitative and quantitative differences in exudation patterns of the different tree species. These reports provide evidence that tree roots, like those of other plants, liberate many organic metabolites into the rhizosphere.

However, in the above investigations the roots of the tree seedlings were not in a mycorrhizal condition. Although differences in exudation patterns between mycorrhizal and nonmycorrhizal roots have not been reported, in the opinion of this author, it is logical to expect differences in exudation patterns between mycorrhizal and nonmycorrhizal roots. This inference is based upon the simple fact that mycorrhizal fungi derive most, if not all, of their required carbohydrates, amino acids, and vitamins from their intimate association between the cortex cells and over the external root surface of their host. Few root exudates could pass through the Hartig net and fungal mantle of mycorrhizae without some absorption and utilization by the fungal symbiont. This suggests that exudates of mycorrhizal roots are (a) those root exudates not utilized by the fungal symbiont, (b) metabolic by-products of the fungal symbiont, or (c) exudates released as a result of the metabolic interaction of the symbiotic partners. If one considers that differences in microbial rhizosphere populations are at least partially due to differences in root exudates, then circumstantial evidence is available indicating differences in exudates between mycorrhizal and nonmycorrhizal roots, since these various root types harbor different rhizosphere populations. (See Chapter 8.)

Although not investigating exudates of mycorrhizae, Lewis and Harley (1965) reported that mycorrhizae of beech (*Fagus sylvatica*) contained quantities of glucose, fructose, sucrose, and trehalose together with the acyclic polyol, mannitol, and two cyclic polyols, *myo*-inositol and an unidentified inositol. Trehalose and mannitol were not extracted from nonmycorrhizal roots of beech showing their presence to be dependent on mycorrhizal infection.

The chemical attraction or chemotaxis of motile zoospores of phycomycetous root pathogens to plant roots is one of the major means by which these fungi initially attack roots (Chang-Ho and Hickman, 1970; Zentmyer, 1970). Numerous studies have revealed that roots, both excised and on intact plants, and chemicals, known to be components of root exudate of these plants, are strongly attractive to zoospores of numerous species of phycomycetes. It is assumed, however, that previous work on chemotaxis has been accomplished on nonmycorrhizal roots, since the roots tested were grown in such a manner (i.e., hydroponics, aseptic culture, or closed root containers) as to limit infection by mycorrhizal fungi.

The only research to date aimed at investigating chemotaxis of zoospores to mycorrhizal and nonmycorrhizal roots is that of Marx and Davey (1969a,b). These workers used intact mycorrhizae formed by several different fungal symbionts and nonmycorrhizal roots on *Pinus echinata* and *P. taeda* seedlings and found that zoospores of *P. cinnamomi* were not strongly attracted to either nonmycorrhizal or mycorrhizal roots. After zoospore encystment on root surfaces, zoospores were observed to germinate faster and more vigorously at the growing tips and the region of cell elongation on nonmycorrhizal roots than on other parts of the root. However, on mycorrhizae, encysted zoospores germinated slowly, and germ tubes elongated slowly, at a rate comparable to that observed on heavily suberized parts of the root. This indicated indirectly to these authors that mycorrhizae were not as chemically stimulating to the zoospores as were nonmycorrhizal, nonsuberized roots. These workers did report chemotaxis of zoospores of *P. cinnamomi* to cut feeder root tips of these pine species. Nonmycorrhizal short and lateral roots and mycorrhizal roots had their apices excised at either 0.1 or 1 mm from the growing tip. Strong chemotaxis was observed on nonmycorrhizal short and lateral roots and on mycorrhizae with 1 mm of their tips removed. As observed before by these authors, zoospores germinated faster and with more vigor on cut surfaces of nonmycorrhizal roots than on mycorrhizal roots.

E. PROTECTIVE MICROBIAL RHIZOSPHERE POPULATIONS

Garrett (1960) described the rhizosphere as the outermost defense of the plant against attack by root pathogens. This zone normally supports a much greater population of microorganisms than is found in nonrhizosphere soil. Several reviews are available concerning the influence of antagonistic rhizosphere populations of many plants which limit pathogenic attack of roots (Katznelson, 1965; Rovira, 1965). Rambelli, in Chapter 8 of this volume, has thoroughly reviewed the research dealing with differences in microbial rhizosphere population of mycorrhizal and nonmycorrhizal roots. Therefore, this section will be appropriately short but hopefully sufficient in content to present the general status of knowledge in this area.

Tribunskaya (1955) found approximately ten times as many fungi in rhizospheres of mycorrhizal pine seedlings as in those of nonmycorrhizal seedlings. She concluded that the various fungal symbionts were responsible for the different microflora of the rhizosphere. Katznelson *et al.* (1962) showed clearly that mycorrhizal roots of yellow birch (*Betula alleghaniensis*) exerted a stimulatory effect on numbers of certain physiological groups of soil bacteria and actinomycetes. Bacteria which grew in a simple chemical medium appeared to be suppressed around mycorrhizae as were total fungal numbers. However, the types of fungi present in the various rhizospheres were different. Fungal genera which contain feeder root pathogens, *Pythium, Fusarium,* and *Cylindrocarpon,* predominated in nonmycorrhizal rhizospheres, while the mycorrhizal roots supported *Mycelium radicis, Penicillium* spp., and other rapidly growing nonpathogenic fungi. *Pythium* and *Fusarium* spp. were completely absent from the rhizospheres of mycorrhizae. These results suggest that the mycorrhizae had an inhibiting effect on pathogens in the root zone. Neal *et al.* (1964), investigated the microbial rhizosphere population of three morphologically distinct mycorrhizae, a white, a gray, and a yellow form as well as the microbial population of suberized roots and nonrhizosphere soil from a *Pseudotsuga menziesii* tree. Each microhabitat contained a distinct microflora. These investigators attributed the differences between mycorrhizal rhizospheres to different associated fungal symbionts. They suggested that the influence of specific mycorrhizal fungi on the rhizosphere may affect the extent of infection of root pathogens and that some mycorrhizae may support a more effective rhizosphere barrier than others. Other workers (Fontana and Luppi, 1966; Foster and Marks, 1967; Oswald and Ferchau, 1968; Neal *et al.*, 1968) have also concluded that microbial populations of mycorrhizal rhizospheres are not only different both qualitatively and quantitatively from those of nonmycorrhizal rhizospheres but that each morphological type of mycorrhiza, each presumably formed by a different fungal symbiont, harbors in its rhizosphere a different microbial population.

The only inference made relating the antagonistic nature of fungal symbionts in mycorrhizal association to population of microorganisms in the rhizosphere is that of Ohara and Hamada (1967). These workers found that bacteria, especially aerobic and heterotrophic types, as well as actinomycetes were strongly inhibited around actively growing mycelium in forest soil or mycorrhizae formed by *Tricholoma matsutake* on *Pinus densiflora.* These microorganisms were found in great abundance in adjacent soil containing neither mycelia nor mycorrhizae formed by this fungus. These authors inferred that antibiotics were the cause for the inhibition.

One may logically surmise, from the above discussion, that since there are differences in microbial rhizosphere populations between mycorrhizal and nonmycorrhizal roots, there are differences also in the microbial competitive potential near these roots. Future research will tell us whether or not these differences are of sufficient ecological significance to influence root pathogen populations in the soil and subsequent development of feeder root disease.

V. Conclusions

The role of ectomycorrhizae as biological deterrents to feeder root infection by pathogens is an important aspect of our understanding of the overall ecological significance of mycorrhizae. When one considers that many forest, tree nursery, and agricultural soils contain significant populations of feeder root pathogens, such as species of *Phytophthora*, *Pythium*, *Fusarium*, and others, capable of causing growth reduction and mortality in mature trees and seedlings (Vaartaja and Bumbieris, 1964; Smith, 1967; Hendrix and Campbell, 1970; Campbell and Hendrix, 1967; Hendrix *et al.*, 1971), then the ecological value of mycorrhizae as deterrents to infections by these pathogens is self-evident. Simply stated, mature trees and seedlings with significant quantities of mycorrhizae growing in soils containing the feeder root pathogens would have little susceptible (nonmycorrhizal) root tissue exposed to attack by the pathogen. These plants will also gain additionally from the well-documented physiological benefit of mycorrhizae. In certain situations, such as in well-fertilized nursery soils containing high populations of feeder root pathogens, the role of mycorrhizae in the control of feeder root disease could be more important than their physiological role.

Based on available information, the protective role of mycorrhizae in root disease appears to be nonsystemic since the presence of a few mycorrhizae on a root system does not ensure control of pathogenic root infections on nonmycorrhizal sections of the same root system. This means, therefore, that the percentage of short roots in the mycorrhizal condition should correlate directly with the degree of control of root infection. A possible exception to this generalization is the involvement of the purported antibiotic mechanism. A fungal symbiont producing in its mycorrhizae a potent antibiotic effective against a feeder root pathogen may afford protection to adjacent nonmycorrhizal roots against this pathogen simply by translocation or diffusion of the antibiotic into the nonmycorrhizal roots (Marx and Davey, 1969a). In this circumstance, the presence of a few mycorrhizae formed by an antibiotic-producing fungal symbiont may be as valuable in control of pathogenic root infections as the presence of considerably more mycorrhizae formed by a non-antibiotic-producing mycorrhizal fungus. This latter circumstance may have more significance than researchers originally thought since it is now known that many species of mycorrhizal fungi are capable of producing antibiotics (Table II).

It is difficult at this stage of our knowledge to determine which of the mechanisms responsible for the resistance of mycorrhizae to pathogenic infection is the most important. Systematic research has shown that the fungal mantle does appear to be a mechanical barrier to penetration by pathogens, but even in the absence of the protective mantle the root cortex cells surrounded by the Hartig net also are resistant. The latter observation strongly suggests a chemical function, probably of host origin, in this resistance phenomenon. In support of this contention, several researchers have demonstrated a chemical function which appears to be involved. The antibiotic mechanism of resistance is apparently functional since normally susceptible roots adjacent to resistant mycorrhizae producing an antibiotic are resistant to pathogenic attack. Several researchers have convincingly demonstrated that mycorrhizae support a microbial rhizosphere population different both qualitatively and quantitatively from those of other mycorrhizae and, more importantly, nonmycorrhizal roots and nonrhizosphere soil. Even though research has not yet proved the value of these different microbial rhizosphere populations in control of feeder root diseases, one can logically conclude that they do have an effect on microbial competition and, thereby, perhaps root pathogens. Circumstantial evidence has also been presented which shows that pathogens are not attracted chemically to mycorrhizae as they are to nonmycorrhizal roots.

In all probability most of the proposed mechanisms of root protection of mycorrhizae are functional at any given time since several appear to be inseparable (i.e., mantle barriers, host origin inhibitors, differences in chemical exudations, etc.). This broad spectrum of defense mechanisms acting in concert assures greater opportunities for biological control of feeder root pathogens by mycorrhizae.

References

Agnihotri, V. P., and Vaartaja, O. (1967). Root exudates from red pine seedlings and their effects on *Pythium ultimum*. *Can. J. Bot.* **45**, 1031.

Akai, S. (1959). Histology of defense in plants. *In* "Plant Pathology" (J. G. Horsfall and A. E. Dimond, eds.), Vol. 1, pp. 391. Academic Press, New York.

Anchel, M. (1952). Acetylenic compounds from fungi. *J. Amer. Chem. Soc.* **74**, 1588.

Anchel, M. (1953). Identification of an antibiotic polyacetylene from *Clitocybe diatreta* as a suberamic acid ene-diyne. *J. Amer. Chem. Soc.* **75**, 4621.

Anchel, M. (1958). Metabolic products of *Clitocybe diatreta*. I. Diatretyne amide and diatretyne nitrile. *Arch. Biochem. Biophys.* **78**, 100.

Anchel, M. (1959). Metabolic products of *Clitocybe diatreta*. III. Characterization of diatretyne 3 as *trans*-10-hydroxy-dec-2-en-4,6,8-triynoic acid. *Arch. Biochem. Biophys.* **85**, 569.

Anchel, M., Silverman, W. B., Valanju, N., and Rogerson, C. T. (1962). Patterns of polyacetylene production. I. The diatretynes. *Mycologia* **54**, 249.

Atkinson, N. (1946). Toadstools and mushrooms as a source of antibacterial substances active against *Mycobacterium phlei* and *Bact. typhosum. Nature (London)* **157**, 441.

Bernard, N. (1911). Sur la function fungicide des bulbes d'ophrydées. *Ann. Sci. Natur.: Bot. Biol. Veg.* **14**, 221.

Bohus, G., Gluz, E. T., and Scheiber, E. (1961). The antibiotic action of higher fungi on resistant bacteria and fungi. *Acta Biol. (Budapest)* **12**, 1.

Boosalis, M. G. (1964). Hyperparasitism. *Annu. Rev. Phytopathol.* **2**, 363.

Brian, P. W. (1957). The ecological significance of antibiotic production. *In* "Microbial Ecology," p. 168. Cambridge Univ. Press, London and New York.

Burström, H. G. (1965). The physiology of plant roots. *In* "Ecology of Soil-borne Plant Pathogens" (K. F. Baker and W. C. Snyder, eds.), Vol. I, p. 154. Univ. of California Press, Berkeley.

Campbell, W. A., and Hendrix, F. F., Jr. (1967). *Pythium* and *Phytophthora* species in forest soils in the southeastern United States. *Plant Dis. Rep.* **51**, 929.

Catalfomo, P., and Trappe, J. M. (1970). Ectomycorrhizal fungi: A phytochemical survey. *Northwest Sci.* **44**, 19.

Chang-Ho, Y., and Hickman, C. J. (1970). Some factors involved in the accumulation of phytomycete zoospores on plant roots. *In* "Root Diseases and Soil-borne Pathogens" (T. A. Toussoun, R. V. Bega, and P. E. Nelson, eds.), pp. 103–109. Univ. of California Press, Berkeley.

Corte, A. (1969). Research on the influence of the mycorrhizal infection on the growth, vigor, and state of health of three *Pinus* species. *Arch. Bot. Biogeogr. Ital.* **45**, 1.

Cruickshank, I. A. M. (1963). Phytoalexins. *Annu. Rev. Phytopathol.* **1**, 351.

Davis, W. C., Wright, E., and Hartley, C. (1942). Diseases of forest tree nursery stock. *Fed. Sec. Agency, Div. Cons. Corp. Forest. Publ.* **9**.

Dickenson, S. (1960). The mechanical ability to breach the host barriers. *In* "Plant Pathology" (J. G. Horsfall and A. E. Dimond, eds.), Vol. 2, pp. 203–232. Academic Press, New York.

Fontana, A., and Luppi, A. M. (1966). Saprophytic fungi isolated from ectotrophic mycorrhizae. *Allionia* **12**, 39.

Foster, R. C., and Marks, G. C. (1966). The fine structure of the mycorrhizae of *Pinus radiata* D. Don. *Aust. J. Biol. Sci.* **19**, 1027.

Foster, R. C., and Marks, G. C. (1967). Observations on the mycorrhizae of forest trees. II. The rhizosphere of *Pinus radiata* D. Don *Aust. J. Biol. Sci.* **20**, 915.

Garrett, S. D. (1960). "Biology of Root-Infecting Fungi." Cambridge Univ. Press, London and New York.

Gaümann, E. (1960). New data on the chemical defense reactions of orchids. *C. R. Acad. Sci.* **250**, 1944.

Gaümann, E., and Kern, H. (1959). On chemical defensive reaction in orchids. *Phytopathol. Z.* **36**, 1.

Gaümann, E., Nuesch, J., and Rimpau, R. H. (1960). Further studies on the chemical defense reaction in orchids. *Phytopathol. Z.* **38**, 274.

Grand, L. F., and Ward, W. W. (1969). An antibiotic detected in conifer foliage and its relation to *Cenococcum graniforme* mycorrhizae. *Forest Sci.* **15**, 286.

Hacskaylo, E. (1971). The role of mycorrhizal associations in the evolution of the higher Basidiomycetes. *In* "Evolution in the Higher Basidiomycetes" (R. H. Petersen, ed.), p. 217. Univ. of Tennessee Press, Knoxville.

Harley, J. L. (1969). "The Biology of Mycorrhiza." Leonard Hill, London.

Hendrix, F. F., Jr., and Campbell, W. A. (1970). Distribution of *Phytophthora* and *Pythium* species in soils in the Continental United States. *Can. J. Bot.* **48**, 377.

Hendrix, F. F., Jr., Campbell, W. A., and Chien, C. Y. (1971). Some phycomycetes indigenous to soils of old growth forests. *Mycologia* **63**, 283.

Hervey, A. H. (1947). A survey of 500 basidiomycetes for antibacterial activity. *Bull. Torrey Bot. Club* **74**, 476.

Hillis, W. E., and Ishikura, N. (1969). The extractives of mycorrhizas and roots of *Pinus radiata* and *Pseudotsuga menziesii*. *Aust. J. Biol. Sci.* **22**, 1425.

Hodges, C. S., Rishbeth, J., and Yde-Andersen, A. (1970). *Fomes annosus*. *Proc. Int. Congr. Fomes annosus, Int. Union Forest Res. Organ., 3rd, 1968*, Sect. 24.

Hutinel, P., and Oddoux, L. (1954). Recherche de l'activité antibiotique chez 100 especes d'Homobasidiomycetes en culture. *C. R. Soc. Biol.* **148**, 1256.

Hyppel, A. (1968a). Antagonistic effects of some soil fungi on *Fomes annosus* in laboratory experiments. *Stud. Forest. Suec.* **64**, 18.

Hyppel, A. (1968b). Effect of *Fomes annosus* on seedlings of *Picea abies* in the presence of *Boletus bovinus*. *Stud. Forest. Suec.* **66**, 16.

Jackson, R. M. (1965). Antibiosis and fungistasis of soil microorganisms. *In* "Ecology of Soil-borne Plant Pathogens" (K. F. Baker and W. C. Snyder, eds.), Vol. I, pp. 363–373. Univ. of California Press, Berkeley.

Katznelson, H. (1965). Nature and importance of the rhizosphere. *In* "Ecology of Soil-borne Plant Pathogens" (K. F. Baker and W. C. Snyder, eds.), Vol. I, pp. 187–209. Univ. of California Press, Berkeley.

Katznelson, H., Rouatte, J. W., and Peterson, E. A. (1962). The rhizosphere effect of mycorrhizal and nonmycorrhizal roots of yellow birch seedlings. *Can. J. Bot.* **40**, 257.

Krupa, S. (1970). Personal communication.

Krupa, S., and Fries, N. (1971). Studies on ectomycorrhizae of pine. I. Production of volatile organic compounds. *Can. J. Bot.* **49**, 1425.

Krywolap, G. N., Grand, L. F., and Casida, L. E., Jr. (1964). The natural occurrence of an antibiotic in the mycorrhizal fungus *Cenococcum graniforme*. *Can. J. Microbiol.* **10**, 323.

Levisohn, I. (1954). Aberrant root infections of pine and spruce seedlings. *New Phytol.* **53**, 284.

Lewis, D. H., and Harley, J. L. (1965). Carbohydrate physiology of mycorrhizal roots of beech. I. Identity of endogenous sugars and utilization of exogenous sugars. *New Phytol.* **64**, 224.

Lorio, P. L. (1966). *Phytophthora cinnamomi* and *Pythium* species associated with loblolly pine decline in Louisiana. *Plant Dis. Rep.* **50**, 596.

MacDougal, D. T., and Dufrenoy, J. (1944). Mycorrhizal symbiosis in *Aplectrum, Corallorhiza*, and *Pinus*. *Plant Physiol.* **19**, 440.

Mandels, M., and Reese, E. T. (1965). Inhibition of cellulases. *Annu. Rev. Phytopathol.* **3**, 85.

Martin, J. T. (1964). Role of cuticle in the defense against plant disease. *Annu. Rev. Phytopathol.* **2**, 81.

Marx, D. H. (1969a). The influence of ectotrophic mycorrhizal fungi on the resistance of pine roots to pathogenic infections. I. Antagonism of mycorrhizal fungi to root pathogenic fungi and soil bacteria. *Phytopathology* **59**, 153.

Marx, D. H. (1969b). The influence of ectotrophic mycorrhizal fungi on the resistance of pine roots to pathogenic infections. II. Production, identification, and biological activity of antibiotics produced by *Leucopaxillus cerealis* var. *piceina*. *Phytopathology* **59**, 411.

Marx, D. H. (1970). The influence of ectotrophic mycorrhizal fungi on the resistance of pine roots to pathogenic infections. V. Resistance of mycorrhizae to infection by vegetative mycelium of *Phytophthora cinnamomi*. *Phytopathology* **60**, 1472.

Marx, D. H. (1973). Growth of ectomycorrhizal and nonmycorrhizal shortleaf pine seedlings in soil with *Phytophthora cinnamomi*. *Phytopathology* (in press).

Marx, D. H., and Bryan, W. C. (1969). *Scleroderma bovista,* an ectotrophic mycorrhizal fungus of pecan. *Phytopathology* **59**, 1128.

Marx, D. H., and Bryan, W. C. (1970). Pure culture synthesis of ectomycorrhizae by *Thelephora terrestris* and *Pisolithus tinctorius* on different conifer hosts. *Can. J. Bot.* **48**, 639.

Marx, D. H., and Davey, C. B. (1969a). The influence of ectotrophic mycorrhizal fungi on the resistance of pine roots to pathogenic infections. III. Resistance of aseptically formed mycorrhizae to infection by *Phytophthora cinnamomi*. *Phytopathology* **59**, 549.

Marx, D. H., and Davey, C. B. (1969b). The influence of ectotrophic mycorrhizal fungi on the resistance of pine roots to pathogenic infections. IV. Resistance of naturally occurring mycorrhizae to infections by *Phytophthora cinnamomi*. *Phytopathology* **59**, 559.

Marx, D. H., Bryan, W. C., and Grand, L. F. (1970). Colonization, isolation, and cultural descriptions of *Thelephora terrestris* and other ectomycorrhizal fungi of shortleaf pine seedlings grown in fumigated soil. *Can. J. Bot.* **48**, 207.

Mathieson, J. (1947). Antibiotics from Victorian basidiomycetes. *Aust. J. Exp. Biol. Med. Sci.* **24**, 57.

Melin, E., Wiken, T., and Öblom, K. (1947). Antibiotic agents in the substrates from cultures of the genus *Marasmius*. *Nature* (*London*) **159**, 840.

Morimoto, M., Iwai, M., and Fukumoto, J. (1954). Antibiotic substances from mycorrhizal fungi. I. Isolation of antibiotic-producing strains. *Kagaku To Kogyo* (*Osaka*) **28**, 111.

Napier, C. J. (1969). The occurrence and seasonal variation of the fine roots and mycorrhizae of loblolly pine (*Pinus taeda* L.) in the West Bay area of Allen Parish, Louisiana. M.S. Thesis, Northwestern State College of Louisiana, Natchitoches.

Neal, J. L., Jr., Bollen, W. B., and Zak, B. (1964). Rhizosphere microflora associated with mycorrhizae of Douglas-fir. *Can. J. Microbiol.* **10**, 259.

Neal, J. L., Jr., Lu, K. C., Bollen, W. B., and Trappe, J. M. (1968). A comparison of rhizosphere microfloras associated with mycorrhizae of red alder and Douglas-fir. *In* "Biology of Alder" (J. M. Trappe, J. F. Franklin, R. F. Tarrant, and G. M. Hansen, eds.), pp. 57–72. USDA Forest Serv. Pac. Northwest Forest and Range Exp. Sta., Portland, Oregon.

Norkrans, B. (1950). Studies in growth and cellulolytic enzymes of *Tricholoma*. *Symb. Bot. Upsal.* **11**, 1.

Ohara, H., and Hamada, M. (1967). Disappearance of bacteria from the zone of active mycorrhizas in *Tricholoma matsutake* (S. Ito et Imai) Singer. *Nature* (*London*) **213**, 528.

Oswald, E. T., and Ferchau, H. A. (1968). Bacterial associations of coniferous mycorrhizae. *Plant Soil* **28**, 187.

Park, J. Y. (1970). Antifungal effect of an ectotrophic mycorrhizal fungus, *Lactarius* sp., associated with basswood seedlings. *Can. J. Microbiol.* **16**, 798.

Powell, W. M., Hendrix, F. F., and Marx, D. H. (1968). Chemical control of feeder root necrosis of pecans caused by *Pythium* species and nematodes. *Plant Dis. Rep.* **52**, 577.

Pratt, B. H. (1971). Isolation of basidiomycetes from Australian Eucalypt forest and assessment of their antagonism to *Phytophthora cinnamomi*. *Trans. Brit. Mycol. Soc.* **56**, 243.

Richard, C., Fortin, J. A., and Fortin, A. (1971). Protective effect of an ectomycorrhizal fungus against the root pathogen *Mycelium radicis atrovirens*. *Can. J. Forest Res.* **1**, 246.

Robbins, W. J., Hervey, A., Davidson, R. W., Ma, R., and Robbins, W. C. (1945). A survey of some wood destroying and other fungi for antibacterial activity. *Bull. Torrey Bot. Club* **72**, 165.

Ross, E. W., and Marx, D. H. (1972). Susceptibility of sand pine to *Phytophthora cinnamomi*. *Phytopathology* **62**, 1197.

Rovira, A. D. (1965). Plant root exudates and their influence upon soil microorganisms. *In* "Ecology of Soil-borne Plant Pathogens" (K. F. Baker and W. C. Snyder, eds.), Vol. I, p. 170. Univ. of California Press, Berkeley.

Rypáček, V. (1960). Die gegenseitigen Beziehungen zwischen Mykorrhizapilzen und holzzerstorenden Pilzen. *In* "Mycorrhizae," pp. 233–40. Fischer, Jena.

Santoro, T., and Casida, L. E., Jr. (1959). Antibiotic production by mycorrhizal fungi. *Bacteriol. Proc.* **59**, 16 (abstr.).

Santoro, T., and Casida, L. E., Jr. (1962). Elaboration of antibiotics by *Boletus luteus* and certain other mycorrhizal fungi. *Can. J. Microbiol.* **8**, 43.

Sasek, V. (1967). The protective effect of mycorrhizal fungi on the host plant. *Proc., Int. Union Forest Res. Organ., 14th, 1967*, Sect. 24, p. 182.

Sasek, V., and Musilek, V. (1967). Cultivation and antibiotic activity of mycorrhizal basidiomycetes. *Folia Microbiol. (Prague)* **12**, 515.

Sasek, V., and Musilek, V. (1968a). Two antibiotic compounds from mycorrhizal basidiomycetes. *Folia Microbiol. (Prague)* **13**, 43.

Sasek, V., and Musilek, V. (1968b). Antibiotic activity of mycorrhizal basidiomycetes and their relation to the host-plant parasites. *Cesk. Mykol.* **22**, 50.

Sevilla-Santos, P., and Encinas, C. J. (1964). The antibacterial activities of aqueous extracts from Philippine basidiomycetes. *Philip. J. Sci.* **93**, 479.

Slankis, V. (1958). Mycorrhiza of forest trees. *In* "First North American Forest Soils Conference" (T. D. Stevens and R. L. Cook, eds.), pp. 130–137. Michigan State Univ. Press, East Lansing.

Slankis, V., Runeckles, V. C., and Krotkov, G. (1964). Metabolites liberated by roots of white pine (*Pinus strobus* L.) seedlings. *Physiol. Plant.* **17**, 301.

Smith, R. S., Jr. (1967). Decline of *Fusarium oxysporum* in the roots of *Pinus lambertiana* seedlings transplanted into forest soils. *Phytopathology* **57**, 1265.

Smith, W. H. (1969). Release of organic materials from the roots of tree seedlings. *Forest Sci.* **15**, 138.

Tomiyama, K. (1963). Physiology and biochemistry of disease resistance of plants. *Annu. Rev. Phytopathol.* **1**, 295.

Trappe, J. M. (1962). Fungus associates of ectotrophic mycorrhizae. *Bot. Rev.* **28**, 538.

Tribunskaya, A. J. (1955). Investigations of the microflora of the rhizosphere of pine seedlings. *Mikrobiologiya* **24**, 188.

Utech, N. M., and Johnson, J. (1950). The inactivation of plant viruses by substances obtained from bacteria and fungi. *Phytopathology* **40**, 247.

Vaartaja, O., and Bumbieris, M. (1964). Abundance of *Pythium* species in nursery soils in South Australia. *Aust. J. Biol. Sci.* **17**, 436.

Vaartaja, O., and Salisbury, P. J. (1965). Mutual effects *in vitro* of microorganisms isolated from tree seedlings. *Forest Sci.* **11**, 160.

Wilcox, H. E. (1968). Morphological studies of the roots of red pine, *Pinus resinosa*. II. Fungal colonization of roots and the development of mycorrhizae. *Amer. J. Bot.* **55**, 688.

Wilkins, W. H. (1946). Investigations into the production of bacteriostatic substances by fungi. *Ann. Appl. Biol.* **33**, 188.

Wilkins, W. H., and Harris, G. C. M. (1944). Investigations into the production of bacteriostatic substances by fungi. VI. Examination of the larger basidiomycetes. *Ann. Appl. Biol.* **31**, 261.

Wingfield, E. B. (1968). Mycotrophy in loblolly pine. I. The role of *Pisolithus tinctorius* and *Rhizoctonia solani* in survival of seedlings. II. Mycorrhiza formation after fungicide treatment. Unpublished Ph.D. Dissertation, Virginia Polytechnic Institute, Blacksburg.

Wright, J. M. (1956a). The production of antibiotics in soil. III. Production of gliotoxin in wheatstraw buried in soil. *Ann. Appl. Biol.* **44**, 461.

Wright, J. M. (1956b). The production of antibiotics in soil. IV. Production of antibiotics in coats of seed sown in soil. *Ann. Appl. Biol.* **44**, 561.

Zak, B. (1964). Role of mycorrhizae in root disease. *Annu. Rev. Phytopathol.* **2**, 377.

Zentmyer, G. A. (1970). Tactic responses of zoospores of *Phytophthora*. In "Root Diseases and Soil-borne Pathogens" (T. A. Toussoun, R. V. Bega, and P. E. Nelson, eds.), pp. 109–111. Univ. of California Press, Berkeley.

CHAPTER 10

Application of Mycorrhizal Symbiosis in Forestry Practice

PEITSA MIKOLA

I. Introduction

A. EARLY FINDINGS ON THE IMPORTANCE OF MYCORRHIZAL INFECTION

The necessity of mycorrhizal association for forest trees was first noticed when experimental plantations of exotic pines in different parts of the world invariably failed, until suitable mycorrhizal fungi had been imported in one way or another (e.g., Kessell, 1927; Roeloffs, 1930; anonymous, 1931; Oliveros, 1932; Clements, 1941; Briscoe, 1959; van

Suchtelen, 1962; Gibson, 1963; Madu, 1967).[1] The Russian scientist
Vysotskii, as early as 1902, suggested using mycorrhizal inoculation
in afforesting grasslands. Likewise, in America the importance of introducing mycorrhizal fungi was detected when afforestation was started
in formerly treeless areas (McComb, 1938; White, 1941).

On the other hand, in some countries cultivation of exotic pines was
started successfully without intentionally importing mycorrhizal fungi.
Thus, there are no historical data on importation of mycorrhizal fungi
to South Africa or New Zealand, both of which today are famous for
their extensive exotic pine plantations. Most likely, however, mycorrhizal
fungi have been imported unintentionally into these countries too, and
in present plantations infection is well established. Long ago the first
pines were introduced into many countries as potted plants with mycorrhizal roots. When such seedlings were planted in natural soil where
mycorrhizal fungi were lacking, the fungi usually spread along with
the roots (White, 1941; Mikola, 1953). In general, grassland soils contain no harmful factors inhibiting mycorrhizal fungi. Many fungi also
spread easily through air in the form of spores or in other ways to such
a degree that special measures are often necessary, in scientific experiments, to protect nonmycorrhizal seedlings in sterilized soil from outside
infection.

B. PROBLEMS IN NATURAL AND MAN-MADE FORESTS

The practical application of mycorrhizal symbiosis to forestry is primarily in introduction of exotic tree species and establishment of man-
made forests. When exotic tree species with mycorrhizae are cultivated
in new areas, several questions should be answered. What species of
mycorrhizal fungi should be introduced with the trees, or is it at all
possible to choose the symbionts for the trees? Furthermore, in which
form should the fungi be imported and how should inoculation be performed? How can importation and inoculation be made in such a way
that the introduction of pathogens is avoided?

The same kinds of problems may occur on a smaller scale in areas
where some mycorrhizal fungi already exist. Forest nurseries are often
established on former agricultural soils or on other sites which have
long been treeless. Although in such cases experience shows that mycorrhizal infection usually takes place without artificial inoculation, it may
be possible to direct favorable mycorrhizal development by introducing

---

[1] The early reports have been reviewed by Hatch (1936) and Rayner (1938), for
instance, and the later ones by Mikola (1970).

new fungal species or by improving soil conditions for the fungi already present.

In natural forests there are few possibilities of applying our present knowledge on ectomycorrhizal symbiosis. Mycorrhizal fungi are present everywhere, and most probably, as the result of competition and long evolution, undisturbed soil already contains those species that are best adapted to prevailing ecological conditions. Under such circumstances introducing new fungal species into the soil has little chance of success. Thus, "99 per cent of all practising foresters will not have to lose any sleep over the problem of mycorrhizal infection" (Wilde, 1944).

In the management of natural forests, however, artificial regeneration through seeding or planting is gaining more and more importance. Mycorrhizal relations also deserve more consideration, particularly when tree seedlings are transplanted from the nursery to natural forest soils.

II. Mycorrhizae in Forest Nurseries

A. Inoculation of Nursery Soil

1. Importance of Inoculation

Early reports on the failure of exotic pines usually came from forest nurseries. Afforestation trials were started by establishing nurseries where seedlings suffered and died, but after introduction of appropriate fungi, healthy and vigorous seedlings with mycorrhizae were produced. Such experiences were reported in areas where native ectotrophic tree species were lacking, particularly in many tropical and subtropical countries. Most of the common tree species of cool and temperate forests belong to the ectotrophs and inoculation is usually not needed there. However, nursery guides for the tropics (e.g., Parry, 1956; Letourneux, 1957) normally give instructions for soil inoculation. The same techniques have also been recommended for establishment of new nurseries in the temperate zone in treeless regions, such as the American prairies (Stoeckeler and Slabauch, 1965) or areas which have long been in agricultural use (Wakeley, 1954; Stoeckeler and Jones, 1957). In the Soviet Union the need for soil inoculation was first observed when nurseries were established for raising tree seedlings for shelterbelt plantations in the steppes. Numerous studies have since been conducted there to find out when the inoculation is needed and when it is unnecessary (see references in Imshenetskii, 1967; Lobanow, 1960; Shemakhanova, 1967). Usually, however, oak shelterbelts in the Soviet Union have been established by direct seeding (cf. Section III,A).

Mycorrhizal infection can easily spread by spores carried by air currents or in other ways. Thus, new nurseries may become infected even without intentional inoculation. The inoculation of the first pine nurseries has proven most necessary on remote tropical islands, such as Trinidad (Lamb, 1956) or Puerto Rico (Briscoe, 1959), where chances of an airborne or other incidental infection are smallest. On continents, however, inoculation of the first pine nurseries has also proven necessary, in general, if these have been established at a great distance from older nurseries or plantations, as is clearly shown by examples from Nigeria and Zambia (see Mikola, 1970).

When it was discovered that exotic pines showed a need to form mycorrhizal associations, an annual inoculation of nursery soil was adopted as a standard routine in several countries and detailed instructions for nursery technique were issued (e.g., May, 1953; Waterer, 1957). If the nursery soil is once inoculated, however, an adequate mycorrhizal population is established and repeated inoculation has no further effect.

If a nursery is extended or a new one established in the vicinity, inoculation of the new section is not quite necessary, for the infection usually comes from the old section through the air, attached to tools, in the roots of transplants, or in other ways. A year or two may elapse, however, before enough fungi have spread from these infection points to cover the whole nursery. Therefore it is advisable to inoculate even new sections of old nurseries, in order to secure a healthy and even growth of the seedlings from the very beginning. This also holds true for new nurseries in the vicinity of old ones or established plantation.

In soils which have long been in agricultural use, mycorrhizal fungi usually are not completely lacking but the population density may be low or the existing fungi may belong to less effective species. If a nursery is established on such soil, introduction of new fungal species may be beneficial. Inoculation alone is not often an effective measure of improving the growth of seedlings and development of mycorrhizae, but, in addition, soil properties, such as acidity, fertility, or organic matter content, must be corrected in order to favor the introduced fungi (Dominik, 1961). In general, correction of soil conditions is a more successful procedure in forest nurseries on agricultural soils than effort to introduce new fungi (Sobotka, 1963).

In order to control weeds and pathogens, partial soil disinfection with steam or chemicals is a common practice in modern nursery management. Then a new question arises: What happens to mycorrhizal fungi with such drastic soil treatment? There is a possibility that they are killed at the same time, and if so, the appropriate fungal population should be returned to the soil by artificial inoculation. Partial soil disinfection usually does not destroy the mycorrhizal population completely, but it can retard commencement of root infection (Hacskaylo and Palmer, 1957; Laiho and Mikola, 1964). Partial sterilization probably destroys the fungi in the uppermost soil layer, but deeper in the soil they survive and infect the roots when the roots reach that layer, or they may grow upward back to the surface layer. Furthermore, airborne reinfection is always possible in forested areas. A considerable delay in commencement of mycorrhizal infection can be harmful and therefore some kind of soil inoculation after partial sterilization may be advisable (cf. Section III,B,1). Pine needles or forest litter serve as an effective inoculum when used for mulching the nursery beds. Some chemical pesticides, however, may cause severe growth disturbances in seedlings and effectively exterminate mycorrhizal fungi; then both inoculation and correction of soil conditions are necessary (Iyer *et al.*, 1971).

Sometimes partial soil disinfection has even encouraged mycorrhizal development in nurseries. This has happened when formaldehyde and allyl alcohol have been used as soil disinfectants. A probable explanation may be that *Trichoderma viride*, a common soil saprophyte, is rather resistant to these chemicals and therefore dominates in the soil after their application. This fungus, in turn, somehow favors the growth of some mycorrhizal fungi (Laiho and Mikola, 1964).

According to present Finnish nursery technique, seedlings are raised on peat substrate in plastic greenhouses. Inoculation is not needed even though the peat, which had been dug from treeless bogs, originally contained no mycorrhizal fungi. Richly mycorrhizal seedlings grow on such a peat substrate, probably because of airborne infection.

2. Techniques of Inoculation

Three types of inocula have been used for mycorrhizal inoculation in nurseries: (a) soil from natural forests, plantations, or old nurseries; (b) mycorrhizal seedlings; and (c) pure cultures of mycorrhizal fungi. Of these inocula, soil has been most widely used in forestry practice. For the first introduction of mycorrhizal fungi, a thin layer of soil taken from an old nursery or an established plantation was usually spread on the top of a nursery bed and mixed thoroughly with the soil beneath. The same method is still commonly used in various parts of the world if plants are raised in open beds on natural soil.

In many nurseries, however, in the tropics and subtropics in particular, it is customary to raise seedlings in various kinds of containers, such as clay pots, wooden boxes, or plastic tubes, or in specially prepared beds ("Swaziland beds"), and soil for them is prepared by mixing various ingredients (Mikola, 1969a). The composition of the soil mixture varies in different nurseries. In East Africa, for example, the "Muguga" mixture (May, 1953) was once widely used. If pines were grown it was customary to add a certain amount, usually 10–20% of volume, of pine soil, i.e., top soil from a pine plantation. A sufficient amount is 10% of the volume, and, if procurement of pine soil is difficult, there is no reason for using larger amounts. Since old nursery soil, which already contains sufficient amounts of mycorrhizal fungi, is generally applied to the new mixture, introduction of new pine soil is no longer necessary and has been discontinued in old nurseries. Different soil mixtures and the use of pine soil for them have been reviewed by Mikola (1970).

Instead of incorporating pine soil into the mixture for transplant beds or containers, seedlings can be inoculated individually by placing about 1 teaspoonful of pine soil at the base of each seedling. This method is cumbersome and time-consuming, but it may be recommended if there is a need for economizing pine soil and inoculation can be timed properly (in tropical nurseries about 1 month after transplanting).

The use of living plants as mycorrhizal inoculum in large-scale nursery practice is uncommon. This method was first applied in Indonesia (Roeloffs, 1930) and is still used there (Becking, 1950; van Alphen de Veer *et al.*, 1954). Vigorous mycorrhizal pine seedlings are planted in seedbeds at 1- to 2-m intervals, and from these mother trees the infection spreads to adjacent seedlings, as can be seen from the green color and improved growth of seedlings. At the time of lifting, some seedlings can be left in beds as mother trees to infect the next crop.

The Indonesian technique of inoculation has also been used on an experimental scale in some other tropical countries. In Indonesia it is said to be the most reliable method for inoculation of *Pinus merkusii*, whereas the use of soil inoculum has resulted in failures (van Alphen de Veer *et al.*, 1954).

The use of pure cultures of mycorrhizal fungi has been repeatedly recommended as the soundest method of inoculation from the scientific point of view and has been commonly used in many experiments. However, because of several difficulties, pure cultures have been used in large-scale nursery practice only in some exceptional cases. First of all, it is not yet known which fungal species should be used under different conditions. Although pure-culture experiments have revealed great physiological differences between various fungi, very little is known so far about the symbiotic efficiency of individual species (cf. Section II,C). Another difficult problem is production of sufficient amounts of mycelia of suitable fungi for mass inoculation of nursery beds.

Ectomycorrhizal fungi, in general, are hard to cultivate. Many probable mycorrhiza-formers have never been isolated into pure culture, and those species which have been isolated, often grow slowly, require specific growth substances and are sensitive to various growth-inhibiting substances (Melin, 1946). For these reasons efforts of pure-culture inoculation have often failed. For pure-culture inoculation to be successful, the introduced fungus must be able to compete with the natural microbial population of the soil, and to achieve this, large amounts of inoculum must be applied and conditions in the soil made favorable for the introduced fungus.

Pure-culture inoculation has reached the stage of practical application in Austria (Moser, 1958b, 1959a, 1962). Prerequisities for successful application of this method are created by the fact that *Boletus plorans* probably is a very suitable symbiont for *Pinus cembra* at high elevations near the timberline, and this fungus is easy to cultivate and grows reasonably well in pure culture. Because this species is most often lacking in the nursery soils of valleys, it is introduced artificially, so that the seedlings may have an efficient symbiont in their roots at the time of transplanting into the field. Since afforestation is conducted on alpine areas above the present timberline, *Boletus plorans* perhaps does not exist in the soil of these planting sites either; thus, it is introduced with the inoculated seedlings.

For mass production of pure-culture inocula, the fungus is first grown in a nutrient solution in Erlenmeyer flasks (Moser, 1958b). The mycelia are transferred from the flasks to the same solution in 10-liter tanks which are aerated intermittently. After a few months of growth in the tanks, the mycelia are transferred to a mixture of vermiculite and peat moistened with the same nutrient solution. If sufficient moisture and good aeration are maintained the mycelia grow throughout the substrate in a few months and the inoculum is ready for use.

Rather similar methods of mass production of fungal cultures for nursery inoculation have been described by Voznyakovskaya and Ryžkova (1958, cited by Shemakhanova, 1967), Sobotka (1955), Shemakhanova (1967), Takacs (1967), and Park (1971). In all these techniques, mycelia are first grown in liquid culture and then mixed into a suitable solid substrate (peat, sawdust, needle litter, wheat grain, or straw), and after several months of incubation the mixture is used for nursery bed inoculation. Because the suitability of different fungi for different tree species and soil conditions is not known, pure cultures of several fungal species are sometimes mixed to make the final inoculum.

According to the Austrian technique, the inoculum is sent from the laboratory to the nursery in plastic bags and used immediately. The dosage is 3–4 liters of inoculum per m^2 of nursery bed. The inoculum is worked lightly into the surface soil and young *Pinus cembra* seedlings are immediately transplanted to the inoculated bed (Moser, 1959a).

Pure-culture inoculum can also be mixed into sterilized soil, and either immediately or after a few weeks of incubation this soil can be applied to nursery beds or used for filling boxes or plastic tubes. Individual inoculation of seedlings grown in pots or plastic tubes has also been practiced on an experimental scale. Results of these methods are so far uncertain.

Besides pure cultures, spores or crushed sporophores of mycorrhizal fungi have been used for inoculating nurseries. Fresh sporophores have been crushed and then mixed into nursery soil in the same way as mycorrhizal soil and pure-culture inocula are used. To increase the infective capacity, crushed sporophores of mycorrhizal fungi have been sometimes added into soil inoculum. Dried sporophores also have been used in some cases when transportation of inoculum for a longer distance has been necessary.

Spores have been used in attempts to inoculate seeds. Although mycorrhizal infection is known to spread easily through the air (Robertson, 1954), efforts using spores for artificial inoculation have not always been successful. Some positive results have been reported (e.g., Klyushnik, 1952; Theodorou, 1971), but the method has not yet reached the stage of practical application.

3. *Shipment and Storage of Mycorrhizal Inocula*

Mycorrhizal infection is usually transmitted from one country or nursery to another in the form of soil inoculum. Because plant protection regulations in many countries prohibit importation of unsterilized soil, great difficulty has been sometimes experienced in procuring mycorrhizal inocula. In fact, in several countries the mycorrhizal infection has been smuggled illegally, or legal procedures have caused considerable delay. Examples of such cases have been described by Mikola (1970).

To keep mycorrhizal fungi alive and in active condition, the time of transportation should be made as short as possible and the soil must be kept moist. According to Letourneux (1957), "nine times out of ten failure occurs if the shipment takes more than eight days." At the present time international air transport is feasible if plant quarantine procedure does not cause undue delay.

There are other reports, however, indicating that mycorrhizal soil can be stored in moist condition for considerably longer times. According to Olatoye (1966) soil which had been stored in plastic bags for 15 months was as effective as freshly collected soil. In the Soviet Union soil for inoculation purposes has been collected in autumn, stored in a moist condition over the winter and applied in the spring with good success (Lobanow, 1960). Some mycorrhizal fungi may survive a long time even in dry soil, probably in the form of spores or sclerotia.

Shemakhanova (1967) has shown that soil which had been stored for 15 months in a dry condition, was as effective as fresh soil. It is possible, however, that in dry soil some fungal species survive, whereas some others, perhaps the most beneficial symbionts, die out. The black mycorrhizal fungus *Cenococcum graniforme* is known to be very drought-resistant (Worley and Hacskaylo, 1959; Trappe, 1962b). Thus, storage of soil in a dry condition may exert a selective influence on the fungal population, and, therefore, it is recommended to keep mycorrhizal soil moist during shipping and storage.

It is also desirable to protect soil inoculum from excessive temperatures, because mycorrhizal fungi, in general, are not very heat-tolerant. In tropical nurseries mycorrhizal soil is usually stored under shelter to protect it against heat and desiccation.

Although the first introduction of ectomycorrhizal fungi into many countries took place in roots of potted plants, living seedlings have seldom been used intentionally for shipping inoculum. New nurseries have often been started with transplants from older nurseries and then, of course, mycorrhizal fungi were carried at the same time. In a few cases it is known, however, that living seedlings have been shipped intentionally from one country to another for mycorrhizal inoculation, because plant quarantine regulations prohibited importing soil inoculum.

Living seedlings provide a good means of preserving mycorrhizal infection. Mother trees which are left in seed beds after lifting the other seedlings transfer the infection to the next crop. This method is particularly practicable under climatic conditions in which a prolonged interval is customary, during the dry season, between lifting seedlings and sowing of the next crop.

Pure cultures of mycorrhizal fungi are shipped and stored in the same way as all the other microbial cultures. Since practical application of pure cultures for mycorrhizal inoculation has, so far, been rare, shipping and exchange of pure cultures of mycorrhizal fungi have taken place for research and experiments only.

Spores of mycorrhizal fungi have been shipped sometimes in the form of dried and crushed sporocarps. Since very little is known so far about the possibilities of using spores for inoculation, information about their storage and survival is also scanty. According to Russian experience, spores of some mycorrhizal fungi can maintain viability for 1 year if they are kept in dry conditions (Lobanow, 1960).

4. Advantages and Drawbacks of Different Methods of Inoculation

Application of mycorrhizal soil is the easiest and simplest method of nursery inoculation. It is also quite a reliable method. Nevertheless, it has several drawbacks.

The bulkiness of soil inoculum can be a serious obstacle in practical work. For nursery inoculation a rather large amount of soil is needed. According to African experience, pine soil must constitute at least 10% of the volume of soil mixture for seedling tubes or Swaziland beds. Procuring pine soil is no problem if there are pine plantations in the neighborhood of the nursery, but if a planting program is started in a new, remote area transportation of sufficient amounts of pine soil for the first nurseries may involve difficulties and high costs.

Soil inoculum contains an indiscriminate mixture of both mycorrhizal fungi and other soil organisms, and perhaps root pathogens and parasites. It is because of the risk of introducing pests that the import of unsterilized soil is prohibited in many countries. Illegal smuggling cannot be accepted. In the few known cases of illegal smuggling, the inoculum has first been mixed with a small amount of soil, in which pine seedlings have been planted. After this soil has been infected throughout, it has been used to inoculate a still greater amount of soil, and the process repeated until a sufficient amount of soil has been available for inoculating nursery beds.

Thus far there are no uniform regulations concerning international exchange of mycorrhizal inocula and necessary precautions. An appropriate system would be desirable to make safe legal import of inocula possible. Fortunately, in no known case has a pest been introduced with mycorrhizal soil inoculum, but the risk still exists.

Soils from natural pine forests and healthy plantations most probably contain a large variety of mycorrhizal fungi, among them species which are effective and well suited for nursery conditions. On the other hand, soil inoculum from an old nursery may contain damping-off fungi, nematodes, weed seeds, and other nursery pests. It is probable that infections were often introduced this way. Soils cannot be disinfected without killing mycorrhizal fungi. Therefore, soil for inoculation should not be taken from old nurseries, and even natural forests or plantations, where *Armillaria mellea, Phytophthora cinnamomi*, or other dangerous pathogens are known to occur, should be avoided.

The application of living seedlings or mother trees for inoculation is quite a reliable method. The species of fungi are usually not known, but if the mother trees are vigorous and healthy and if inspection reveals good mycorrhizal structure in their root systems, the conclusion can be made that efficient fungi are present. If the spread of infection from the mother trees is slow, only 40–50 cm/year, some unfavorable soil properties, such as excessively high pH, are indicated. This can be corrected by appropriate soil treatment, e.g., acidification. Infection usually spreads faster in sterilized soil; therefore soil sterilization is recommended before planting mother trees.

Nursery inoculation with mycorrhizal soil, if successful, usually results in faster and more even infection than is accomplished with mother trees. Inoculation with mother trees, however, has been successful even in cases where application of mycorrhizal soil has failed (van Alphen de Veer *et al.*, 1954).

When a new nursery is established, the first crop often consists of seed-lings transplanted from an older nursery. This is an easy method of in-fecting the new nursery. Likewise, the first field plantations in some areas have been established with imported seedlings. With regard to shipping living plants for long distances, the same objections as to ship-ping soil are applicable. Seedlings are perhaps not quite as heavy and bulky as soil but they need more care, e.g., irrigation, during the trans-portation. Pests and diseases, especially needle diseases, can be carried with seedlings as well as with soil inoculum. Importation of living plants is often more strictly controlled than that of unsterilized soil, and illegal smuggling is more difficult. For the above reasons, living plants have sel-dom been used particularly for transmission of mycorrhizal infection. To reduce the risk of carrying pests, various possibilities are available. For example, seedlings can be washed or treated with insecticides and trans-planted into sterilized soil before shipping; the shoots can even be treated with fungicides.

Theoretically, pure cultures and spores would be the best inocula; therefore their use is often recommended (Bakshi, 1967). Pure cultures need little space and are easy to handle, and there is no risk of introduc-ing pests. So far, however, because of technical difficulties and unknown factors involved, pure-culture inoculation has been used very little in nursery practice on a field scale.

One drawback is the slow growth of mycorrhizal fungi in pure culture. Production of sufficient amounts of mycelia for field inoculation is cum-bersome and time-consuming. It is not possible under ordinary nursery conditions, and for this reason the inoculum must be produced in a lab-oratory and shipped to nurseries for use. The whole procedure is rather expensive.

Furthermore, experience has shown that pure-culture inoculation is an unreliable method. The fungus often dies immediately after inoculation; probably because of competition with other soil fungi, mycorrhizal in-fection does not take place. The chances of success can be improved by soil sterilization before inoculation and by using a rather large amount of inoculum. If large amounts of pure-culture inoculum are needed new problems of procurement and handling may appear. The rate of applica-tion of 3 liters of pure culture inoculum per m^2, for instance, as is cus-tomary in the Austrian nurseries, corresponds to 30 m^3/hectare.

The greatest obstacle for the use of pure cultures is, however, the lack of information about which fungal species should be used. According to the present state of knowledge concerning the efficiency of different species, use of a mixture of pure cultures of several species may be recommended (Takacs, 1967). It is not known, however, whether the species that gain dominance in the mixture are also the best mycorrhizal symbionts. Furthermore, there is a great number of potential mycorrhiza-formers which have never been isolated into pure culture (cf. Modess, 1941; Trappe, 1962a).

The same factors which restrict the use of pure cultures for inoculation on a nursery scale, also apply to the use of spores. Spores would have many advantages. They need little space, they are easy to send from one country to another, and the risk of disease transmission is eliminated. There is the slight disadvantage, however, that spores can only be collected during a brief period of the year, which may be a less favorable season for inoculation, and therefore spores might have to be stored for a rather long time.

If a sufficient knowledge about spore viability and symbiotic efficiency of different fungal species can be obtained and reliable methods of inoculation developed, then spore inoculation may become a standard technique. It must be kept in mind, however, that there may also be effective mycorrhizal fungi which form no sporophores or whose spores for some reasons are difficult to obtain.

B. Promotion of Mycorrhizal Development

In the natural range of ectotrophic tree species hardly any soils are completely devoid of mycorrhizal fungi. The poor development of mycorrhizae, and at the same time seedlings, in nurseries is probably most often not due to the absence of appropriate fungi but to adverse soil conditions which affect the seedlings either directly or indirectly through mycorrhizal fungi. Since close correlation exists between the good growth and healthy condition of seedlings and the abundance and morphology of mycorrhizae (e.g., Göbl, 1967), the suitability of nursery soils can be evaluated by examining mycorrhizal relations, and soil management in nurseries should aim at improving conditions for mycorrhizal fungi. Although an improvement in the quality of nursery stock is the ultimate goal of soil amendment, this aim is partly achieved through activity of mycorrhizal fungi. Mycorrhizae are sensitive to alterations in soil conditions and, consequently, the effect of soil amendment is perhaps first detectable in mycorrhizal relations.

Adjustment of soil conditions is often an indispensable prerequisite for successful mycorrhizal inoculation (Göbl and Platzer, 1967). Fertilization, acidification, and addition of organic matter are the most common measures used for promoting mycorrhizal development in forest nurseries.

McComb and Griffith (1946) have shown that mycorrhizal development in a prairie soil nursery could be promoted with phosphorus fertilization. White pine (*Pinus strobus*) seedlings which without inoculation and fertilization were chlorotic and nonmycorrhizal, started vigorous growth and developed numerous mycorrhizae as a result of a moderate phosphorus fertilization. Thus, the soil contained some mycorrhiza-forming fungi but they were in an inactive condition and were activated by phosphorus. Still better growth and mycorrhizal development, however, was achieved with inoculation, i.e., by introducing new fungal species which established a symbiotic relationship with the pine without additional phosphorus. Douglas fir (*Pseudotsuga menziesii*), however, behaved differently in the same experiment. Phosphorus fertilization improved the growth of seedlings, but no mycorrhizae developed without inoculation. The conclusion can be made that some fungal species originally present in the soil were able, when activated with phosphorus, to form mycorrhizae with white pine but not with Douglas fir, or perhaps pine roots were activated somehow by phosphorus to establish symbiosis more readily with local fungi.

The development of mycorrhizae in nurseries can also be promoted by the addition of other nutrients. When there is a severe shortage of some nutrients, usually N, P, or K, the number of mycorrhizae is small and their structure abnormal, but with moderate and balanced supply of the major nutrient, optimum development is achieved (Hatch, 1937; Björkman, 1942, 1949, 1956). Excessive fertilization, with nitrogen in particular, also reduces the intensity of mycorrhiza formation, which may be harmful for the survival of the seedlings after planting in the field (cf. Section III,B,1 and Chapter 9).

Most ectomycorrhizal fungi are acidophilic, their pH optimum usually ranging from 4.0 to 6.0 (Modess, 1941). Since forest nurseries often have been established on former agricultural soils which have been previously limed, acidification may be an essential measure to improve soil conditions for mycorrhizal fungi. Acidification alone may be effective or it can be used in conjunction with inoculation. Tserling (1960), for instance, showed that soil inoculation alone was not sufficient to establish mycorrhizal association on larch in an alkaline chernozem soil and that acidification was also necessary. Likewise, Moser (1959a) stressed the importance of a suitable pH as well as sufficient nutrient levels and the presence of micronutrients for successful mycorrhizal inoculation.

Sufficient irrigation and good aeration are also essential for mycorrhizae and for successful inoculation in particular.

Several investigators have reported on the favorable effect of organic matter on mycorrhizae (e.g., Rayner, 1936, 1939; Rubtov, 1964). Addition of organic matter alone has promoted the formation of mycorrhizae. Furthermore, mycorrhizal inoculation is often associated with a liberal application of organic matter, e.g., if forest humus is used as soil inoculum, or if mycelia for pure-culture inoculation are grown on a peat or forest litter substrate.

Peat, forest humus, and compost are the most commonly used organic soil amendments in nurseries. It is, in general, hard to decide in each case which is the effective factor promoting the growth of seedlings. Thus, forest humus is often added to nursery soil to inoculate it with mycorrhizal fungi and other natural forest soil microbes, but the favorable effect on seedlings of such treatment may be due to other factors as well, such as adding macro- or micronutrients or growth substances, or to the influence of humus on the pH or physical properties (aeration or water-holding capacity) of the nursery soil. Peat and forest humus are usually more acid than the original nursery soil and therefore they may lower pH, thus making the soil more suitable for mycorrhizal fungi.

Björkman (1954, 1956), on the other hand, has shown that vigorous and richly mycorrhizal plants can be raised in a forest soil nursery with very low organic matter content if other conditions (fertilization, irrigation, aeration) are favorable.

C. EFFECTIVENESS OF DIFFERENT SPECIES OF FUNGI

It is well known, that different species and races of *Rhizobium* vary greatly in their ability to fix atmospheric nitrogen. Corresponding differences may occur between the symbiotic efficiency of different mycorrhizal fungi. Consequently, for inoculation with either spores or pure cultures it would be most important to know which fungal species should be used with different tree species and ecological conditions.

Since the early studies of Melin (1923, 1925) numerous pure-culture experiments have revealed great differences between mycorrhizal fungi with respect to their ability to use various carbon and nitrogen compounds, growth substance requirements, and their relation to pH, temperature, and oxygen supply (e.g., Modess, 1941; Norkrans, 1950; Moser, 1958a, 1959b; Hacskaylo *et al.*, 1965; Pachlewski, 1967b; Lundeberg, 1970). These results have been supplemented by field observations (e.g., Modess, 1941). Nevertheless, very little is known about the relative effectiveness of different fungi as mycorrhizal symbionts. Of particular interest is the question of whether some species can do better than others in their ability to utilize and transmit to the host the organic nitrogen of humus compounds or difficultly soluble phosphorus and potassium minerals.

Rosendahl (1942) first showed that different species of mycorrhizal fungi may vary in capacity to dissolve phosphorus and potassium from rocks. *Boletus felleus* promoted the growth of pine seedlings more than did *B. granulatus* and *Amanita muscaria*. In this experiment, however, mycorrhizae were only formed by *Boletus felleus*. In the pot experiments of Bowen and Theodorou (1967) seedlings inoculated with *Boletus granulatus* and *Rhizopogon luteolus* utilized rock phosphate more effectively than those inoculated with *Boletus luteus* and *Cenococcum graniforme*, which, in turn, were not superior to the uninoculated controls. The growth of pine seedlings was consistently promoted more by *Boletus granulatus* and *Rhizopogon luteolus* than by *Boletus luteus* and *Cenococcum graniforme* (Theodorou and Bowen, 1970). In other experiments, in contrast, *B. luteus* was a more beneficial symbiont for pine than *B. granulatus* (Krangauz, 1967). Although none of the fungi studied by Lundeberg (1970) were able to mobilize organically bound soil nitrogen, considerable differences appeared in their ability to promote or suppress growth of the host plant (*P. sylvestris*). Great differences have also been noticed between different strains of the same species, e.g., in *Boletus scaber* (Levisohn, 1959) and *Rhizopogon luteolus* (Theodorou and Bowen, 1970). Some species may occur in nature as both mycorrhizal and nonmycorrhizal races, e.g., *Boletus subtomentosus* (Lundeberg, 1970) and *Paxillus involutus* (Laiho, 1970). Of particular interest is the unexpected observation of Lundeberg (1970) that a strain of *Boletus subtomentosus*, which was not able to form mycorrhizae on pine, was most effective in promoting the growth of pine seedlings. The favorable influence of mycorrhizal fungi on the growth of young trees even without mycorrhizal association was also demonstrated in an earlier experiment (Levisohn, 1956) (see Chapter 8).

Although it has been shown that great differences exist between different fungal species in their influence on the growth of host plants, the results have been somewhat inconsistent, and, therefore, this knowledge has so far little applicability in nursery practice. Thus, for instance, inoculation with *Paxillus involutus* greatly improved growth and survival of pine seedlings in the field (Laiho, 1970), whereas in the experiment of Lundeberg (1967), the same fungus suppressed growth. Likewise the common mycorrhizal fungus *Cenococcum graniforme*, which is very drought-tolerant and therefore has been particularly recommended for inoculation in dry areas (Trappe, 1964), has inhibited growth of seedlings in some

TABLE I

THE RELATIVE EFFECT OF DIFFERENT FUNGI ON GROWTH OF PINE SEEDLINGS[a]

Reference:	Hacskaylo and Vozzo (1971)		Theodorou and Bowen (1970)		Shemakhanova (1967)			Krangauz (1967)		Björkman (1970)	Lamb and Richards (1971)	
Tree species:	P. caribaea		P. radiata		P. silvestris			P. silvestris		P. silvestris	P. radiata	P. elliottii
Type of experiment:	Field Nursery		Field		Nursery			Pot		Pot	Pot	
Measurement:	Height		Height		Dry weight			Dry weight		Dry weight	Dry weight	
Age of seedlings:	40 weeks	43 weeks	15 months	43 months	1 year [b]	[c]	[d]	1 year	2 years	1 year	8 months	
Boletus bovinus					166	134	117			93		
B. cothurnatus	151	120										
B. granulatus			144	133				115	90		153	608
B. luteus			122	114	105	99	99	112	244			
B. subtomentosus		119								144		
Cenococcum graniforme	125	168			231	127	118				209	581
Corticium bicolor			141	129								
Rhizopogon luteolus 1			143	124								
Rhizopogon luteolus 2												
Rh. roseolus	131	131									160	535
Scleroderma verrucosum					170	103	109					
Unidentified isol. P. radiata											296	231
Unidentified isol. P. elliottii											256	1080
Unidentified isol. P. radiata											324	200
Unidentified isol. P. radiata											497	427
Soil inoculum	141	121						139	157			

[a] 100 = uninoculated control.
[b] Sandy soil.
[c] Sandy soil, low humus content.
[d] Sandy soil, high humus content.

experiments (Lundeberg, 1970) and has been the most beneficial symbiont in others (Shemakhanova, 1967). The results of some such comparative experiments are summarized in Table I.

The varying responses of different fungi suggest that symbiotic efficiency is determined not only by the species and race of the fungus but also depends on the host species and environmental conditions (Levisohn, 1960). Thus, of the fungi tested by Krangauz (1967), *Boletus luteus* was the best symbiont for pine, whereas growth of oak was promoted most by *Hebeloma crustuliniforme*. According to Shemakhanova (1967), *Boletus luteus, B. bovinus, Scleroderma verrucosum,* and *Cenococcum graniforme* are the most suitable symbionts for pine and *Boletus subtomentosus, Amanita muscaria,* and *A. rubescens* are recommended for *Quercus*. Probably *Boletus edulis* also is a good symbiont for oak (Runov, 1967). Rayner and Neilson-Jones (1944) recommend *Boletus bovinus* as a good symbiont for pine especially in poor heath soils.

A consequence of the different temperature requirements of different mycorrhizal fungi is that some species are more adapted to cold and others to warm conditions. *Boletus plorans* is an example of a suitable species for cold habitats, whereas *Pisolithus tinctorius* probably prefers warm sites and therefore is common in the tropics (Schramm, 1966; Mikola, 1969b; Marx et al., 1970).

The influence of soil fertility on the effect of different fungi is illustrated by an experiment of Mikola (1967). An unidentified ectendomycorrhizal fungus suppressed growth of *P. sylvestris* seedlings on unfertilized peat substrate when compared to uninoculated control seedlings, whereas on fertilized peat the same fungus promoted growth even more than a mixed forest humus inoculum (Table II). Furthermore, the experiments of Shemakhanova (1967) indicate that the relative efficiency of different fungi may depend greatly on the humus content of the soil (Table I).

Because of the particular requirements of individual fungal species, a

TABLE II

THE LENGTH (CM) OF THE TOP SHOOTS OF 2-YEAR-OLD PINE SEEDLINGS AS INFLUENCED BY DIFFERENT INOCULATION AND FERTILIZATION TREATMENTS[a]

	Not fertilized	Fertilized
Control, without inoculation	7.59	12.20
Inoculated with pure culture of the ectendotrophic fungus	5.51	14.05
Inoculated with a forest humus suspension	10.12	12.86

[a] From Mikola (1967).

mixed population, i.e., soil inoculum, has often been superior to any pure culture. A mixture of pure cultures of several species has sometimes been recommended for inoculation. Unidentified isolates from pine mycorrhizae have often proved much superior to pure cultures isolated from sporophores of known ectomycorrhizal fungi (Lamb and Richards, 1971; cf. Table I). Thus, there is insufficient information to give detailed instructions about fungal species which should be used for inoculation under different conditions. On the basis of experiments and practical experience, however, the following general recommendations may be suggested.

In all probability, *Rhizopogon* spp. (*R. luteolus* and *R. roseolus*) are favorable symbionts, at least for pines. Their sporophores are very common in nursery beds and young plantations in different parts of the world. Thus, soil conditions in nurseries are evidently favorable for them, which is an indispensable prerequisite for successful inoculation. They also thrive well in different forest soils (e.g., Pachlewski, 1967a; Mikola, 1969b). Crushed sporophores of *Rhizopogon* spp., mixed with soil, have been used in nursery practice for soil inoculation (Kessell and Stoate, 1938); however, the success of this treatment is unknown.

Boletus spp. belong to the best known and most intensively studied ectomycorrhizal fungi. Often numerous fruiting bodies of *B. granulatus* and *B. luteus* are seen in exotic pine plantations; therefore inoculation of nursery beds with these fungi has been suggested. However, the success of such practice is unknown. Sporophores of *Boleti* are uncommon in nursery beds, whereas those of *Rhizopogon* spp. are very common. According to Adams (1951), *Rhizopogon* spp. are the most common fungi in nurseries, whereas they are replaced by *Boleti* after transplanting into the field. *Boletus plorans* is probably the best symbiont of *Pinus cembra* for afforestation of subalpine areas, and, according to Moser (1959a), its inoculation into nursery soil is feasible. Likewise, *Boletus elegans* apparently is a very suitable symbiont for *Larix* spp. As a rule, the symbiosis between *Boletus elegans* and *Larix* spp. is easily established and no inoculation is usually necessary; infection by fungus, however, can be promoted by proper management of nursery soils.

According to Laiho (1970), *Paxillus involutus* is a suitable symbiont of trees in fertile soils and can be recommended in the afforestation of former agricultural land. So far possibilities for its inoculation on a nursery scale are unknown.

The unidentified ectendotrophic fungus, which often prevails naturally in nurseries on former agricultural soils (Mikola, 1965; Laiho, 1965; Wilcox, 1971) seems to thrive well in relatively fertile soils. If it is not present in a nursery, inoculation with it is easy.

Probably some other unidentified species also may be suitable for

nursery inoculation. At least under experimental conditions, mycelia isolated from mycorrhizae have sometimes proved superior to known species
(Lamb and Richards, 1971). It would be particularly important to find
fungi suitable for inoculation under alkaline conditions, e.g., for *Pinus
halepensis* nurseries.

III. Mycorrhizae at the Establishment of Forest Stands

A. DIRECT SEEDING

When direct seeding is practiced in previously forested areas, mycorrhizae raise no problems. The soil always contains suitable mycorrhizal
fungi and more spores come from nearby forests, and, therefore, mycorrhizae develop regularly immediately after seed germination. The fungal
population usually is well adapted to local conditions, and introduction
of new species has hardly any chance of success. On the other hand, there
are problems if direct seeding is practiced in regions where forests of the
same or related ectotrophic species are lacking.

Direct seeding is considered in Australia where shortage of labor,
limited nursery capacity, and other reasons prevent the execution of extensive afforestation programs by planting only. In such cases broadcast
seeding from an airplane would be a practicable solution, if the presence
of mycorrhizal fungi could be secured at the same time. Mixing seed
with mycorrhizal soil is possible and has proven successful in experiments, but it is impractical because of the large amount of soil needed
(Forrest, 1966). Inoculation of pine seed with fungal spores has been
successful on an experimental scale (Theodorou, 1971; several Soviet
investigators, cf. Shemakhanova, 1967, p. 215), so that feasible methods
for field use might be developed.

Mycorrhizal inoculation of tree seed has so far been practiced most
widely in the Soviet Union for establishing oak shelterbelts in forest–
steppe zones (cf. Lobanow, 1960; Shemakhanova, 1967). Acorns are
usually sown in holes, and the mycorrhizal soil put into the same holes.
The soils of the forest–steppe zone are not completely devoid of mycorrhizal fungi but inoculation greatly increases survival and early growth
of the seedlings. The purpose for inoculation is both to introduce new
fungi into the soil and to improve conditions for existing fungi.

To be effective, a rather large amount of soil inoculum is required.
Shemakhanova (1967) recommends 2–15 gm/acorn, corresponding to
15–100 gm/seeding hole with 6–8 acorns or up to 2 m³/hectare. Even
considerably larger amounts, 250–500 gm per hole, have been used
(Krasovskaya and Smirnova, 1950, cited in Shemakhanova, 1967). Suc-

cessful inoculation of acorns for field sowing with pure cultures of mycor-rhizal fungi has also been reported (Runov, 1967).

As was mentioned previously, soil inoculation is not always sufficient to guarantee development of mycorrhizae and survival of the plants, and it must sometimes be supplemented with other measures, such as irrigation of phosphorus fertilization (Michowitsch, 1963). Irrigation or fertilization has sometimes had the same effect even without inoculation. Thus, several Soviet scientists consider inoculation unnecessary since the same effect can be achieved by cheaper and less laborious methods (see refer-ences in Lobanow, 1960; Shemakhanova, 1967). Whether the inoculation is beneficial or unnecessary probably depends on soil and climatic condi-tions. After a critical review of the literature, Lobanow (1960, p. 279) came to the conclusion that inoculation is advisable, although not neces-sary, in chernozem soils, but absolutely necessary in chestnut soils. Any-way, the afforestation of steppes by direct seeding requires that mycor-rhizal development be encouraged by inoculation, soil amendment, or both.

B. Planting of Nursery-Grown Seedlings

1. Quality of Planting Stock

When a seedling is transplanted from the nursery into the field its en-vironmental conditions are changed greatly, and adaptation to this radical change may be difficult. To reduce trauma, the change should be as small as possible and the seedling should be resistant to adverse condi-tions of the new habitat. Normally, tree seedlings in natural forest soils are mycorrhizal and, accordingly, the seedlings in the nursery should have a mycorrhizal structure that corresponds to the conditions of the planting site. Vigorous and healthy-looking nonmycorrhizal seedlings can be raised in the nursery with heavy fertilization and sufficient water-ing. However, the survival of such seedlings is low. Nonmycorrhizal or weakly mycorrhizal seedlings are less resistant to drought and pathogenic organisms than seedlings with well-developed mycorrhizae.

If the soil at the planting site contains a natural population of mycor-rhizal fungi, even nonmycorrhizal seedlings may recover. Their adap-tation to new conditions and development of mycorrhizae, however, take time which is reflected as a period of retarded growth. Björkman (1944) first suggested that the well-known stagnation of spruce seedlings after planting was due to such an adaptation and the formation of proper kinds of mycorrhizae. Later, Björkman (1953, 1954, 1956) showed that survival and early growth of seedlings after planting depended not only on their size or healthy appearance, but also on a balanced fertilization in the nursery and mycorrhizal structure of the root systems. Accordingly, when judging the quality of planting stock, mycorrhizal relations must be considered. The abundance and structure of mycorrhizae must be included in the characteristics which are used in grading nursery stock. In general, when seedlings remain in the nursery for several years, as is customary in cool and temperate climates, they normally are mycorrhizal at the time of lifting. On the other hand, if seedlings are kept in the nursery for less than 1 year, as is a common practice for subtropical pines, or if soil sterilization and heavy fertilization are practiced, both of which retard mycorrhizal development, there is a risk that the seedlings are still nonmycorrhizal at the time of lifting, which results in poor survival after planting in the field (Jorgensen and Shoulders, 1967).

Because microscopic root examination is not possible during routine grading of nursery stock, evaluation of the mycorrhizal state must be based on visible macroscopic characteristics. In pines dichotomous branching of short roots is usually closely correlated with mycorrhizal development. The thickness and color of short roots also depend on mycorrhizal relations. Thick, light-colored, and richly branched short roots indicate healthy and vigorous mycorrhizal development. For practical purposes Göbl (1967) distinguishes between favorable and unfavorable mycorrhizal types based on branching and the color and type of the mantle. Likewise Dominik (1956) and Levisohn (1963) consider some morphological types of mycorrhizae more beneficial to the host tree than others.

2. Soil Conditions in the Planting Site

When a seedling is planted in a natural forest soil or in some other soil where mycorrhizal fungi are present, the later development of mycorrhizae depends totally on soil conditions and local fungal population. The presence and structure of mycorrhizae at the time of planting mainly influence survival and early growth of seedlings. Thus, even though nonmycorrhizal seedlings, usually, rapidly acquire mycorrhizae in the field, inoculation or other measures to promote mycorrhizal development in the nursery may be justified in order to reduce mortality or to avoid a stagnation period after planting. This is demonstrated by Theodorou and Bowen (1970). Although mycorrhizal fungi apparently were present in the soil and uninoculated seedlings rapidly developed mycorrhizae, difference in growth between inoculated and uninoculated seedlings was still detectable 32 months after planting. In the experiment of Mikola (1967), uninoculated seedlings were infected immediately after planting but the difference in growth leveled off only 3 or 4 years later.

Because ecological conditions and, consequently, fungal populations

in the nursery and in the field are often very different, local fungi gradually replace the fungal species the seedlings carried in their roots from
the nursery. This change seems to take place easily and without serious
disturbances to the seedlings (Mikola, 1965). A short period of growth
stagnation, however, eventually may occur owing to the change of
mycorrhizal population. Seedlings grown in a forest soil nursery usually
have a lower mortality and shorter stagnation period than seedlings
raised in former agricultural soil (Björkman, 1956, 1962). On the other
hand, seedlings raised in a nursery on agricultural soil may be most
suitable for afforestation of abandoned fields (Laiho, 1967).

As a whole, possibilities for selecting fungal symbionts in the nursery
with respect to future planting sites are very small so far. Moser's (1962)
success at afforesting alpine areas is a promising exception in this respect.
More important than selecting fungal species is creating suitable physical
and chemical soil conditions, both in the nursery and planting site, for
mycorrhizal development (Moser, 1963; Sobotka, 1963).

IV. Other Aspects in Forestry

A. SITE EVALUATION

Since the number and morphology of mycorrhizae depend on ecological conditions in the soil, site quality also can be studied by observing
the mycorrhizal relations. In the boreal forests, as well as elsewhere in
corresponding conditions where podsol is the dominant soil type, optimum mycorrhizal development (numerous, richly branched, thick, and
usually light-colored mycorrhizae) indicate healthy soil conditions,
whereas slender, dark-colored short roots are signs of some deficiencies,
such as shortage of nutrients or water-logged conditions (Björkman,
1942; Mikola and Laiho, 1962; Göbl, 1965).

Weak mycorrhizal development can also be an indication of a very
high nutrient content of the soil (Björkman, 1942, 1949). In such cases,
however, rich ground vegetation also indicates high fertility.

B. EFFECT OF SILVICULTURAL PRACTICES ON MYCORRHIZAE

If mycorrhizal studies show that defects occur in the soil, improvement can be attempted with appropriate soil amelioration. The effect of
such measures may be detected in the structure of mycorrhizae even before any change in growth of trees appears. Thus, sporophores of some
mycorrhizal fungi are particularly numerous in fertilized stands (Laiho,
1970). If the soil is severely deficient in a certain nutrient, such as phos-

phorus, the first visible effect of the addition of this nutrient is often a flush of sporophores of mycorrhizal fungi.

In water-logged soils, such as peat bogs, mycorrhizae occur only near the soil surface and even there are often dark and slender, indicating poor site conditions (Heikurainen, 1955). Artificial drainage lowers the ground water table and improves aeration, changes which are immediately reflected in the structure of mycorrhizae and their vertical distribution (Paavilainen, 1966). The production of sporophores increases at the same time. Thus, the need for both fertilization and drainage can be studied through examination of mycorrhizae. Referring to the close correlation between nutrient supply and mycorrhizal development, Björkman (1970) has suggested that perhaps forest fertilization could be replaced by some other method of activating mycorrhizal fungi.

Because mycorrhizal fungi are essential for their hosts, it is necessary to know what measures are harmful for these vital organisms. Of primary importance is the question of how mycorrhizal fungi are affected by clear cutting, i.e., the removal of the higher symbionts from the system. Romell (1930) observed that sporophores of certain fungal species never grew after clear cutting in areas where they were regularly found before the cutting; the same observation has been made by several later investigators. However, although sporophores are absent, the fungi may survive in the soil for several decades, as is shown by immediate mycorrhizal infection of seedlings if the area is reseeded. Some mycorrhizal fungi, such as *Hebeloma crustuliniforme, Paxillus involutus,* and *Rhizopogon* spp., are able to fruit around young seedlings, even in nursery beds, whereas some other species form sporophores only in closed stands.

Accordingly, clear cutting does not seriously harm mycorrhizal fungi unless the area remains treeless for a very long time. Even prolonged agricultural use or absence of trees for other reasons does not exterminate all ectomycorrhizal fungi. Some species possibly disappear while others survive. Controlled burning of slash, which often follows clear cutting, is a more drastic treatment; the heat may kill a part of the fungal population, and, in addition, the pH and nutrient relations in soil are altered. The heat, however, penetrates only a few centimeters and fungi which are deeper in the soil survive without injury. An experiment by Mikola *et al.* (1964) showed that intensive burning can somewhat delay the commencement of mycorrhizal infection, but not to a harmful degree.

Weed killers or other pesticides have not been found to exert a harmful influence on mycorrhizal fungi when applied at the rates which are customary in forestry practice. The harmful effect of industrial air pollutants on mycorrhizae has been observed in severely polluted areas (Sobotka, 1968).

406 Peitsa Mikola

C. Protection of Mycorrhizal Fungi

Ectomycorrhizal fungi survive persistently where they once have been established. Therefore mycorrhizal fungi, in general, do not need any particular protection. Sometimes apprehension has been expressed that collecting mushrooms, i.e., sporophores of ectomycorrhizal fungi, would be harmful to tree growth. There is no reason for such fear since collecting sporocarps has no adverse effect on the vegetative mycelia in the soil or on the mycorrhizae. On the contrary, if the sporocarps of mycorrhizal fungi, which often are extremely numerous in forest plantations, are harvested and utilized, this can considerably improve the profitability of artificial afforestation (Mikola, 1969b).

Protecting rare species of fungi as well as of other plants is a task of the conservationists. Regarding rare species of ectomycorrhizal fungi, their protection is best achieved by site protection, i.e., keeping the known habitats of such fungi forested by their probable hosts.

References

Adams, A. J. S. (1951). A forest nursery for *Pinus radiata* at Mt. Burr in the southeast of South Australia. *Aust. Forest.* **15**, 47.

Anonymous. (1931). Establishing pines. Preliminary observations on the effect of soil inoculation. *Rhodesia Agr. J.* **28**, 185.

Bakshi, B. K. (1967). Mycorrhiza—its role in man-made forests. *Doc. FAO World Symp. Man-Made Forests, 1967* Vol. 2, p. 1031.

Becking, J. H. (1950). Der Anbau von *Pinus merkusii* in den Tropen. *Schweiz. Z. Forstw.* **101**, 181.

Björkman, E. (1942). Über die Bedingungen der Mykorrhizabildung bei Kiefer and Fichte. *Symb. Bot. Upsal.* **6**, No. 2.

Björkman, E. (1944). Om skogsplanteringens markbiologiska förutsättningar. (Forest planting and soil biology.) *Sv. Skogsvårdsfören. Tidskr.* **1944**, 333.

Björkman, E. (1949). The ecological significance of ectotrophic mycorrhizal association of forest trees. *Sv. Bot. Tidskr.* **43**, 223.

Björkman, E. (1953). Om orsakerna av granens tillväxtsvårigheter efter plantering i nordsvensk skogsmark. (Factors arresting early growth of the spruce after plantation in northern Sweden.) *Norrlands Skogsvårdsförb. Tidskr.* **1953**, 285.

Björkman, E. (1954). Betydelsen av gödsling i skogsträdsplantskolor för plantornas första utveckling i skogsmarken. *Norrlands Skogsvårdsförb. Tidskr.* **1954**, 543.

Björkman, E. (1956). Über die Natur der Mykorrhizabildung unter besonderer Berücksichtigung der Waldbäume und die Anwendung in der forstlichen Praxis. *Forstwiss. Zentralbl.* **75**, 265.

Björkman, E. (1962). The influence of ectotrophic mycorrhiza on the development of forest tree plants after planting. *Proc. Int. Union Forest Res. Organ., 13th, 1961* Part 2, Vol. 1, Sect. 24-1.

Björkman, E. (1970). Mycorrhiza and tree nutrition in poor forest soils. *Stud. Forest. Suec.* **83**, 1.

Bowen, G. D., and Theodorou, C. (1967). Studies on phosphate uptake by mycor-rhizas. *Proc. Int. Union Forest Res. Organ., 14th, 1967* Vol. V, p. 116.

Briscoe, C. B. (1959). Early results of mycorrhizal inoculation of pine in Puerto Rico. *Carib. Forest.* 20, 73.

Clements, J. B. (1941). The introduction of pines into Nyasaland. *Nyasaland Agr. Quart. J.* 1, 5.

Dominik, T. (1956). Projekt nowego podziatu mikoryz ektotroficznych oparty na cechach morfologiczno-anatomicznych. *Rocz. Nauk Roln. Les.* 14, 223.

Dominik, T. (1961). Badania nad przeszczepianiem mikrobocenoz gleb leśnych na tereny rolne. (Experiments with inoculation of agricultural land with microbial cenosis from forest soils.) *Prace Inst. Bad. Lesn.* No. 210, p. 103.

Forrest, W. G. (1966). Mycorrhiza problems in pine establishment. *Proc. Refresh. School Forest., Univ. N. Engl., Armidale, N.S.W.* Stencil.

Gibson, I. A. S. (1963). Eine Mitteilung über die Kiefernmykorrhiza in den Wäldern Kenias. *In* "Mykorrhiza" (W. Rawald and H. Lyr, eds.), p. 49. Fischer, Jena.

Göbl, F. (1965). Mykorrhizauntersuchungen in einem subalpinen Fichtenwald. *Mitt. Forstl. Bundes-Versuchsanst. Mariabrunn* 66, 173.

Göbl, F. (1967). Mykorrhiza-Typen in Pflanzgärten. *Proc. Int. Union Forest Res. Organ., 14th, 1967* Vol. V, p. 60.

Göbl, F., and Platzer, H. (1967). Düngung und Mykorrhizabildung bei Zirbenjung-pflanzen. *Mitt. Forstl. Bundes-Versuchsanst. Wien* 74, 1.

Hacskaylo, E., and Palmer, J. G. (1957). Effect of several biocides on growth and incidence of mycorrhizae in field plots. *Plant Dis. Rep.* 41, 354.

Hacskaylo, E., and Vozzo, J. A. (1971). Inoculation of *Pinus caribaea* with ectomy-corrhizal fungi in Puerto Rico. *Forest Sci.* 17, 239.

Hacskaylo, E., Palmer, J. G., and Vozzo, J. A. (1965). Effect of temperature on growth and respiration of ectotrophic mycorrhizal fungi. *Mycologia* 57, 748.

Hatch, A. B. (1936). The role of mycorrhizae in afforestation. *J. Forest.* 34, 22.

Hatch, A. B. (1937). The physical basis of mycotrophy in *Pinus. Black Rock Forest Bull.* 6.

Heikurainen, L. (1955). Der Wurzelaufbau der Kiefernbestände auf Reisermoor-böden und seine Beeinflussung durch die Entwässerung. *Acta Forest. Fenn.* 65, No. 3.

Imshenetskii, A. A., ed. (1967). "Mycotrophy in Plants." Isr. Program Sci. Transl., Jerusalem. (Original in Russian, *Izd. Akad. Nauk SSSR*, Moscow, 1955.)

Iyer, J., Lipas, E., and Chesters, G. (1971). Correction of mycotrophic deficiencies of tree nursery stock produced on biocide-treated soils. *In* "Mycorrhizae" (E. Hacskaylo, ed.), USDA Forest Serv. Misc. Publ. No. 1189, p. 233. US Govt. Printing Office, Washington, D. C.

Jorgensen, J. R., and Shoulders, E. (1967). Mycorrhizal root development vital to survival of slash pine nursery stock. *Tree Plant. Notes* 18, 7.

Kessell, S. L. (1927). Soil organisms. The dependence of certain pine species on a biological soil factor. *Emp. Forest J.* 6, 70.

Kessell, S. L., and Stoate, T. N. (1938). Pine nutrition. An account of investigations and experiments in connexion with the growth of exotic conifers in Western Australian plantations. *Forest Bull., Perth, W.A.* No. 15.

Klyushnik, P. J. (1952). Mycorrhizal fungi of oak. *Les. Khoz.* 5, 63.

Krangauz, R. A. (1967). Pure cultures of potential mycorrhiza formers with oak and pine examined in pot cultures. *In* "Mycotrophy in Plants" (A. A. Imshenet-skii, ed.), p. 246. Isr. Program Sci. Transl., Jerusalem.

Krasovskaya, I. V., and Smirnova, I. D. (1950). The use of mycorrhiza in planting of oak acorns in the arid conditions of the Saratov region. *Les i Step* No. 2, p. 29.

Laiho, O. (1965). Further studies on the ectendotrophic mycorrhiza. *Acta Forest. Fenn.* **79**, No. 3.

Laiho, O. (1967). Field experiments with ectendotrophic Scotch pine seedlings. *Proc. Int. Union Forest Res. Organ., 14th, 1967* Vol. V, p. 149.

Laiho, O. (1970). *Paxillus involutus* as a mycorrhizal symbiont of forest trees. *Acta Forest. Fenn.* **106**, 1.

Laiho, O., and Mikola, P. (1964). Studies on the effect of some eradicants on mycorrhizal development in forest nurseries. *Acta Forest. Fenn.* **77**, No. 2.

Lamb, A. F. A. (1956). "Exotic Forest Trees in Trinidad and Tobago." Govt. Printing Office, Trinidad and Tobago.

Lamb, R. J., and Richards, B. N. (1971). Effect of mycorrhizal fungi on the growth and nutrient status of slash and radiata pine seedlings. *Aust. Forest.* **35**, 1.

Letourneux, C. (1957). Tree planting practices in tropical Asia. *FAO Forest. Develop. Pap.* **11**, 1.

Levisohn, I. (1956). Growth stimulation of forest-tree seedlings by the activity of free-living mycorrhizal mycelia. *Forestry* **29**, 53.

Levisohn, I. (1959). Strain differentiation in a root-infecting fungus. *Nature (London)* **183**, 1065.

Levisohn, I. (1960). Physiological and ecological factors influencing the effect of mycorrhizal inoculation. *New Phytol.* **59**, 42.

Levisohn, I. (1963). Über Mykorrhizen und Pseudomykorrhizen. *In* "Mykorrhiza" (W. Rawald and H. Lyr, eds.), p. 27. Fischer, Jena.

Lobanow, N. W. (1960). "Mykotrophie der Holzflanzen" (transl. by W. Rawald). Deut. Verlag Wiss., Berlin. (Original in Russian, Moscow, 1953.)

Lundeberg, G. (1967). Raw humus as the nitrogen source for pine seedlings. *Proc. Int. Union Forest Res. Organ., 14th, 1967* Vol. V, p. 112.

Lundeberg, G. (1970). Utilisation of various nitrogen sources, in particular bound soil nitrogen, by mycorrhizal fungi. *Stud. Forest. Suec.* **79**, 1.

McComb, A. L. (1938). The relation between mycorrhizae and the development and nutrient absorption of pine seedlings in a prairie nursery. *J. Forest.* **36**, 1148.

McComb, A. L., and Griffith, J. E. (1946). Growth stimulation and phosphorus absorption of mycorrhizal and non-mycorrhizal northern white pine and Douglas fir seedlings in relation to fertilization treatment. *Plant Physiol.* **21**, 11.

Madu, M. (1967). The biology of ectotrophic mycorrhiza with reference to the growth of pines in Nigeria. *Obeche, J. Tree Club, Univ. Ibadan* **1**, 9.

Marx, D. H., Bryan, W. C., and Davey, C. B. (1970). Influence of temperature on aseptic synthesis of ectomycorrhizae by *Thelephora terrestris* and *Pisolithus tinctorius* on loblolly pine. *Forest Sci.* **16**, 424.

May, W. B. (1953). Nursery notes to remember. *East Afr. Agr. Forest. Res. Organ., For. Technol. Note* No. 1.

Melin, E. (1923). Experimentelle Untersuchungen über die Konstitution und Ökologie der Mykorrhizen von *Pinus silvestris* L. und *Picea Abies* (L). Karst. *Mykol. Untersuch.* **2**, 73.

Melin, E. (1925). "Untersuchungen über die Bedeutung der Baummykorrhiza." Fischer, Jena.

Melin, E. (1946). Der Einfluss von Waldstreuextrakten auf das Wackstum von

Bodenpilzen, mit besonderer Berücksichtigung der Wurzelpilze von Waldbäumen. *Symb. Bot. Upsal.* **8**, No. 3.

Michowitsch, A. J. (1963). Untersuchungen zur Infizierung der Eiche mit Mykorrhizapilzen in der Waldsteppe und in der Halbwüste. *In* "Mykorrhiza" (W. Rawald and H. Lyr, eds.), p. 441. Fischer, Jena.

Mikola, P. (1953). An experiment on the invasion of mycorrhizal fungi into prairie soil. *Karstenia* **2**, 33.

Mikola, P. (1965). Studies on the ectendotrophic mycorrhiza of pine. *Acta Forest. Fenn.* **79**, No. 2.

Mikola, P. (1967). The effect of mycorrhizal inoculation on the growth and root respiration of Scotch pine seedlings. *Proc. Int. Union Forest Res. Organ., 14th, 1967* Vol. V, p. 100.

Mikola, P. (1969a). Comparative observation on the nursery technique in different parts of the world. *Acta Forest. Fenn.* **98**, 1.

Mikola, P. (1969b). Mycorrhizal fungi of exotic forest plantations. *Karstenia* **10**, 169.

Mikola, P. (1970). Mycorrhizal inoculation in afforestation. *Int. Rev. Forest Res.* **3**, 123.

Mikola, P., and Laiho, O. (1962). Mycorrhizal relations in the raw humus layer of northern spruce forests. *Commun. Inst. Forest. Fenn.* **55**, No. 18.

Mikola, P., Laiho, O., Erikäinen, J., and Kuvaja, K. (1964). The effect of slash burning on the commencement of mycorrhizal association. *Acta Forest. Fenn.* **77**, No. 3.

Modess, O. (1941). Zur Kenntniss der Mykorrhizabildner von Kiefer und Fichte. *Symb. Bot. Upsal.* **5**, No. 1.

Moser, M. (1958a). Der Einfluss tiefer Temperaturen auf das Wachstum und die Lebenstätigkeit höherer Pilze mit besonderer Berücksichtigung von Mykorrhizapilzen. *Sydowia* **12**, 386.

Moser, M. (1958b). Die künsliche Mykorrhizaimpfung von Forstpflanzen. II. Die Torfstreukultur von Mykorrhizapilzen. *Forstwiss. Zentralbl.* **77**, 257.

Moser, M. (1959a). Die künstliche Mykorrhizaimpfung an Forstpflanzen. III. Die Impfmethodik im Forstgarten. *Forstwiss. Zentralbl.* **78**, 193.

Moser, M. (1959b). Beiträge zur Kenntnis der Wuchsstoffbeziehungen im Bereich ectotropher Mykorrhizen 1. *Arch. Mikrobiol.* **34**, 251.

Moser, M. (1962). Soziologische und ökologische Fragen der Mykorrhiza-Induzierung. *Proc. Int. Union Forest Res. Organ., 13th, 1961* Part 2, Vol. 1, Sect. 24-2.

Moser, M. (1963). Die Bedeutung der Mykorrhiza bei Aufforstung unter besonderer Berücksichtigung von Hochlagen. *In* "Mykorrhiza" (W. Rawald and H. Lyr, eds.), p. 407. Fischer, Jena.

Norkrans, B. (1950). Studies in growth and cellulolytic enzymes of *Tricholoma*. *Symb. Bot. Upsal.* **11**, No. 1.

Olatoye, S. T. (1966). A report of mycorrhizal investigations. (Investigation 317.) *Tech. Note, Dep. For. Res., Nigeria* No. 33.

Oliveros, S. (1932). Effect of soil inoculation on the growth of Benguet pine. *Makiling Echo* **11**, 205.

Paavilainen, E. (1966). On the effect of drainage on root systems of Scots pine on peat soils. *Commun. Inst. Forest Fenn.* **66**, No. 1.

Pachlewski, R. (1967a). Studies on mycorrhizal fungi of pine (*Pinus silvestris* L.)— *Lactarius rufus* (Scop. ex Fr.) Fr. and *Rhizopogon luteolus* Fr. and North. under natural conditions and in pure culture. *Proc. Int. Union Forest Res. Organ., 14th, 1967* Vol. V, p. 12.

Pachlewski, R. (1967b). "Investigations of Pure Culture of Mycorrhizal Fungi of Pine." Forest Res. Inst., Warsaw.

Park, J. Y. (1971). Preparation of mycorrhizal grain spawn and its practical feasibility in artificial inoculation. *In* "Mycorrhizae" (E. Hacskaylo, ed.), USDA Forest Serv. Misc. Publ. No. 1189, p. 239. US Govt. Printing Office, Washington, D. C.

Parry, M. S. (1956). Tree planting practices in tropical Africa. *FAO Forest. Develop. Pap.* **8,** 1.

Rayner, M. C. (1936). The mycorrhizal habit in relation to forestry. II. Organic composts and the growth of young trees. *Forestry* **10,** 1.

Rayner, M. C. (1938). The use of soil or humus inocula in nurseries and plantations. *Emp. Forest J.* **17,** 236.

Rayner, M. C. (1939). The mycorrhizal habit in relation to forestry. III. Organic composts and the growth of young trees. *Forestry* **13,** 19.

Rayner, M. C., and Neilson-Jones, W. (1944). "Problems in Tree Nutrition." Faber & Faber, London.

Robertson, N. F. (1954). Studies on the mycorrhiza of *Pinus silvestris.* I. The pattern of development of mycorrhizal roots and its significance for experimental studies. *New Phytol.* **53,** 253.

Roeloffs, J. W. (1930). Over kunstmatige Verjonging van *Pinus Merkusii* Jungh. et de Vr. en *Pinus Khasya* Royle. *Tectona* **23,** 874.

Romell, L.-G. (1930). Blodriskan, en granens följesvamp. *Sv. Bot. Tidskr.* **24,** 524.

Rosendahl, R. O. (1942). The effect of mycorrhizal and non-mycorrhizal fungi on the availability of difficultly-soluble potassium and phosphorus. *Soil Sci. Soc. Amer., Proc.* **7,** 477.

Rubtov, S. (1964). Humusul de larice-stimulator puternik al cresterii puietilor de larice in pepiniere. *Rev. Padurilor* **69,** 279.

Runov, E. V. (1967). Experimental introduction of mycorrhizas into oak sowings in arid steppe. *In* "Mycotrophy in Plants" (A. A. Imshenetskii, ed.), p. 174. Isr. Program Sci. Transl., Jerusalem.

Schramm, J. R. (1966). Plant colonization studies on black wastes from anthracite mining in Pennsylvania. *Trans. Amer. Phil. Soc.* [N.S.] **56,** No. 1.

Shemakhanova, N. M. (1967). "Mycotrophy of Woody Plants." Isr. Program Sci. Transl., Jerusalem. (Original in Russian, Izd. Akad. Nauk SSSR Moscow, 1962.)

Sobotka, A. (1955). Umělá mykorrhizace sadebniho materialu. *Sb. Cesk. Cesk. Akad. Zemeved. Lesn.* **28,** 67.

Sobotka, A. (1963). Die praktische Anwendung der Mykorrhiza bei der Aufforstung. *In* "Mykorrhiza" (W. Rawald and H. Lyr, eds.), p. 461. Fischer, Jena.

Sobotka, A. (1968). Wurzeln von *Picea excelsa* L. unter dem Einfluss der Industrieexhalate im Gebiet des Erzgebirges in der ČSSR. *Immissionen und Waldzönosen. Cesk. Akad. Ved. Ustav pro Tvorlu a Ochr. Kran., Praha, 1968,* p. 45.

Stoeckeler, J. H., and Jones, G. W. (1957). Forest nursery practice in the Lake States. *U. S., Dep. Agr., Agr. Handb.* **110.**

Stoeckeler, J. H., and Slabauch, P. E. (1965). Conifer nursery practice in the prairie-plains. *U. S., Dep. Agr., Agr. Handb.* **279.**

Takacs, E. A. (1967). Produccion de cultivos puros de hongos micorrhizógenos en el Centro Nacional de Investigaciones Agropecuarias, Castelar. *IDIA Suppl. For.* **4,** 83.

Theodorou, C. (1971). Inoculation of mycorrhizal fungi into soil by spore inoculation of seed. *Aust. Forest.* **35**, 23.

Theodorou, C., and Bowen, G. D. (1970). Mycorrhizal responses of radiata pine in experiments with different fungi. *Aust. Forest.* **34**, 183.

Trappe, J. M. (1962a). Fungus associates of ectotrophic mycorrhizae. *Bot. Rev.* **28**, 538.

Trappe, J.M. (1962b). *Cenococcum graniforme*—its distribution, ecology, mycorrhiza formation, and inherent variation. Unpublished Thesis, University of Washington, Seattle.

Trappe, J. M. (1964). Mycorrhizal hosts and distribution of *Cenococcum graniforme*. *Llyodia* **27**, 100.

Tserling, G. I. (1960). Mycorrhiza formation with larch in chernozems of the Volga region and methods of its stimulation. *Mikrobiologiya* **29**, 401.

van Alphen de Veer, E. J. *et al.* (1954). Artificial regeneration of *Pinus merkusii*. *Proc. World Forest Congr., 4th, 1954* Vol. III, p. 565.

van Suchtelen, M. J. (1962). Mykorrhiza bij *Pinus* spp. in de tropen. *Meded. Landbouwhogesch. Opzoekingssta. Staat Gent* **27**, 1104.

Voznyakovskaya, Y. M., and Ryžkova, A. S. (1958). Mass cultures of mycorrhiza-forming fungi in laboratory conditions. *Tr. Vses. Nauch.-Issled. Inst. Sel'skokhoz. Mikrobiol.* **15**, 164.

Vysotskii, G. N. (1902). Mycorrhiza of oak and pine seedlings. *Lesoprom. Vestn.* **29**, 504.

Wakeley, P. C. (1954). Planting the southern pines. *U. S., Dep. Agr., Agr. Monogr.* **18**.

Waterer, R. R. (1957). Nursery practice for conifers in Kenya. *Technol. Order, Kenya For. Dep.* No. 17.

White, D. P. (1941). Prairie soil as a medium for tree growth. *Ecology* **22**, 398.

Wilcox, H. E. (1971). Morphology of ectendomycorrhizae in *Pinus resinosa*. *In* "Mycorrhizae" (E. Hacskaylo, ed.), USDA Forest Serv. Misc. Publ. No. 1189, p. 54. US Govt. Printing Office, Washington, D. C.

Wilde, S. A. (1944). Mycorrhizae and silviculture. *J. Forest.* **42**, 290.

Worley, J. F., and Hacskaylo, E. (1959). The effect of available soil moisture on the mycorrhizal association of Virginia pine. *Forest Sci.* **10**, 214.

Glossary

Actinomycetes (lit. ray fungi), very small, usually 1.0 μm in diameter, gram-positive, with well-developed hyphae.

Antagonism, usually refers to an association that is injurious to one or more of the organisms in the association.

Antibiosis, antagonism occurring between two organisms resulting in the inhibition or destruction of one by the other.

Ascomycetes (lit. the sac fungi), a large group with about 15,000 spp. Many are parasitic on higher plants. Very few members of this group are known to form mycorrhizae.

Basidiocarp, the fruiting body of the basidiomycetes.

Basidiomycetes, a large group of fungi, with about 1500 spp. The characteristic structure of this group is the basidium. Most known mycorrhiza-forming fungi are found in subgroup, Homobasidiomycetidae.

Basidiospore, a spore produced from a basidium. Usually a basidium produces four, one-celled basidiospores which are discharged violently and carried by air currents.

Carpophore, sometimes used to describe the whole fruiting body of the higher fungi. More usually refers to the stalk of the sporocarp.

Chlamydospore, a thick-walled, resting spore formed asexually on the vegetative hyphae.

Clamp connexion (or connection), formed only in some basidiomycetes during cell division of the hyphae. It is a short, backwardly directed hypha formed at the septum that is used to distribute the two daughter nuclei formed during nuclear division.

Concolorous, the same color.

Coralloid, coral-like usually caused by repetitious, bifurcate branching of the mycorrhizal root.

Core, usually that part of the root that is not invaded by the mycorrhizal fungus. It includes the central stele and part of the cortex.

Cystidium, (pl. cystidia), a sterile, pale-colored end of a hyphal element in the mantle of the mycorrhiza. Usually conical or prism shaped.

Dendritic, tree-like, much branched.

Dolipore, a thick-walled, barrel-like pore found in the cross wall separating two hyphal cells. It has a complicated structure and is characteristic of basidiomycetous fungi.

Ectendomycorrhiza (pl. ectendomycorrhizae, ectendomycorrhizas, syn. ectendotrophic mycorrhiza), a mycorrhizal association wherein the mycorrhizal fungus occupies the intercellular spaces of the cortex of the root and penetrates some (or all) of the adjacent cortical cells.

Ectomycorrhiza (pl. ectomycorrhizae, ectomycorrhizas, Syn. ectotrophic mycorrhiza), a mycorrhizal association wherein the mycorrhizal fungus is confined exclusively to the intercellular spaces of the cortex of the root, unlike ectendomycorrhiza.

Endomycorrhiza (pl. endomycorrhizae, endomycorrhizas, Syn. endotrophic mycorrhiza), a mycorrhizal association where the fungus is confined exclusively to the intracellular spaces of cortical cells of the host root.

Fungistatic, not permitting fungal growth.

Gasteromycetes, a subclass of the basidiomycetes containing about 700 spp. Many of these are associated with tree roots and are suspected of forming mycorrhizal associations.

Hartig net, named after Robert Hartig, refers to the intercellular hyphae of mycorrhizae. They have a characteristic close-linked, beaded appearance in cross sections of the root.

Haustorium, a special absorptive hypha that invades a living cell.

Hymenomycetes, a subclass of the basidiomycetes where the basidium develops in an open fruiting body. This group contains many mycorrhizal fungi.

Hypogeous, developing in the earth.

Mantle, the mat of hyphae that develops on the surface of the mycorrhizal root. The structure and color of the mantle are highly variable.

Mycelium, a mass of hyphae.

Mycorrhiza (pl. mycorrhizae, mycorrhizas), a symbiotic association between a non-pathogenic (or weakly pathogenic) fungus and living, primary cortical cells of a root.

Mycorrhizosphere, the rhizosphere of a mycorrhizal root.

Mycotrophic, having mycorrhizae.

Nodule, a rounded mass of densely packed mycorrhizal roots.

Oidium (pl. oidia), spermatium (nonmotile gamete) formed on hyphal branches or flat-ended asexual spores formed by the breaking up of a hypha.

Peritrophic, mycelium surrounding the surface of the root.

Phytoalexin, coined by K. O. Müller, a fungi-toxic substance produced in a plant cell in response to infection or damage.

Prosenchyma (syn. plectenchyma), a thick tissue formed by hyphae twisting and adhering together. Also refers to various tissues of higher plants composed of elongated cells mostly with little or no protoplasm and including tissues specialized for conduction and support.

Pseudomycorrhiza, a mycorrhizal association in which the fungal partner shows pathogenic characteristics.

Pseudoparenchyma, a mass of closely interwoven hyphae that appear like parenchyma in prepared sections.

Ramiform, much branched.

Resupinate, being reversed in position, usually where the hymenium is borne on the upper surface.

Rhizomorph, a densely packed mass of fungal hyphae that looks like a tree root.

Rhizoplane, the region of the soil in immediate contact with the surface of the root.

Rhizosphere, coined by Hiltner in 1904, and defined as that region of the soil surrounding the root in which the microflora are influenced by the root.

Rind, the outer layers of the fruiting body of a fungus.

Saprophyte, an organism that obtains its food material from dead organic matter.

Satellitism, the dependence for existence of a smaller organism on a larger one.

Sclerotium (pl. sclerotia), a firm, rounded, often hard, mass of hyphae devoid of spores that forms a resting stage.

Septal pore, a pore in the cross wall of a fungal hypha connecting two contiguous cells. In the basidiomycetes these have a complicated structure.

Seta (pl. setae), a stiff hair, usually thick-walled and dark-colored.

Spermatosphere, that region of the soil around a seed in which the microorganisms are influenced by the seed, especially as it germinates.

Spermosphere, see Spermatosphere.

Spore, a reproductive body, usually containing one to few cells.

Sporocarp, any fruiting body that produces spores.

Sporophore, a structure that either produces spores or supports a spore-bearing organ.

Stipitate, stalked.

Symbiosis, the living together (and benefit derived therefrom) of two unlike organisms.

Synenchyma, adhering parenchyma.

Tannin layer, the outermost layers of cortical cells in a mycorrhiza that are impregnated with tannins. Some of these cells are true cortical cells and some have been derived from the root cap. The latter are not sloughed but preserved and retained by the mycorrhizal fungus.

Thigmotropism (syn. thigmotaxis), the response of an organism to touch or contact.

Tuberculate, provided with tubercles, usually small, rounded swellings.

Vesicular–arbuscular, usually applies to endomycorrhizae where the hyphae inside the cortical cells are either coiled or finely divided into haustorial branches.

Author Index

Numbers in italics refer to the pages on which the complete references are listed.

417

Subject Index

A

ABA (Abscisic acid), 278, 281
Abortion, 253, 261
Abscisic acid, *see* ABA
Abscission, 195, 367, *see also* Abortion
Absorbing power, 172–174
Absorption, 82, 87, 95, 116, 120, 125, 127, 318, 373, *see also* Active uptake, Passive uptake
  auxin, 286
  exudates, 126
  nutrients, 13, 45, 79, 108, 142, 151–154, 156–161, 169–188, 193, 196, 211, 225, 227, 267, 299, 325, 353
  oxygen, 226, 311
  sugars, 223, 224
  water, 79, 93, 98, 151, 152, 161, 174, 182, 183, 187, 193
Accumulation of minerals, 171
Acetic acid, 119, 122, 123, 128, 321
Acetoin, 122
Acetone, 328
Acetylene, 169
Acetylene reduction test, 168, 169
Acidification, 392, 395, *see also* pH
Acidity, 386, *see also* pH
cis-Aconitic acid, 119, 318
Actinomycetes, 14, 299, 301–303, 310, 314, 316, 325, 328, 329, 335, 357, 375
Activators, 111–113
Active transport, 192
Active uptake, 159, 167, 168
Adenosine diphosphate, *see* ADP
Adenosine triphosphate, *see* ATP
ADP, 227

Adsorption, 158, 159, 161, 167, 170, 184, 186
Aeration, 12, 93, 99, 100, 138, 245, 354, 396, 405
Afforestation, 86, 96, 98, 100, 384, 389, 400–402, 404
Agar, 52, 64, 110, 113, 114, 130, 241, 304, 314, 339, 340, 365, *see also* Nutrient agar, Potato-dextrose agar
Aging, 13, 34, 72, 73, 138, 139, 161, 188, 251, 301, *see also* Senescence
A-horizon, 93
Air-layering, 117
Air pollution, *see* Pollution
α-Alanine, 118, 123, 124, 129, 240, 278, 319, 320
β-Alanine, 118, 123, 129, 319, 320
Alcohol, 219–221, *see also* specific alcohols
Aldose, 219
Algae, 160, 188, 191
Alkali metals, 170
Alkalinity, *see* pH
Allelopathy, 107
Allyl alcohol, 387
Amides, 118, 124, 125, 128, 168, 278, 304, 318, 319, 372, *see also* specific amides
Amination, 168
Amino acids, 112, 113, 116–118, 120–126, 128, 129, 133, 144, 168, 187, 211, 243, 247, 251, 267, 276, 278, 279, 285, 287, 305–309, 311, 318–320, 325, 335, 340, 372, 373, *see also* specific amino acids
o-Aminobenzoic acid, 240
p-Aminobenzoic acid, 319, 321
α-Aminoadipic acid, 319

Weed killers, *see* Herbicides
Wilting, 135, *see also* Water deficits
Wind, 7
Wood rot, 179

X

Xanthophylls, 281
Xylanase, 221
Xylem, 7, *see also* Primary xylem,
 Protoxylem, Vessels
Xylose, 119, 216, 217, 318, 319

Y

Yeast, 112, 114, 122, 306, 307

Z

Zeatin, 241
Zeatin ribonucleotide, 241
Zinc, 95, 154, 156, 158, 170, 176, *see
 also* ^{65}Zn
Zone of apposition, 34
Zoospores, 364, 366, 368, 374
^{65}Zn, 188

66